MULTIPHASE TRANSPORT AND PARTICULATE PHENOMENA

MULTIPHASE TRANSPORT AND PARTICULATE PHENOMENA

VOLUME 2

Edited by

T. Nejat Veziroğlu
Clean Energy Research Institute
University of Miami

○HEMISPHERE PUBLISHING CORPORATION
A member of the Taylor & Francis Group

New York　　Washington　　Philadelphia　　London

MULTIPHASE TRANSPORT AND PARTICULATE PHENOMENA: Volume 2

Cover design by Sharon DePass.

1 2 3 4 5 6 7 8 9 0   E B E B   8 9 8 7 6 5 4 3 2 1 0 9

Library of Congress Cataloging-in-Publication Data

Multiphase transport and particulate phenomena / edited by T. Nejat
  Veziroğlu.
    p.   cm.
   Papers presented at the 5th Miami International Symposium on
Multiphase Transport and Particulate Phenomena.

   1. Multiphase flow—Congresses. 2. Heat—Transmission—
Congresses. 3. Mass transfer—Congresses. I. Veziroğlu, T.
Nejat. II. Miami International Symposium on Multiphase Transport
and Particulate Phenomena (5th)
TA357.M85   1990
620.1'064—dc20
                         80-19776
                           CIP

ISBN 1-56032-026-5 (set)
ISBN 1-56032-031-1 (Vol. 2)

Contents

TWO–PHASE FLOW PRESSURE DROPS

MULTIPHASE TRANSIENTS

MULTIPHASE FLOW INSTABILITIES

LOCA AND OTHER ACCIDENT PHENOMENA

ENERGY/THERMAL APPLICATIONS

Preface

Multiphase transport and particulate phenomena applications are found in a wide range of engineering systems, such as heat exchangers, boilers, evaporators, condensers, boiling water reactors, pressurized water reactors, particle separators, contamination detectors and controllers, filters, slurry transporters, and fluidized beds. Over the past three decades, problems in two-phase flow heat transfer and instabilities have challenged many investigators. Instabilities can induce boiling crises, disturb control systems, and/or cause mechanical damage. It is thus important to be able to predict the conditions under which a two-phase system will perform reliably. At the same time, the importance of the particulate phenomena science and other technology is growing in other areas, including particle sizing, liquid particle interactions, mechanics of suspensions and emulsions, and sedimentation.

Due to recent energy and environmental crises, many other multiphase transport and particulate phenomena problems have also become important. Some of these include the modeling of the loss-of-coolant accident in pressurized water nuclear reactors, scaling up of fluidized bed reactors for converting coal to clean gaseous and liquid fuels, design of heat exchangers for liquified natural gas and liquified petroleum gas, control of microcontamination, effect of aerosols, and air pollution control.

The Fifth Miami International Symposium on Multiphase Transport and Particulate Phenomena continued the tradition established by its four predecessors. It provided a high-level international platform in pleasant surroundings for the presentation of the latest research results and for the exchange of ideas on the important topics of multiphase transport and particulate phenomena.

The lectures and papers presented at the symposium and prepared in accordance with the paper guidelines, have been divided by their subject matter into 16 parts and 3 volumes. The reader should be advised that it was difficult to classify specifically some of the papers where there was an overlap in subject matter. In such cases, we tried to make the best possible choice. This three-volume set should serve as a valuable reference, covering the latest developments in the growing areas of multiphase transport and particulate science and technology.

T. Nejat Veziroğlu

Acknowledgments

The Organizing Committee gratefully acknowledges the assistance and cooperation of the International Association for Hydrogen Energy, the International Atomic Energy Agency, the International Solar Energy Society, the Florida International University, the Florida Solar Energy Center, and the Department of Mechanical Engineering, University of Miami.

We also wish to extend sincere appreciation to the keynote speaker, William L. Grosshandler, Thermal Sciences and Engineering Program, National Science Foundation, Washington, D.C., and to the banquet speaker, Robert T. Lahey, Nuclear Engineering Department, Rensselaer Polytechnic Institute, Troy, New York.

Special thanks are due to our authors and lecturers, who have provided the substance of the proceedings.

And last, but not least, our debt of gratitude is owed to session developers, chairpersons, and co-chairpersons for the organization and execution of the technical sessions. In acknowledgment we list these session officials on the following pages.

Organizing Committee

Organizing Committee

K. Akyüzlü
University of New Orleans
New Orleans, LA, USA

S. G. Bankoff
Northwestern University
Chicago, IL, USA

K. J. Bell
Oklahoma State University
Stillwater, OK, USA

A. E. Bergles
Rensselaer Polytechnic Institute
Troy, NY, USA

J. S. Chang
McMaster University
Ontario, Canada

X. Chen
Xian Jiaotong University
Xian, China

A. Duyar
Florida Atlantic University
Boca Raton, FL, USA

M. Gashgari
King Abdulaziz University
Saudi Arabia

D. Geldart
University of Bradford
Bradford, UK

D. Gidaspow
Illinois Institute of Technology
Chicago, IL, USA

N. Güven
Texas Tech University
Lubbock, TX, USA

S. A. Hoenig
University of Arizona
Tucson, AZ, USA

H. Hoffman
Oak Ridge National
Laboratory, USA

W. S. Janna
University of New Orleans
New Orleans, LA, USA

K. Johannsen
Berlin Technical University
Berlin, FRG

V. Kakabadze
Academy of Sciences
Tbilisi, USSR

S. Kakaç
University of Miami
Coral Gables, FL, USA

R. T. Lahey
Rensselaer Polytechnic Institute
Troy, NY, USA

S. S. Lee
University of Miami
Coral Gables, FL, USA

R. Lyczkowski
Argonne National Laboratory
Argonne, IL, USA

A. Mertol
Science Applications
International Corp.
Los Altos, CA, USA

A. S. Mujumdar
McGill University
Canada

V. J. Novick
EGG&G Idaho, Inc.
Seattle, WA, USA

M. N. Ozisik
North Carolina State University
Raleigh, NC, USA

M. R. Parker
University of Salford
Salford, UK

R. W. Peters
Purdue University
West Lafayette, IN, USA

J. T. Pytlinski
University of Puerto Rico
Puerto Rico

P. Ramakrishnan
Indian Institute of Technology
Madras, India

T. Raunemaa
University of Kuopio
Helsinki, Finland

T. M. Romberg
CISRO
Sutherland, Australia

C. Schweiger
University of Duisburg
Duisburg, FRG

N. Selcuk
Middle East Technical University
Turkey

J. S. Sheffield (Co-Chairperson)
University of Missouri
Rolla, MO, USA

S. Sideman
Israel Institute of Technology
Haifa, Israel

C. W. Snoek
Atomic Energy of Canada, Ltd.
Chalk River, Canada

I. K. Stephan
University of Stuttgart
Stuttgart, FRG

R. L. Sterling
University of Minnesota
Minneapolis, MN, USA

Y. K. Tan
South China Institute of
Technology
Guangzhou, China

Y. Ueno
University of Missouri
Rolla, MO, USA

T. N. Veziroğlu (Chairperson)
University of Miami
Coral Cables, FL, USA

J. H. Vincent
Institute of Occupational Medicine
Edinburgh, UK

J. Weisman
University of Cincinnati
Cincinnati, OH, USA

A. A. Zkauskas
Mokslu Akapemija
Vilnius, USSR

STAFF

Executive Secretary   Lucille Walter

Conference Coordinators   Javonne Gelineau
Aymara Schmidt

Graduate Assistants	L. Kazi
	N. Lutfi
	T. Özgökmen
	M. Padki
	T. Tekindur
Undergraduate Assistants	M. Akcin
	C. Blaisure
	K. Cerretti
	V. Yankowski

TWO-PHASE FLOW PRESSURE DROPS

A Study of Two-Phase Pressure Drop of Wye-Type Branching Pipes

D. F. CHE and Z. H. LIN
Department of Energy and Power Engineering
Xi'an Jiaotong University
Xi'an, PRC

Abstract

On an air-water test loop the two-phase pressure drop characteristics of 5 kinds of wye-type branching pipes have been investigated. Based upon the experimental data, the gas-liquid two-phase loss coefficients of these branching pipes are presented by using the homogeneous analytical model, these coefficients can be applied to predict the pressure drops of the wye-type branching pipes in pipe systems.

1. INTRODUCTION

The pipe systems composed of branching pipes in which gas-liquid two-phase mixtures flow are widely used in industry and engineering processes. The redistribution of the two-phase mixture in the branching pipe is very important to the safe operation of equipments. For example, it may directly influence the working saftey of large-capacity steam generators or boilers. The distribution characteristics of two-phase mixtures in branching pipes is governed mainly by the two-phase pressure drop characteristics, therefore, it is neccessary to study the two-phase pressure loss characteristics of branching pipes. However, up to now the gas-liquid two-phase pressure loss characteristics of branching pipes are studied insufficiently.

This paper presents the research results of the two-phase pressure losses of 5 types of branching pipes that may be used in large-capacity once-through boilers [Fig. 1]. The tested fluid is air-water mixture. Based upon the experimental data, the gas-liquid two-phase loss coefficients of the branching pipes are obtained by using the homogeneous analytical model.

2. DESCRIPTION OF TEST LOOP AND SECTION

The scheme of the test loop is shown in Fig. 2.

The air-water two-phase mixture is formed by mixing the air produced by the air compressor and the water provided by the pump. The two-phase mixtures distributed by the branching pipe is directed to the separator by the branches. Air and water is separated in the separator. The separated air flows out from the top of the separator, its flowrate is measured by an orifice flow meter. The separated water is discharged from the bottom of the separator, its flowrate is measured directly by weighing the discharged water.

During experiment all the branches are arranged vertically. The pressure of the system is slightly above the atmospheric pressure and the temperature of the air-water two-phase mixture is near to the atmospheric temperature. all the pressure drops and the absolute pressures are measured by U-type differential manometers. The manometric fluid is water or mercury, depending upon the magnitude of the pressure difference.

The inner diameters of all the legs of the branching pipes are 18.8 mm, and all the branching pipes are made of carbon steel.

The tested parameters are as follows:

the inlet air flowrate and water flowrate, the air and the water flowrates in each branch, static pressure drop between the inlet of the branching pipe and the outlet of each branch.

The air flowrate is from 0.0025 to 0.018 kg/s, the water flowrate ranges from 0.03 to 0.2 kg/s.

3. DEFINITION OF TWO-PHASE LOSS COEFFICIENT

Assume the air-water two-phase mixture flows homogeneously in the pipes and the friction between the wall and the fluid is neglected.

For type (a) of the tested branching pipes as shown in Fig. 1, the energy equation between point 0 and point 1 of the first branch can be represented as follows,

$$P_0 + \rho_{m0}U_{m0}^2/2 = P_1 + \rho_{m1}U_{m1}^2/2 + \rho_{m0}gL_0\sin\theta + \rho_{m1}gL_1 + \zeta_1\rho_{m0}U_{m0}^2/2$$

therefore, the loss coefficient of the first branch

$$\zeta_1 = (P_0 - P_1 + \rho_{m0}U_{m0}^2/2 - \rho_{m1}U_{m1}^2/2 - \rho_{m0}gL_o\sin\theta - \rho_{m1}gL_1)/(\rho_{m0}U_{m0}^2/2)$$

where
$$\rho_{m0} = \rho_l/(1+x_0(\rho_l/\rho_g-1)) \ ; \qquad \rho_{m1} = \rho_l/(1+x_1(\rho_l/\rho_g-1))$$

$$x_0 = W_{g0}/(W_{g0}+W_{10}) \qquad ; \qquad x_1 = W_{g1}/(W_{g1}+W_{11})$$

$$U_{m0} = (W_{10}+W_{g0})/(\rho_{m0}A) \ ; \qquad U_{m1} = (W_{11}+W_{g1})/(\rho_{m1}A)$$

Similarly, the loss coefficient of the second branch of type (a) can be expressed,

$$\zeta_2 = (P_0 - P_2 + \rho_{m0}U_{m0}^2/2 - \rho_{m2}U_{m2}^2/2 - \rho_{m0}gL_0\sin\theta - \rho_{m2}gL_2 - \rho_{m2}gH tg\theta)/(\rho_{m0}U_{m0}^2/2)$$

By using the same method the loss coefficient of the first and the second branches for type (b) and (c) can be obtained as follows,

$$\zeta_1 = (P_0 - P_1 + \rho_{m0}U_{m0}^2/2 - \rho_{m1}U_{m1}^2/2 - \rho_{m0}g(L_0\sin\theta+L) - \rho_{m1}gL_1)/(\rho_{m0}U_{m0}^2/2)$$

$$\zeta_2 = (P_0 - P_2 + \rho_{m0}U_{m0}^2/2 - \rho_{m2}U_{m2}^2/2 - \rho_{m0}g(L_0\sin\theta+L) - \rho_{m2}gL_2)/(\rho_{m0}U_{m0}^2/2)$$

The loss coefficients of the second and the third branch of type (d) are as follows,

$$\zeta_2 = (P_0 - P_2 + \rho_{m0}U_{m0}^2/2 - \rho_{m2}U_{m2}^2/2 - \rho_{m0}gL_0\sin\theta - \rho_{m12}gHtg\theta - \rho_{m2}gL_2)/(\rho_{m0}U_{m0}^2/2)$$

$$\zeta_3 = (P_0 - P_3 + \rho_{m0}U_{m0}^2/2 - \rho_{m3}U_{m3}^2/2 - \rho_{m0}gL_0\sin\theta - \rho_{m12}gHtg\theta - \rho_{m3}gHtg\theta - \rho_{m3}gL_3)/(\rho_{m0}U_{m0}^2/2)$$

where
$$\rho_{12} = \rho_l/(1+x_{12}(\rho_l/\rho_g-1)) \ ; \quad x_{m12} = (W_{g1}+W_{g2})/(W_{g1}+W_{g2}+W_{11}+W_{12})$$

ζ_1, ζ_2 of type (e) are the same as those of type (d).

The loss coefficients of the third and the fourth branch of type (e) are as follows,

$$\zeta_3 = (P_0 - P_3 + \rho_{m0}U_{m0}^2/2 - \rho_{m3}U_{m3}^2/2 - \rho_{m0}gL_0\sin\theta - \rho_{m12}gHtg\theta - \rho_{m23}gHtg\theta - \rho_{m3}gL_4)/(\rho_{m0}U_{m0}^2/2)$$

$$\zeta_4 = (P_0 - P_4 + \rho_{m0}U_{m0}^2/2 - \rho_{m4}U_{m4}^2/2 - \rho_{m0}gL_0\sin\theta - \rho_{m12}gHtg\theta - \rho_{m23}gHtg\theta - \rho_{m4}gHtg\theta$$
$$- \rho_{m4}gL_4)/(\rho_{m0}U_{m0}^2/2)$$

$$\rho_{23} = \rho_l/(1+x_{23}(\rho_l/\rho_g-1)) \ ; \quad x_{23} = (W_{g2}+W_{g3})/(W_{g2}+W_{g3}+W_{12}+W_{13})$$

4. RESULTS AND DISCCUSIONS

The gas-liquid two-phase loss coefficients defined as above can be obtained experimentally if P_0-P_i (i=1,4), W_{gi}, W_{1i} (i=0,4) are measured.

The experimental results of ζ_i of type (a)--type (e) branching pipes are shown respectively in Fig.3--Fig.7. The abscissa expresses the kinetic head of the inlet two-phase mixture of the branching pipe, i.e., $H_0 = \rho_{m0}U_{m0}^2/2$.

It can be seen from these figures the two-phase loss coefficients can be considered approximately equal to different constants in the range of the tested parameters.

For type (a), ζ_1 and ζ_2 are approximately equal to 0.95 and 0.88 respectively. For type (b), ζ_1, ζ_3 are nearly equal to 1.5. For type (c), ζ_1 and ζ_2 are both equal to 1.7. For type (d), ζ_1, ζ_2 and ζ_3 are equal to 0.95, 0.925 and 0.84 respectively. For type (e), ζ_1, ζ_2, ζ_3 and ζ_4 are equal to 1.1, 1.1, 0.95 and 0.9 respectively.

By comparison one may find that ζ_1 of type (b) is smaller than that of type (c). This expresses the two-phase loss coefficients of branching pipes are influenced by the inclined angle θ of the inlet leg. The smaller the angle θ, the larger the two-phase loss coefficient. However, the quantative relationship between ζ_i and θ has not been obtained yet, it will be conducted in further research work.

5. CONCLUSIONS

The gas-liquid two-phase loss coefficients of five types of branching pipes have been investigated on air-water test loop, following conclusions can be drawn:

(1) ζ_i of each branching pipe can be considered invariant in the range of the tested parameters.

(2) The following two-phase loss coefficient values are recommended to be used in the tested range:

For type (a) $\zeta_1 = 0.95$, $\zeta_2 = 0.88$.

For type (b) $\zeta_1 = \zeta_2 = 1.5$.

For type (c) $\zeta_1 = \zeta_2 = 1.7$.

For type (d) $\zeta_1 = 0.95$, $\zeta_2 = 0.925$, $\zeta_3 = 0.84$.

For type (e) $\zeta_1 = \zeta_2 = 1.1$, $\zeta_3 = 0.95$, $\zeta_4 = 0.9$.

(3) The maximum deviation of the prediction of pressure drops is less than 30% if the two-phase loss coefficient values above is used.

(4) The gas-liquid two-phase loss coefficients are influenced by the inlet leg of the branching pipe to the horizon. The smaller the inclined angle θ, the larger the two-phase loss coefficient ζ.

6. NOMENCLATURE

P -- pressure N/m^2, W -- mass flowrate kg/s,   U -- velocity m/s
ρ -- density kg/m^3, x -- quality ,   A -- cross-sectional area m^2
g -- gravitional acceleration m/s^2, L -- length m, θ -- inclined angle degree
ζ -- loss coefficient ,   H -- length   m

Subscript
l -- liquid phase,   g -- gas phase,   m -- homogeneous,   0 -- inlet leg
i (1,4) -- ith branch

Acknoledgement

The authors would like to express their deep appreciation for the financial support provided by National Natural Science Foundation of China and Shanghai Boiler Works.

5

REFFERENCES

1. D.F.Che, Master Degree Thesis, Xi'an Jiaotong University, 1986
2. F.Frass, Two-phase Mixture In Branching Pipes, VGB, Vol. 10, 1978, p729.
3. Z.H.Lin, D.F.Che, X.J.Chen, Two-Phase Distribution In Branching Pipes, 4th Miami International Symposium On Multi-Phase Transport & Particulate Phenomena, 15-17 December, 1986.

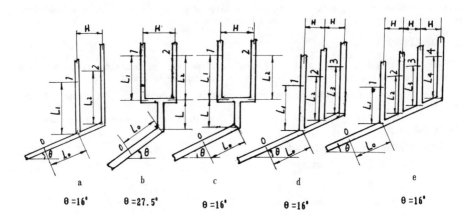

a $\theta = 16°$

b $\theta = 27.5°$

c $\theta = 16°$

d $\theta = 16°$

e $\theta = 16°$

Fig.1   Studied Branching Pipes

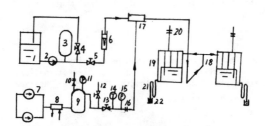

1 - water tank 2 - pump 3 - stabilizer
4, 12 - by-pass valve 5, 13 - regulating valve
6 - variable area flowmeter 7 - air compressor
8 - cooler 9 - air tank 10 - safety valve
11, 15 pressure guage 14 - thermometer
16, 20 - air orifice flowmeter 17 - mixer
18 - test pipe system 19 - gas and liquid separation tank
21 - drain tube 22 - pot

Fig.2   Test Loop

6

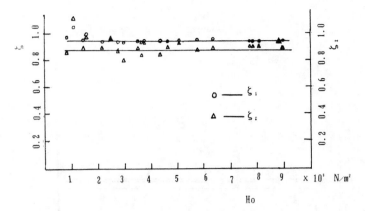

Fig.3 ζ_1 and ζ_2 of Type (a)

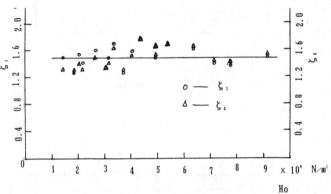

Fig.4 ζ_1 and ζ_2 of Type (b)

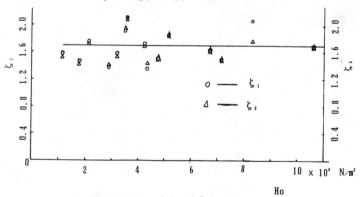

Fig.5 ζ_1 and ζ_2 of Type (c)

7

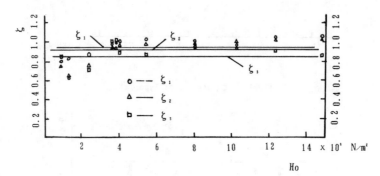

Fig.6 ζ_1 , ζ_2 and ζ_3 of Type (d)

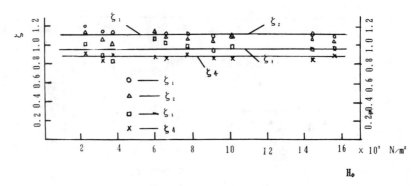

Fig.7 ζ_i (i=1,4) of Type (e)

8

Two-Phase Pressure Drop in Bends during Gas-Liquid Flow

Z. H. LIN and J. LI
Department of Energy and Power Engineering
Xi'an Jiaotong University
Xi'an, PRC

Abstract

In this paper, an experimental study of two-phase pressure drop in eight bends during gas-liquid flow has been conducted. The influence of the ratio of bend radius to pipe diameter, plane of bend, gas mass quality and working pressure have been discussed. Based on the experimental data, a predicting correlation is established. Analytical predictions are compared with test data and the predicting results of other correlations.

1. INTRODUCTION

Bends are most common pipe components in pipe systems. The ability to reliably predict the two-phase pressure drop in bend is important for hydraulic calculation of many engineering systems with gas-liquid two-phase flows, yet there is only limited information on the pressure drop in bends during gas-liquid flows. Some authors suggested to use the homogeneous flow model to calculate the two-phase pressure drop of a horizontal bend [1] [2], others intended to extended the correlation obtained from steam-water data into the gas-liquid region [3]. However, experimental studies and theoretical work have not been widely conducted.

In this paper, an experimental study of two-phase pressure drop in eight bends during air-water flow has been conducted. The geometry parameters of tested bends are as follows: inner pipe diameter, 26 and 51 mm; the ratio of bend radius to pipe diameter, $R/d = 3.08 \sim 6.54$; angle of bend, 90 and 180 degree respectively; plane of bend, vertical and horizontal. The tested flow parameters are: air flow rate $0.999 \sim 1.167$ kg/min, water flow rate $3.625 \sim 27.150$ kg/min; air mass quality $0.035 \sim 0.2$; air pressure $2.5 \sim 3$ bar.

Based on the experimental data of this paper, a predicting correlation is established. Analytical predictions are compared with test data and the predicting results of other correlations.

2. EXPERIMENTS

The experimental loop is shown in Fig. 1. The compressed air produced by the air compressor was mixed up with the water pumped by the pump to form a gas-liquid two-phase flow in the mixer. Before entering the mixer, the air flow rate was measured by an orifice and the water flow rate was measured by a variable area flow meter. After the mixer, the two-phase fluid passed the tested bend and flowed out through a drain tube. The pressure drops of the bend were measured by U tube differential manometers (Fig. 2).

The pressure drop of a horizontal bend $\triangle p_b$ was determined as follows:

$$\triangle p_b = \triangle P - \triangle P_1 - \triangle P_2 \qquad (1)$$

The pressure drop of a bend with a horizontal inlet and a vertical down outlet or a bend with a vertical down inlet and a horizontal outlet was determined according to following expression:

$$\triangle p_b = \triangle P - \triangle P_1 - \triangle P_2 - \gamma \cdot H \qquad (2)$$

9

TABLE I : a=f(R/d)

R/d	3	4	5	6	7
a	0	0.7	1.3	1.8	2

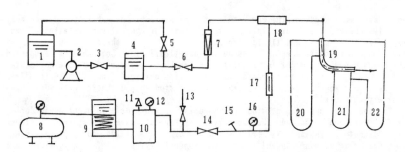

1 - water tank 2 - pump 3, 6, 14 - regulating valves
4 - stabilizer 5, 13 - by-pass valve 7 - variable area
flow meter 8 - air compressor 9 - cooler 10 - air tank
11 - safety valve 12, 16 - pressure gauges 15 - thermometer
17 - orifice 18 - mixer 19 - test bend 20, 21, 22 - U tube
differential manometers

Fig.1   Test loop

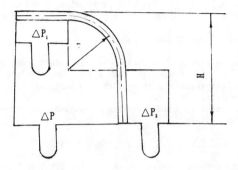

Fig. 2 The arrangement of differential pressure manometer

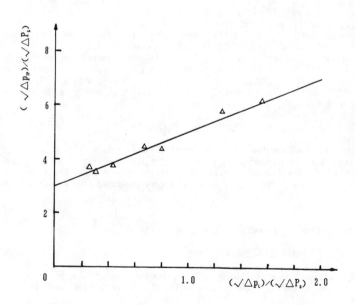

Fig. 3 Experimental data of a 90° bend with R/d=3.08
(p=2.5~3 bar)

where $\gamma_\bullet$ — the specific weight of the air-water mixture.
           H — the height of the vertical tube section as shown in Fig. 2.
    $\triangle p, \triangle p_1, \triangle p_2$— measured differential pressures of differential manometers as shown in Fig. 2.

Four kinds of 90° bends and two kinds of 180° bends were tested. For 90° bends, the ratios of bend radius to pipe inner diameter R/d were, $R/d=80/26=3.08$; $R/d=160/26=6.15$; $R/d=175/51=3.43$ and $R/d=276/51=5.41$. For 180° bends, $R/d=170/26=6.54$, $R/d=265/51=5.20$.

For 90° bend with $R/d=3.08$, tests of a bend with vertical down inlet and horizontal outlet or a bend with vertical up inlet and horizontal outlet were also conducted.

3. EXPERIMENTAL DATA TREATMENT AND COMPARISON

Reported conrrelations for two-phase flow were mainly from two flow models, the homogeneous flow model and the separated flow model. However, in reality, on account of the complexity of the gas-liquid two-phase flow in a bend, the flow in a bend is neither a homogeneous nor a separated one.

For a wholly separated flow, under certain assumptions, the pressure drop of a bend can be expressed as follows [4],

$$(\sqrt{\triangle P_{tr}})/(\sqrt{\triangle P_6})=(\sqrt{\triangle P_t})/(\sqrt{\triangle P_6})+1 \qquad (3)$$

where $\triangle p_{tr}$ — the pressure drop across a bend for gas-liquid two-phase flow;
        $\triangle p_t$ — the pressure drop across a bend for gas phase flow alone;
        $\triangle p_6$ — the pressure drop across a bend for liquid phase flow alone.

For a wholly homogeneous flow the expression of the pressure drop of a bend is [4],

$$\triangle p_{tr}/\triangle P_6=1+x(\gamma_t/\gamma_6-1) \qquad (4)$$

where $\triangle p_6$— the pressure drop across a bend assuming total flow to be liquid;
        x— mass quality of gas phase;
    γ_t, γ_6— specific weight of liquid phase and gas phase.

Experimental data were first correlated in the form of Martinelli parameters ($\sqrt{\triangle p_{tr}}/(\sqrt{\triangle P_6})$ and $(\sqrt{\triangle p_t})/(\sqrt{\triangle P_6})$) to check whether the flow is a wholly separated one. The parameter $(\sqrt{\triangle p_{tr}})/(\sqrt{\triangle P_6})$ was plotted against $(\sqrt{\triangle p_t})/(\sqrt{\triangle P_6})$ at different R/d. One of them ($R/d=3.08$) is shown in Fig. 3.

Experiments show $(\sqrt{\triangle p_{tr}})/(\sqrt{\triangle P_6})$ varies approximately linearly with $(\sqrt{\triangle p_t})/(\sqrt{\triangle P_6})$, and in the tested range the following relationship can be obtained for the 90° bend with $R/d=3.08$,

$$(\sqrt{\triangle P_{tr}})/(\sqrt{\triangle P_6})=2(\sqrt{\triangle P_t})/(\sqrt{\triangle P_6})+3 \qquad (5)$$

This expresses that the two-phase flow in the bend is not a wholly separated one.

Equation (4) was recommended to predict the pressure drop of a two-phase flow across a bend [1] [2]. The following correlation obtained from steam-water data was also suggested to predict the pressure drop of a gas-liquid two-phase flow across a bend [3],

$$\triangle p_{tr}/\triangle P_6=1+(\gamma_t/\gamma_6-1)(Bx(1-x)+x^2) \qquad (6)$$

where B— coefficient calculated from equation(7)

$$B=1+2.2/(K_\phi(2+R/d)) \qquad (7)$$

where K_ϕ— pressure drop coefficient of single phase flow.

For the sake of making comparison, equation (5) can be derived and rearranged as follows,

$$(\sqrt{\triangle p_{tp}})/(\sqrt{\triangle P_0})=2(1-x)+3x(\sqrt{\gamma_l})/(\sqrt{\gamma_g}) \qquad (8)$$

In Fig.4, correlation (8) is plotted and compared with correlations (3),(4) and (6) as well as with current experimental data of horizontal 90° bend with $R/d=3.08$.

As can be seen from Fig.4, the current experimental results fall within the range of the listed predictive equations. Among these equations, the proposed predictive correlation appears to agree best with the experimental data.

4. DISCUSSIONS

Fig.4 indicates that the two-phase flow in a bend is neither a homogeneous one nor a separated one. Equation (6) may agree with Sekoda's air-water data in the small quality range $x < 0.04$ [3] but deos not fit well for the current data in the quality range $x=0.04\sim0.20$.

Equation (8) only fit for the range from $0.04\sim0.20$ and for a bend around $R/d=3$ within the tested region.

For the sake of obtaining a more general correlation, more experimental studies and research work have to be done in the region of $x=0.2\sim1.0$ and higher pressure.

The current experiments also indicates that the R/d is a influential factor of the two-phase flow pressure drop of a bend. Fig.5 expresses the experimental results for bends with different R/d.

Fig.5 expresses the largeer the R/d, the smaller the value of $(\sqrt{\triangle p_{tp}})/(\sqrt{\triangle P_0})$. For R/d within $3\sim7$, the predicting correlation (8) can be expressed as follows:

$$(\sqrt{\triangle p_{tp}})/(\sqrt{\triangle P_0})=2(1-x)+3x(\sqrt{\gamma_l})/(\sqrt{\gamma_g})-a \qquad (9)$$

where   a — coefficient of R/d, can be obtained from table I.

The influence of plane of bend on the two-phase flow pressure drop has alse been studies for 90° bends with $R/d=3.08$. Experiments express that a horizontal bend, a bend with a vertical down inlet and a horizontal outlet or a bend with vertical up inlet and a horizontal outlet give essentially the same pressure drop (Fig.6).

The influence of working pressure p has not been researched in detail and this will be done in further studies. Generally speaking, $\triangle p_{tp}/\triangle P_0$ decreases with the increase of pressure.

The experimental data of two 180° bends ($R/d=170/26=6.54$ and $R/d=265/51=5.20$) showed similar results. For the same mass quality and R/d, the $(\sqrt{\triangle p_{tp}})/(\sqrt{\triangle P_0})$ of a 180° bend is smaller than that of a 90° bend. within the experimental range of this paper, the difference of $(\sqrt{\triangle p_{tp}})/(\sqrt{\triangle P_0})$ is about 4.

The comparison of the experimental data between a 180° bend with $R/d=5.20$ and a 90° bend with $R/d=5.41$ is shown in Fig.7.

5. CONCLUSIONS

(1) The present experiments express that the gas-liquid two-phase flow across a bend is a complex flow problem. For the sake of obtaining a general correlation for predicting the two-phase flow pressure drop of a bend, more studies have to be conducted, especially work in the range of high mass quality $x=0.2\sim1.0$ and higher pressure.

(2) Within the tested range of this paper equation (9) is recommended for predicting the pressure drop of a 90° bend with $R/d=3\sim7$.

(3) Experiments shows the value of $(\sqrt{\triangle p_{tp}})/(\sqrt{\triangle P_0})$ of a 180° bend is smaller than that of a 90° bend with the same R/d. The difference between them is about 4.

(4) $(\sqrt{\triangle p_{tp}})/(\sqrt{\triangle P_0})$ values of a horizontal bend, a bend with a vertical down inlet and a horizontal outlet or a bend with a vertical up inlet and a horizontal outlet are essentially the same, therefor equation (9) can be used for all of these bends.

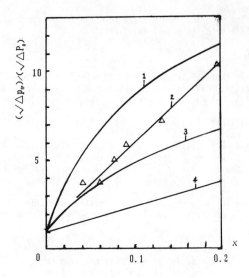

1 - equation (6)   2 - equation (8)   3 - equation (4)   4 - equation (3)

Fig. 4   Comparison of Experimental Data with various correlatios
(R/d=3.08, horizontal 90° bend

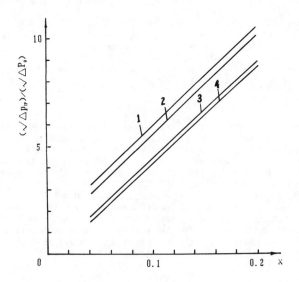

1 - R/d=3.08   2 - R/d=3.43   3 - R/d=5.41   4 - R/d=6.15

Fig.5 Experimental rusults for bends with differen R/d

14

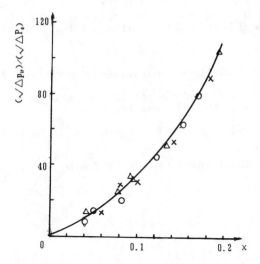

△ horizontal bend; × bend with a vertical up inlet and a horizontal outlet; ○ bend with a vertical down inlet and a horizontal outlet.

Fig.6   Experimantal results for 90° bends with different inlets and outlets (R/d=3.08)

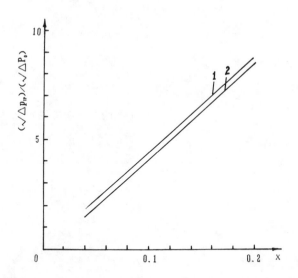

1 - a 90° bend with R/d=5.41   2 - a 180° bend with R/d=5.20

Fig.7   The comparison of the expeimental data between a 180° bend and a 90° bend

Acknowledgements

The authors of this paper would like to express their thanks to the National Natural Science Fundation of China for the financial support of this research.

REFERENCES

1. The Hydraulic Calculation of Boiler (Standard Method), Energy Press, USSR, 1978.

2. The Hydraulic Calculation Method of Utility Boiler, JB/Z201-83, Approved by the Ministing of Machinery Industry of China, PRC, 1983.

3. D.Chisholm, Two-phase pressure Drop in Bends, Int. J. Multiphase Flow, Vol. 6. p363~367, 1980.

4. Z.H,Lin, Gas-Liquid Flows and Boiling Heat Transfer, Xi' an Jiaotong University Press, 1987.

Particle Trajectory Modeling of Mixing in Confined Turbulent Two-Phase Flows

C. P. CHEN
Department of Mechanical Engineering
The University of Alabama in Huntsville
Huntsville, Alabama 35899, USA

A. A. ADENIJI-FASHOLA
Fluid Dynamics Branch
NASA—Marshall Space Flight Center
Huntsville, Alabama 35812, USA

Abstract

In this study, the approach of modeling the particulate phase flow by calculating particle trajectories in a fluid continuum is adopted. A SIMPLE-type algorithm is used in calculating the fluid flow field and equations for the turbulent kinetic energy and its dissipation rate are solved along with the continuity and momentum equations in order to characterize the turbulent velocity and length scales in the fluid phase.

The very important phenomena of particle dispersion by the fluid turbulence and the modulation of the turbulence by the particles are considered. In addition, the effect on the particle motion of the transverse force due to the interaction between the velocity shear existing in the fluid and the velocity slip between the particle and the fluid has been included. This is due to the fact that the usual assumption of the operation of the slip and shear Reynolds numbers within the Stokes regime is not always satisfied for many practical gas–particle flows.

1. INTRODUCTION

Turbulent fluid-particle flows are encountered in numerous technological applications in the chemical process industry as well as in various energy-related industries. The presence of a dispersed phase made up of small, light particles in some turbulent flows further complicates the already complex situation encountered in single-phase turbulent flows. However, the need to optimize the performance of the system components in which turbulent fluid-particle flows take place makes it necessary to continue the quest for a deeper understanding of the fundamental problems involved. Besides, the various interacting complex phenomena encountered in the modeling of this class of flows provide them with an intrinsic challenge to the fluid flow researcher.

The two common approaches adopted for the modeling of two-phase flows are the homogeneous and the separated flow models. The former is applicable to situations in which the mean slip between the phases is negligible and for which the design parameters of interest are of the bulk variety such as the pressure drop or mass fluxes. In situations where more detailed information about the intra- or inter-phase behavior is of interest, or where there is substantial segregation of the phases, the separated two-phase models are invariably preferred. For such flows, an additional decision has to be made with regard to the scheme for the description of the dispersed phase - whether to adopt an Eulerian or a Lagrangian approach. Important considerations necessary for making such a decision include the concentration of the dispersed phase which influences the mean separation distance between particles. The relative magnitude of this length scale as well as the particle size and the length microscale of the underlying turbulence in the continuous phase serve as a guide as to whether the dispersed phase can be treated as a continuum and thus described using the Eulerian approach or whether a Lagrangian description of the dispersed phase will be more appropriate.

In the present study, we present a discussion of turbulent fluid-particle flow modeling in which the continuous phase is described using the continuum Eulerian approach while a Lagrangian description is adopted for the dispersed phase which is assumed to be dilute. We restrict ourselves to confined flows and thus include a discussion of the treatment of solid boundaries within the Eulerian-Lagrangian framework. Also, the details of the single-particle hydrodynamics as they influence the two-phase flow simulation obtained are presented.

2.   FORMULATION

In this section, we present the governing equations for the Eulerian-Lagrangian description of the turbulent gas-particle flows. A Lagrangian description of the particulate phase is considered more appropriate if the mean separation between particles is much greater than the particle size but much less than the Kolmogorov microscale of the continuous phase turbulence, as discussed by Chen and Wood [1]. The assumptions regarding particulate phase concentration and other conditions specific to the turbulent gas-particle flow situation being investigated in the present study are presented in the next section.

The turbulent flow equations are derived in the usual fashion by considering the velocity vector components of the flow to be composed of a mean and a fluctuating component and employing time averaging to the continuous phase. The alternative approach of defining the continuous phase velocity vectors as comprising convective and diffusive components was adopted by Smoot and Smith [2], and Fletcher [3] and, more recently, by Hwang and Chang [4] and Nagano and Kim [5].

<u>Governing Equations</u>

In the following, for the sake of brevity, only the final, time-averaged equations containing up to the second-order terms are presented. The governing conservation equations for the continuous phase mean turbulent motion obtained by applying the Reynolds decomposition and time averaging of the instantaneous continuity and momentum equations are written [6,7,8] as follows:

$$\frac{\partial}{\partial x_j} \left(\rho U_j \right) = 0, \tag{1}$$

$$\frac{\partial}{\partial x_j} \left(\rho U_i U_j \right) = - \frac{\partial}{\partial x_i} P + \frac{\partial}{\partial x_j} \left(\mu \frac{\partial}{\partial x_j} U_i \right)$$

$$- \frac{\partial}{\partial x_j} \left(\rho \overline{u_i' u_j'} \right) + \overline{F}_{pi} \tag{2}$$

where

$$u_i = U_i + u'_i . \tag{3}$$

19

The term F_{pi} on the RHS of equation (2) is the momentum interface interaction term. Closure for the Reynold stresses in the gas phase is achieved using the k-ε turbulence model. The transport equations for k, the turbulence kinetic energy and ε, its dissipation rate are written as follows:

$$\frac{\partial}{\partial x_j}\left(\rho U_j k\right) - \frac{\partial}{\partial x_j}\left(\frac{\mu_t}{\sigma_k}\frac{\partial}{\partial x_j}k\right) + \rho G_k - \rho\epsilon + \rho\overline{u_j F_{pj}}, \tag{4}$$

and

$$\frac{\partial}{\partial x_j}\left(\rho U_j \epsilon\right) - \frac{\partial}{\partial x_j}\left(\frac{\mu_t}{\sigma_\epsilon}\frac{\partial}{\partial x_j}\epsilon\right) + \rho\frac{\epsilon}{k}\left[C_1 G_k - C_2\epsilon\right]$$
$$+2\mu_t\,\overline{\frac{\partial u_i'}{\partial x_j}\frac{\partial F_{pi}'}{\partial x_j}} \tag{5}$$

where, in the axi-symmetric co-ordinate system, ρG_k, the turbulence kinetic energy production, is given by

$$\rho G_k = \mu_t\left[2\left\{\left(\frac{\partial}{\partial x}U_x\right)^2 + \left(\frac{\partial}{\partial r}U_r\right)^2 + \left(\frac{U_r}{r}\right)^2\right\} + \left(\frac{\partial}{\partial x}U_r + \frac{\partial}{\partial r}U_x\right)^2\right] \tag{6}$$

The last term on the RHS of each of equations (4) and (5) are extra terms specific to two-phase turbulent flows and responsible for the modulation effect of the particulate phase on the continuous phase turbulence. The derivation of these forms of the k-ε equations of turbulence can be found in Shuen et al. [9] while specific modeling of the terms has been done by Chen and Wood [10].

In order to describe the dispersed phase, the trajectories of a large number of particles are tracked and the source terms which account for the fluid-particle interaction are computed for each computational cell traversed by each particle, following the particle-source-in (PSI) cell scheme of Crowe et al. [11]. The number of particles tracked must be sufficient to ensure a statistically stationary solution for the overall particulate flow field.

The starting point for the particle equation of motion is the Basset-Boussinesq-Oseen (B-B-O) equation

$$m_p \frac{dv_i}{dt} = 3\pi \mu d_p \left(u_i - v_i\right) + \frac{\rho_f}{\rho_s} m_p \frac{du_i}{dt} + \frac{1}{2} \frac{\rho_f}{\rho_s} m_p \left(\frac{du_i}{dt} - \frac{dv_i}{dt}\right)$$
$$\text{(I)} \qquad\qquad \text{(II)} \qquad\qquad \text{(III)} \qquad\qquad\qquad \text{(IV)}$$

$$+ \frac{3}{2} d_p^2 \sqrt{\pi \rho_f \mu} \int_{t_o}^{t} \left[\left(\frac{du_i}{d\tau} - \frac{dv_i}{d\tau}\right) \Big/ \sqrt{(t-\tau)}\right] d\tau + F_e \qquad\qquad (7)$$
$$\text{(V)} \qquad\qquad\qquad\qquad\qquad \text{(VI)}$$

where the terms are

I   -   Particle Inertia
II   -   Stokes Viscous Resistance
III   -   Pressure Gradient from Fluid Acceleration
IV   -   Virtual Mass
V   -   Basset History
VI   -   External Potential Force

Corssin and Lumley [12] resolved some inconsistencies in the B-B-O equation as presented by Tchen [13] by emphasizing the role of the pressure gradient of the undisturbed flow in contributing also to the net fluid force experienced by the particle. Maxey and Riley [14] later re-examined the forces due to non-uniform flow by providing a rational derivation of the equation of motion of a small sphere having a relative motion of low Reynolds number. Both the slip Reynolds number, $d_p |u_i - v_i|/\nu$ and the shear Reynolds number $(d_p^2/\nu)U/L$, where U is the mean fluid flow velocity and L is a characteristic dimension of the flow domain, are assumed to satisfy the conditions

$$[\, d_p |u_i - v_i|/\nu \,] \ll 1 \qquad \text{and} \qquad [\,(d_p^2/\nu)U/L\,] \ll 1$$

For such flows in which the particle inertia term in equation (7) is not negligible, the condition $\rho_s \gg \rho_f$, if satisfied, renders terms III, IV and V negligible. This assumption has been made by most investigators that implement particle trajectory schemes and will be made here.

The other important assumption that is usually made in the implementation of particle tracking schemes is that the slip and shear Reynolds numbers are within the Stokes regime. This assumption is not always justified. The details of the derivation of the transverse force on the particle that results from the interaction of the slip between the particle and the fluid and the shear in the mean fluid flow have been presented by Saffman [15]. The slip-shear side force of Saffman is given by

$$F_L = 81.2 \mu v_{rel} d_p^2 \, (\xi/\nu)^{1/2} \qquad\qquad (8)$$

where v_{rel} is the particle velocity relative to the local undisturbed velocity of the suspending fluid and ξ is the magnitude of the velocity gradient.

The effect of the slip-shear transverse force of Saffman is included in the present study. The aerodynamic effect of the departure of the slip Reynolds number from the Stokes regime is also included through a modification of the empirical Stokes drag relation to account for the case $[d_p \, | \, u_i - v_i \, | \, /\nu] > 1$. This later action of accounting for the viscous drag outside of the Stokes regime has also been implemented in previous work such as those of Durst et al. [16], Mostafa and Mongia [17] and Shuen et al. [9,18]. However, the former effect of the Saffman lift force has, to the knowledge of the present authors, always been neglected.

The simplified form of the B-B-O equation in which the particle inertia, viscous drag and buoyancy terms are retained and for which the assumption $\rho_s/\rho_f \gg 1$ is made has been used in the present study and is given in the simplified form as

$$\frac{dv_i}{dt} = \frac{(u_i - v_i)}{\tau} + g.$$

(9)

where each of u_i and v_i is made up of a mean and a fluctuating component. The particle relaxation time, τ, is related to the particle aerodynamic response time, t^*, and is given by

$$\tau = t^* / f$$

(10)

where f is a drag correction factor which accounts for the relative motion between the particle and the gas being within or outside of the Stokes regime as follows:

$$f = (C_D Re_p/24) = \begin{cases} 1 & \text{for } Re_p = |U_i - V_i| \, d_p/\nu \le 1 \\ 1 + 0.15 Re_p^{0.687} & \text{for } Re_p > 1 \end{cases}$$

(11)

The particle aerodynamic response time, t^*, is defined as

$$t^* = \rho_s d_p^2 / 18\mu .$$

(12)

Fluid-Particle Interface Interaction

The interface interaction between the phases is accounted for, in general, by obtaining the mass, momentum and energy source terms due to a particle for each computational cell visited, as the particle traverses the fluid continuum. These additional source terms due to the fluid-particle interaction are introduced into the continuous phase equations during the next super-iteration cycle. A sufficiently smooth particle source spatial distribution is assured through the use of a large number of computational particles which are representative of groups of particles having similar attributes of starting location, size, velocity and temperature. In the present investigation, the fluid and particle temperatures are assumed to be equal and unchanging. Hence, the fluid phase and particle energy equations are not included in the formulation.

The momentum source term has its origin in the drag force exerted on the particle by the fluid. Following Durst et al. [16], the particle momentum source term for an arbitrary control volume for particles of a single size group, α, and starting location, β, is given by Adeniji-Fashola et al. [19] as

$$F_{pi}^{\alpha,\beta} = (\dot{m}_p^{\alpha,\beta}/\tau) \int_{t_{in}}^{t_{out}} (u_i - v_i^{\alpha,\beta}) \, dt \tag{13}$$

where $\dot{m}_p^{\alpha,\beta}$ is the mass flow rate of particles of size group α and from starting location β; t_{in} and t_{out} are the times when the particle enters and leaves the control volume respectively. It is to be noted that

$$\dot{m}_p^{\alpha,\beta} = \dot{m}_p f_\alpha f_\beta = m_\alpha \dot{N}^{\alpha,\beta} \tag{14}$$

where $\dot{N}_p^{\alpha,\beta}$ is the particle number flow rate corresponding to $\dot{m}_p^{\alpha,\beta}$ and is given by

$$\dot{N}_p^{\alpha,\beta} = (6\dot{m}_p/\pi\rho_s d_p^3) \, f_\alpha f_\beta \tag{15}$$

and $\dot{m}_p$ is the particulate phase mass flow rate.

The total momentum source due to all the particles that traverse the control volume is finally obtained as a summation over all such particles

$$F_{pi}^{cell} = \sum_\alpha^{cell} \sum_\beta^{cell} F_{pi}^{\alpha,\beta} \tag{16}$$

23

The initiating solution of the momentum conservation equations for the gas phase is obtained for zero interface source terms. Subsequent super-iterations on these equations are later carried out with the appropriate source terms included and source term under-relaxation factors introduced as needed.

Particle-Turbulence Interaction

In turbulent gas-solid flows, the continuous and dispersed phases interact with each other at both the mean and fluctuational levels. At the fluctuational level, the solid particles experience a dispersion due to the action of the turbulence field while the turbulence field in turn experiences a modulation effect due to the particles. This latter effect of turbulence modulation has been reported even at relatively low loading levels of solid particles [20,21]. The turbulence modulation effect is manifested as extra dissipation terms in the transport equations for the turbulent kinetic energy, k and its dissipation rate, ϵ, as given in equations (4) and (5) respectively.

The hierarchy of turbulence closure models for single-phase flows range from simple mixing-length type zero-equation models to second order mean Reynold stress models in which transport equations are written for each component of the turbulent flux vectors and stress tensors. Extending the single-phase turbulence models to also account adequately for turbulence-particle interactions has been reported in the recent literature [22,23]. The present modeling approach is based on the two-equation k-ϵ turbulence model. The turbulent stresses of equation (2) are modeled using the Boussinesq gradient diffusion equation

$$u_i'u_j' = - \nu_t [\ 1/2\ (\partial U_i/\partial x_j + \partial U_j/\partial x_i)\ ] + 2/3\ k\delta_{ij} \tag{17}$$

where the eddy viscosity of the carrier phase is expressed as

$$\nu_t = C_\mu k^2 / \epsilon \tag{18}$$

In order to account for dispersion effects, a predetermined number of representative computational particles are tracked through a continuous succession of turbulent eddies superimposed upon the mean flow of the continuous phase as shown schematically in Figure 1.

Theoretically, knowledge of the full time history of the turbulent flow, obtained by direct simulation is required. However, since this is as yet unrealistic for most flow situations encountered in practice, the turbulence is simulated by means of a stochastic process. The mean values of the velocity components are determined from equations (1) and (2). The turbulence information required to evaluate the fluctuational component of the fluid velocity is obtained from the turbulence kinetic energy field obtained from the most recent super-iteration on the fluid phase equations. The approach adopted in the present study follows after that outlined by Gosman and Ioannides [24] and Shuen et al. [9,18]. It is assumed that the turbulence field is isotropic and that the fluctuational component of the fluid velocity is given by a Gaussian distribution whose variance is given by

$$\sigma_{ii}^{2} = 2k/3 \tag{19}$$

The fluctuational component of the velocity at any required location is then obtained by randomly sampling the distribution. The assumptions of isotropic turbulence and Gaussian velocity distribution above may not be realistic for most flow situations which demonstrate considerable anisotropy. These assumptions do not, however, cloud the particulate phase spatial distribution characteristics which represent the major interest of the present study.

A particle interacts with different eddies along its trajectory, depending on the relative magnitudes of the time macroscale (eddy lifetime) distribution of the turbulence field and the transit time of the particle within each eddy. If the characteristic size of an eddy is assumed to be its dissipation length scale, l_e where

$$l_e = C_{\mu}^{3/4} k^{3/2} / \epsilon , \tag{20}$$

the eddy lifetime is then estimated as

$$t_e = (3/2)^{1/2} \beta C_{\mu} k / \epsilon . \tag{21}$$

Kallio and Stock [25] have deduced from experimental measurements that β, in equation (21), assumes values between 0.15 and 2.00. Faeth and co-workers [9,18,26] on the other hand used the definition

$$t_e = l_e / (2k/3)^{1/2} , \tag{22}$$

which corresponds to a value of $\beta = 1.8257$ in equation (21). Gosman and Ioannides [24] used $t_e = l_e / |u'|$ where u' is the randomly sampled velocity fluctuation value from the assumed distribution.

An estimate for the transit time of a particle within an eddy is obtained from the solution of a linearized form of the equation of motion of the particle as

$$t_{tr} = -\tau \ln [1.0 - l_e/\tau | u_i - v_i |] \qquad (23)$$

The interaction time, t_{int}, between a particle and an eddy is taken as

$$t_{int} = \min [t_{tr}, t_e] \qquad (24)$$

as suggested by Gosman and Ioannides [24].

3. NUMERICAL SCHEME

The set of governing partial differential equations subject to the appropriate boundary conditions is solved for the fluid phase using the finite-volume SIMPLE algorithm of Patankar [27]. A staggered grid scheme in which the momentum vector cells are displaced from the scalar cells, as outlined in Patankar [27] is adopted. Following Durst et al. [16], the above scheme for solving the fluid phase equations is coupled with a particle-tracking scheme for the dispersed phase.

First, the "clean" fluid flow field is obtained by solving the continuous phase governing equations for zero particulate phase source terms. Next, particle trajectories are computed for a predetermined number of representative particles such that a statistically stationary solution is obtained for the overall particle flow field. The particle trajectories are obtained from the solution of the non-linear ordinary differential equations of motion subject to the currently existing continuous fluid flow field. A fourth order Runge-Kutta algorithm is used for this purpose. During the calculation of a particle's trajectory, the sources of momentum, kinetic energy and its dissipation rate, are accumulated for each computational cell of the flow domain. These source terms are then used in the next super-iteration of the continuous phase flow field until convergence is attained.

In the present study, a variable integration time step scheme is used. An upper bound on the time step through any computational cell is imposed based on the estimated particle residence time for that cell and with the particle being constrained to undergo a predetermined minimum number of integration steps, usually four, within the cell. This constraint eliminates the possibility of particles overshooting one or more cells due, for example, to a sudden reduction in cell dimensions, and the consequent unevenness in the particle source distribution and, possibly, divergence of the super-iterations that could result from such overshoots.

In order to ensure the continuity of the source term fields and hence a globally converged solution for the stochastic simulation procedure under discussion, a minimum of 10,000 computational particles, uniformly distributed in physical space at the inlet plane of the computational domain are used. This is in contrast to the deterministic scheme of Durst et al. [16] in which a maximum of 140 computational particles were introduced at 14 grid nodes.

It is also pertinent to point out that for the particle trajectory calculations, the fluid properties used are those linearly interpolated to the particle's current location using the particle's four nearest neighbors. This results in second order accuracy [28] and is to be contrasted with the more common approach of assuming the fluid properties to be uniform over the cell within which the particle is currently located.

Boundary Conditions

The definition of a unique fluid flow problem is completed through the specification of the boundary conditions after the governing differential equations have been outlined and the appropriate closure of these equations effected. The specific fluid flow problem investigated is the gas-particle upflow within a vertical cylindrical pipe. This simple geometry has been selected as most appropriate for the study of particulate phase distribution in a fluid-particle flow since it does not experience the significant flow complexities that are a necessary feature of flows in complex geometries. The experimental work of Tsuji et al. [29] and the computational work of Durst et al. [16] serve as the benchmarks for the evaluation of the performance of the present formulation.

Inlet Plane:
The specification of the inlet plane boundary conditions (b.c's) is very important as this determines the subsequent evolution of the flow as has been stressed by Sturgess et al. [30] and Westphal and Johnston [31]. The ideal specification for the flow variables at the inlet are the measured values. A complete set of measured values at the inlet is, however, seldom available. In the absence of the desired detailed experimental information, even uniform profiles are commonly specified by fluid flow researchers for the axial velocity at the inlet plane. The 1/7th power law turbulent velocity profile is used in the present study. The turbulent kinetic energy is assumed to be a percentage (usually between 1 and 5 percent) of the inlet flow mean kinetic energy. The kinetic energy dissipation rate at the inlet is the obtained from

$$\epsilon = C_\mu^{3/4} k^{3/2} / l_d \qquad (25)$$

27

where l_d, the dissipation length scale, is specified as a percentage of the characteristic length at the inlet.

Exit Plane:
    At the exit plane, the usual boundary condition imposed for any flow variable, ϕ, is $\partial\phi/\partial n = 0$ where n is the normal to the exit plane. This condition is generally valid if the extent of the computational domain in the primary flow direction is sufficient to ensure fully-developed flow conditions at the exit plane. Particle trajectory computations are discontinued for a computational particle once the particle exits from the computational domain through an open boundary. The computational extent for the present investigation is 60 diameters.

Solid Boundary:
    For the continuous phase, the "wall function" approach of Launder and Spalding [32] is used to avoid the excessively fine grids required to resolve the low Reynolds number viscous phenomena very near the solid boundaries. However, the conventional equilibrium log-law profiles for the mean velocity and turbulent kinetic energy have to be modified to account for the pressure of a suspended dispersed phase [33]. A modified set of wall functions is derived based on the Monin-Obukhov similarity theory for the analogous stably-stratified atmospheric boundary layer, summarized as follows:
the value of the mean velocity of the continuous phase parallel to the solid boundary at the first grid point away from the wall and located above the viscous sublayer, for which $y^+ \approx 50$, is described by

$$(U/u^*) = (1/K) [\ln y^+] + \beta (\rho_p/\rho_f) Ri_f \qquad (26)$$

with $\beta = 5$ [34].

    During the motion of computational particles through the flow domain particles that reach the wall, in general, either adhere to it as observed in erosion problems [35] or collide with the wall and get "reflected" back into the flow domain. In this contribution, the particles are assumed to make perfectly elastic collisions with the wall. The details of the effect of the high level of shear in the vicinity of the wall on the particle trajectories are presented in the next section.

Symmetry axis:
    For the fluid phase Eulerian equations, the condition $\partial\phi/\partial n = 0$, where n is the normal to the symmetry axis, is applied. The velocity component normal to the symmetry axis is, however, made to vanish at the axis.

28

A particle that attempts to cross the symmetry axis is "reflected" back into the computational domain in order to ensure continuity of the particulate phase. The reflection boundary condition is, however, very similar to the perfect slip b.c. applied at a solid wall. The effect of an alternative scheme for handling particles attempting to cross the symmetry axis are also discussed in detail in the next section.

4. RESULTS AND DISCUSSION

The particle trajectory scheme outlined above was applied to the prediction of the particle-laden upflow of air in a vertical cylinder reported by Tsuji et al. [29]. A pipe of 30 mm inner diameter was used for the experimental runs and the particles ranged in size between 200 μm and 3 mm. The test particles used by Tsuji et al. and which are of interest in the present study are polystyrene spheres having a density of 1020 kg/m^3. The mean air velocity used in the experiments ranged between 8 and 20 m/s and the loading ratio w defined as the particle-to-air mass-flow-rate ratio was up to 5. Asymmetric structure due to gravity, which is inevitable in a horizontal pipe, was avoided by selecting a vertical flow geometry.

The results presented in the following are preliminary. However, the clarification which they provide regarding the performance of the particle trajectory scheme in the prediction of such perticle-laden flows, especially with regard to the appropriate boundary conditions for the particulate phase, are very encouraging. Figure 2 shows the radial profile of the axial velocity between the air and the particulate phase for the experimental data of Tsuji et al. as well as the predicted values using the particle trajectory formulation. The mean particle size and loading ratio are 200 μm and 2.1 respectively. It is seen that the experimental velocity profiles are predicted fairly accurately in the central 60 percent of the pipe. However, there is considerable underprediction closer to the wall.

The modulation effect of the particulate phase on the continuous phase turbulence is presented in Figure 3. It is seen that while the predicted modulation effect on the turbulence intensity is of the right order of magnitude, the trend of the radial profiles of the experimental data and predicted results do not coincide. This may be a reflection on the turbulence modulation model [10] used which was based on correct order of magnitude bounds for the turbulence modulation. Also the assumption of isotropic turbulence, especially in the high wall shear region, is considered to be partly responsible for the mismatch between the predictions and the experimental data.

An aspect of the prediction of confined turbulent fluid-particle flows which has received little attention in the literature is that of the particulate phase distribution

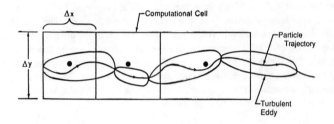

Fig. 1. - Particle Trajectory Through a Succession of Turbulent Eddies and
Computational Cells

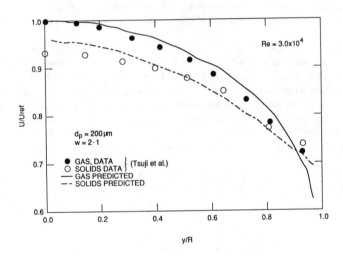

Fig. 2. - Radial Profiles of Axial Velocity in Vertical Pipe Upflow at x = 58.9D

within the flow domain. In the present study, the particulate phase is uniformly distributed at the inlet plane of the pipe to exhibit a constant particulate phase mass flux in the radial direction at this location. However, a considerable particle redistribution takes place by the time the exit plane of the pipe is reached as shown in the radial profile of the normalized mass flux of Figure 4. Such a redistribution of the particulate phase was not observed in the deterministic model of Durst et al. [16] or indeed in the present study when the stochastic option flag was turned off. It soon became apparent that the turbulent dispersion effect, coupled with the "reflection" boundary condition at the symmetry axis combined to yield the observed redistribution in the predicted mass flux radial profiles.

The prediction of Figure 4 was obtained without the inclusion of the Saffman "lift" force. The much more uniform mass flux radial profile of Figure 5 was obtained with the Saffman lift force included in determining the turbulence influenced motion of the particles. Considerable particle aggregation is, however, still observed to be present in the vicinity of the pipe symmetry axis. This, we believe to be as a result of some inadequacy in the specification of the symmetry boundary condition for the particle trajectories. In order to test this hypothesis, a tentative boundary condition in which any particle that reaches the symmetry axis was moved back into the bulk flow was implemented. The resulting mass flux radial profiles are shown in Figure 6. In Figure 6a, particles that reached the symmetry axis were moved to a location near the R/2 point where R is the pipe radius while in Figure 6b, the particles were relocated to a position near the pipe wall. The particle aggregation near the symmetry axis is seen to be considerably reduced in both profiles of Figure 6 while showing little sensitivity for the actual relocation point of the particles. This observation indicates that the nature of the boundary condition employed for particle trajectories in axi-symmetric flow simulations needs further clarification.

In order to obtain the very smooth profiles for the particulate phase variables presented above, 25,000 particle trajectories were computed for each experimental flow data. The use of 10,000 particles were, however, observed to provide sufficiently smooth profiles during the preliminary phase of the investigation. A typical 25,000-particle run on an IBM 3084 took about 1100 minutes while a similar run on the Cray XMP-416 required about 225 minutes. The corresponding times for typical 10,000-particle runs are 440 minutes on the IBM 3084 and 90 minutes on the Cray XMP-416 respectively. Typically five super iterations were required for each of the runs in order to achieve source term convergence.

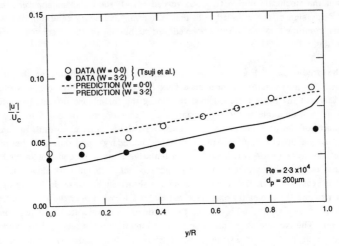

Fig. 3. - Modulation Effect of Particles on the Turbulence in the Continuous
at x = 58.9D

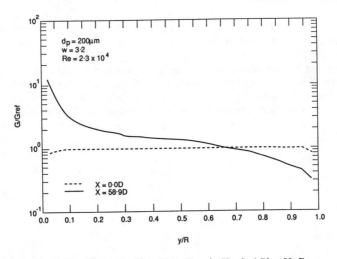

Fig. 4. - Radial Profiles of Particle Mass Flux in Vertical Pipe Upflow

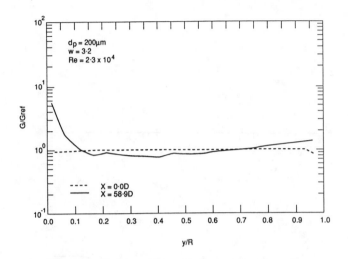

Fig. 5. - Radial Profiles of Particle Mass Flux in Vertical Pipe Upflow -
Saffman Lift Included

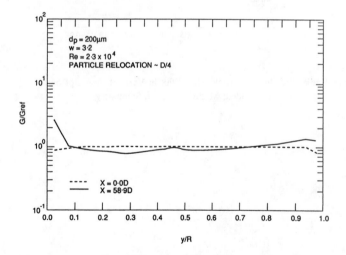

Fig. 6a. - Radial Profiles of Particle Mass Flux in Vertical Pipe Upflow -
Saffman Lift Included; Modified b.c. on Symmetry Axis.

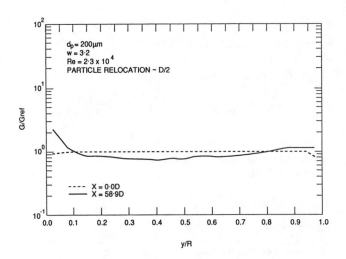

Fig. 6b. - Radial Profiles of Particle Mass Flux in Vertical Pipe Upflow -
Saffman Lift Included; Modified b.c. on Symmetry Axis.

NOMENCLATURE

C_D	drag coefficient
C_μ, C_1, C_2	turbulent model constants
D	pipe diamater
d_p	particle diameter
F_{pi}	interface interaction force
f_α	fraction of particles belonging to diameter group α
f_β	fraction of particles leaving from starting location β
G	particulate mass flux
g	gravitational acceleration
k	continuous phase turbulence kinetic energy
l_e	eddy size
$\dot{m}_p$	particle phase mass flow rate
m_α	mass of particle belonging to diameter group α
$\dot{N}_p$	particle number flow rate
P	pressure
r	radial direction
Re	flow Reynolds number
Re_p	particle slip Reynolds number
t	time
t^*	particle aerodynamic response time
t_e	eddy lifetime
u_i	fluid phase velocity components
v_i	particulate phase velocity components
w	particulate loading ratio
x	axial direction
x_i	general co-ordinate direction
δ_{ij}	Kronecker delta
ϵ	dissipation rate of turbulent kinetic energy
μ, ν	molecular and kinematic viscosity of fluid
ν_t	eddy viscosity of continuous phase
ρ_f	continuous phase density
ρ_p	particulate phase density $= \rho_s \theta$
ρ_s	particle material density
θ	volume fraction of particles
τ	particle relaxation time

REFERENCES

1. Chen, C. P. and Wood, P. E., "A Turbulence Closure Model for Dilute Gas Particle Flows," Can. J. Ch. E. 63, pp. 349-360, 1985.

2. Smoot, L. D. and Smith, P. J., Coal Combustion and Gasification, Plenum Press, New York, 1985.

3. Fletcher, T. H., "A Two-Dimensional Model for Coal Gasification and Combustion," Ph.D. Thesis, Brigham Young University, 1983.

4. Hwang, C. J. and Chang, G. C., "Numerical Study of Gas-Particle Flow in a Solid Rocket Nozzle," AIAA J., 26, pp.682-689, 1988.

5. Nagano, Y. and Kim, C., "A Two-Equation Model for Heat Transport in Wall Turbulent Shear Flows," J. Heat Transfer, 110, pp. 583-589, 1988.

6. Hinze, J. O., "Turbulent Fluid and Particle Interaction," Prog. Heat & Mass Transfer, 6, pp. 433-452, 1972.

7. Dukowicz, J. K., "A Paricle-Fluid Numerical Model for Liquid Sprays," J. Comp. Phys. 35, pp.229-253, 1980.

8. Drew, D. A., "Mathematical Modeling of Two-Phase Flows," Ann. Rev. Fluid Mech., 15, pp. 261-291, 1983.

9. Shuen, J-S., Solomon, A. S. P., Zhang, Q-F. and Faeth, G. M., "A Theoretical and Experimental Study of Turbulent Particle-Laden Jets," NASA CR 168293, 1983.

10. Chen, C. P. and Wood, P. E., "Turbulence Closure Modeling of the Dilute Gas- Particle Axisymmetric Jet," AIChE J. 32, pp. 163-166, 1986.

11. Crowe, C. T., Sharma, M. P. and Stock, D. E., "The Particle-Source-In Cell (PSI-Cell) Model for Gas-Droplet Flows," J. Fluid Engrg. 99, pp. 325-332, 1977.

12. Corrsin, S. and Lumley, J., "On the Equation of Motion for a Particle in Turbulent Fluid," App. Sci. Res. A, 6, p. 114, 1956.

13. Tchen, C. M., "Mean Value and Correlation Problems Connected With the Motion of Small Particles Suspended in a Turbulent Fluid," Ph.D. Thesis, Delft, 1947.

14. Maxey, M. R. and Riley, J. J., "Equation of Motion for a Small, Rigid Sphere in a Non-Uniform Flow," Phys. Fl., 26, p. 883, 1983.

15. Saffman, P. G., "The Lift on a Small Sphere in a Slow Shear Flow,", J. Fluid Mech., 22, pp. 385-400, 1965.

16. Durst, F., Milojevic, D. and Schonung, B., "Eulerian and Lagrangian

Predictions of Particulate Two-Phase Flows: A Numerical Study" Appl. Math. Model., 8, pp. 101-115, 1984.

17. Mostafa, A. A. and Mongia, H. C., "Eulerian and Lagrangian Predictions of Turbulent Evaporating Sprays," AIAA Paper 86-0452, 1986.

18. Shuen, J-S., Solomon, A. S. P., Zhang, Q-F. and Faeth, G. M., "Structure of Particle-Laden Jets: Measurements and Predictions," AIAA J., 23, pp. 396-404, 1985.

19. Adeniji-Fashola, A. A., Chen, C. P. and Schafer, C. F., "Numerical Predictions of Two-Phase Gas-Particle Flows Using Eulerian and Lagrangian Schemes," NASA Technical Paper, in preparation, 1988.

20. Al-Taweel, A. M. and Landau, J., "Turbulence Modulation in Two-Phase Jets," Int. J. Multiphase Flows, 3, p. 341, 1977.

21. Modarress, D., Tan, H. and Elghobashi, S., "Two-Component LDA Measurement in a Two-Phase Turbulent Jet," AIAA J. 22, pp. 624-630, 1984.

22. Chen, C. P. and Wood, P. E., "Turbulence Closure Modeling of Two-Phase Flows," Chem. Eng. Commun. 29, pp. 291-310, 1984.

23. Pourahmadi, F. and Humphrey, J. A. C., "Modeling Solid-Fluid Turbulent Flows With Application to Predicting Erosive Wear," Physicochem. Hydro., 4, p. 191, 1983.

24. Gosman, A. D. and Ioannides, E., "Aspects of Computer Simulation of Liquid Fueled Combustors," AIAA Paper 81-0323, 1981.

25. Kallio, G. A. and Stock, D. E., "Turbulent Particle Dispersion: A Comparison Between Lagrangian and Eulerian Model Approaches," Gas-Solid Flows, ASME FED Vol. 35, pp. 23-34, 1986.

26. Faeth, G. M., "Recent Advances in Modeling Particle Transport Properties and Dispersion in Turbulent Flow," Proc. ASME-JSME Therm. Engrg. Conf. 2, ASME, p. 517, 1983.

27. Patankar, S. V., Numerical Heat Transfer and Fluid Flow, Hemisphere Pub. Corp./McGraw-Hill Book Co., 1980.

28. Sirignano, W. A., "Fuel Droplet Vaporization and Spray Combustion Theory," Prog. Energy. Comb. Sci., 9, p. 291, 1983.

29. Tsuji, Y., Morikawa, Y. and Shiomi, H., "LDV Measurements of an Air-Solid Two-Phase Flow in a Vertical Pipe," J. Fl. Mech., 139, pp. 417-434, 1984.

30. Sturgess, G. J., Syed, S. A. and McManus, K. R., "Importance of Inlet Boundary Conditions for Numerical Simulation of Combustor Flows," AIAA Paper 83-1263, 1983.

31. Westphal, R. V. and Johnston, J. P., "Effect of Initial Conditions on Turbulent Reattachment Downstream of a Backward-Facing Step," AIAA J., 22, p. 1727, 1984.

32. Launder, B. E. and Spalding, D. B., "The Numerical Computation of Turbulent Flows," Comp. Meth. in Appl. Mech. and Eng., 3, pp. 269-289, 1974.

33. Kramer, T. J. and Depew, C. A., "Experimentally Determined Mean Flow Characteristics of Gas-Solid Suspensions," J. Basic Eng., pp. 492-499, June, 1972.

34. Arya, S. P. S., "Parametric Relationships for the Atmospheric Boundary Layer," Boundary Layer Meteorology, 30, pp. 57-73, 1984.

35. Dosanjh, S. and Humphrey, J. A. C., "The Influence of Turbulence on Erosion by a Particle-Laden Fluid Jet," LBL - 17247, Lawrence Berkeley Laboratory, U. Cal., 1984.

Gas-Liquid Two-Phase Flow in Dividing Horizontal Tubes

L. LIGHTSTONE and JEN-SHIH CHANG
Department of Engineering Physics
McMaster University
Hamilton, Ontario
Canada L8S 4M1

Abstract

The time averaged void fraction and pressure drop behaviour of horizontal air-water two-phase flows is studied experimentally and numerically for 2 cm inner diameter tubes with various flow dividing junctions at its end. The time average void and pressure drop behaviour along the channel is simulated using a two fluid separated flow model. The results show that two-phase behaviour (flow regime, void fraction and pressure drop) is strongly affected by the presence of a flow division in the system. These effects extend far upstream of the junction for low momentum flows and far downstream for high momentum flows. Both the numerical and experimental results show that there occurs a large increase in void just downstream of the junction owing to the halving of the fluid volume flow rates and the liquid deceleration. In general, across the junction, there is an increase in voidage because of the flow division. Furthermore, the presence of junction may produce either pressure recovery or significant pressure loss at the flow dividing site with pressure loss always occurring for the symmetric Tee.

1   INTRODUCTION

Two-phase flow is becoming increasingly important in many areas of engineering design such as gas distribution lines, heat exchanger analysis, and power plant heat transport systems. While complete analytical formulations of the equations describing these flows have been developed their complexity makes them exceptionally cumbersome. As a result many researchers have spent their time producing "constitutive laws" describing two-phase systems. However, it still remains for many of the above-mentioned constitutive laws to be tested in the more complex situations such as the accelerating conditions which arise in a flow dividing system. This work examines a particular two-phase system and attempts an numerical and experimental analysis to successfully describe the phenomena observed in the flow dividing and non-dividing configurations.

Single-phase flows in dividing junctions have been studied very early on in the engineering literature [1,2]. The standard approach is to consider the pressure drop in terms of a Bernoulli type equation

$$P_1 + \rho \frac{u_1^2}{2} = P_2 + \rho \frac{u_2^2}{2} + \Omega\rho \frac{u_1^2}{2} \qquad (1.1)$$

L. Lightstone present address is at Atlantis Scientific, Ottawa, Ontario, Canada K2C 0P9.

where P and u are the fluid pressure and speed, and 1 and 2 correspond to upstream and downstream of the "flow perturbation". Here, the irreversible work done on the fluid is represented by a loss coefficient Ω which multiplies a kinetic energy term. These loss coefficients have been well tabulated.

The single-phase problem of inviscid irrotational junction flow was considered by Modi et al. [2] and analytical solutions were arrived at using the method of conformal mapping. Recently computer studies of three-dimensional single-phase turbulent and laminar flows in flow dividing Tee junctions have been conducted by Spedding [3,4,5].

When considering a two-phase flow in a dividing junction two additional complications must be accounted for which do not arise in the single-phase case. They are:

a)   change in void fraction near junction site;
b)   change in flow regime structure near junction site.

Very few studies have been done on horizontal two-phase dividing flow. These studies [6,7,8,9,10,11,12,19,20,21] have not considered any effect of flow regime change nor have they examined void fraction behaviour but have confined themselves to the study of pressure drop and flow quality division.

A flow dividing junction may be viewed, in a one-dimensional sense, as a sudden expansion because of the increase in area available for the flow. In single-phase analysis an expression for the loss coefficient of a sudden expansion was derived by considering a momentum balance on a control volume surrounding the junction and assuming that the pressure before and after the expansion was derived by considering a momentum balance on a control volume surrounding the junction and assuming that the pressure before and after the expansion acted on the area downstream of the junction. This was experimentally shown to yield the correct value for the loss coefficient [13].

$$\Omega = \left(1 - \frac{A_1}{A_2}\right)^2 \tag{1.2}$$

The two-phase analogy of the above was postulated by Romie [14] and is given below

$$A_1 P_1 + \rho_g Q_{g_1} u_{g_1} + \rho_\ell Q_{\ell_1} U_{\ell_1} = A_2 P_2 + \rho_g Q_{g_2} u_{g_2} + \rho_\ell Q_{\ell_2} U_{\ell_2} \tag{1.3}$$

The "Romie" equation was experimentally tested by Lottes [14] and shown to provide a reasonably good method for pressure drop evaluation.

To construct the equations describing the time averaged two-phase dividing flow, the two fluids are assumed to be stratified. This assumption is considered valid as discussed previously. Before and after the junction it is assumed that there is little change in the void fraction. In these regions a simple two fluid separated flow model is applied. At the junction it is assumed that there is a significant change in void fraction. The momentum equations are constructed by considering control volumes surrounding each fluid at the junction. The irreversible losses which occur at the junction location are considered in two parts:

a)   Losses due to a sudden expansion.
b)   Losses due to the turning of the fluid.

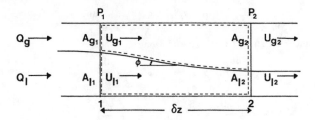

Fig. 1   Control volumes at junction site.

Loss (a) is incorporated into the equations through Romie's formulation. The modelling of loss (b) follows the single-phase analysis of incorporating a loss factor (Ω) multiplying a kinetic energy term. It is assumed that $\Omega = \Omega(\theta)$ where θ is the junction angle and that $\Omega = 0$ for $\theta = 0$. In other words, if the system is a true sudden expansion there should be no "fluid turning" losses. In addition to the above loss terms, a special momentum exchange term is included at the junction to account for enhanced fluid-fluid interactions there. The flow regime structure and fluid friction terms are taken into accou
nt through the use of the constitutive laws which are developed for the straight pipe system.

2.   MOMENTUM ANALYSIS OF TWO-PHASE DIVIDING FLOW

2.1   Momentum Model for Two-Phase Dividing Flow

This section is concerned with the modelling of the pressure drop and void fraction throughout the flow dividing system. The flows are considered incompressible and time averaged and thus may be described as both stratified and time independent. Before and after the junction, it is assumed that there is little change in void fraction. The flows in these regions may therefore be described by Newton's law for each phase j.

$$\rho_j \, u_j \, \frac{du_j}{dz} = -\left(\frac{dP}{dz}\right)_j + F_j \tag{2.1}$$

Here a one-dimensional analysis is considered and the steady state assumption has been applied.

For the liquid body forces consist of the wall and interfacial friction, and gravity forces. In the gas phase gravity forces are neglected and only the wall and interfacial friction is accounted for. These friction terms contain the flow regime information. It is also assumed, as with the majority of the analysis done in this text, that the pressure drops in each phase are equal.

Writing the equations of motion for each phase in full

$$\rho_g u_g \frac{du_g}{dz} = -\frac{dP}{dz} - \frac{\tau_g S_g}{A_g} - \frac{\tau_i S_i}{A_g} \tag{2.2}$$

41

$$\rho_\ell u_\ell \frac{du_\ell}{dz} = -\frac{dP}{dz} - \frac{\tau_\ell S_\ell}{A_\ell} - \frac{\tau_i S_i}{A_\ell} - \rho_\ell g \frac{dh_\ell}{dz} \tag{2.3}$$

Subtracting equation 2.2 from 2.3 yields the following

$$\rho_\ell u_\ell \frac{du_\ell}{dz} - \rho_g u_g \frac{du_g}{dz} = \Psi - \rho_\ell g \frac{dh_\ell}{dz} \tag{2.4}$$

where

$$\Psi = -\frac{\tau_\ell S_\ell}{A_\ell} + \frac{\tau_g S_g}{A_g} + \tau_i S_i \left(\frac{1}{A_g} + \frac{1}{A_\ell} \right) \tag{2.5}$$

The flow dividing junction is considered in a one-dimensional sense. To develop the momentum equations at the junction site consider two control volumes around each fluid as shown by the dotted lines in Figure 1. The junction is assumed to expand, in a direction into the page, to twice the area of the upstream pipe.

Conservation of linear-momentum over a control volume CV bounded by a control surface CS may be written as

$$\Sigma \text{ Forces} = \frac{d(m\underline{u})}{dt} = \frac{\partial}{\partial t} \int_{cv} \rho \underline{u} dV + \int_{cs} \rho \underline{u} (\underline{u} \cdot d\underline{A}) \tag{2.6}$$

For our analysis, applying the one-dimensional steady state assumptions, equation 2.6 may be rewritten for a given phase j

$$A_{1_j} P_{1_j} e - A_{2_j} P_{2_j} + g \left(h_{1_j} A_{1_j} - h_{2_j} \frac{A_{2_j}}{2} \right) - f_j \delta z = \rho_j Q_j (u_{2_j} - u_{1_j}) \tag{2.7}$$

where f represents the friction terms

$$f = \tau S \Big|_w \pm \tau S \Big|_i \cos\phi \tag{2.8}$$

with

$$\tau S = \frac{1}{\delta z} \int_{cs} \tau(z) S(z) \, dz \simeq \tau S \Big|_2 \tag{2.9}$$

The "w" corresponds to wall friction and the "i" to interfacial friction. The $\pm$ depends on the phase under consideration. Furthermore it is assumed that the ϕ is small so that $\cos\phi = 1.0$. Note also that the gravitation term at location 2 is divided by 2. This assumes that the "hydrostatic" forces upstream of the junction act only on the downstream projected area ($A_1/2$) and that the remainder of the "hydrostatic" force exerted downstream of the junction is "supported" by the wall of the sudden expansion.

In this analysis it is assumed that the irreversible losses (with the exception of friction) occur solely at the junction site. These losses are considered in two parts:

a)   sudden expansion losses;
b)   turning losses.

In addition to the above, there is also the possibility of enhanced momentum exchange at the junction because of the fluid-fluid interaction there. This will be accounted for by considering a turning-momentum-exchange term.

Neglecting gravitational effects on the gas phase and accounting for the sudden expansion loss by assuming that P_1 acts on A_2, the momentum equations at the junction become

42

$$A_{2_g} P_1 - A_{2_g} P_2 - (\tau_{g2} S_{g2} + \tau_{i2} S_{i2}) \delta z = \rho_g Q_g (u_{g2} - u_{g1}) + \Omega_g(\theta) + \Omega(\theta) \tag{2.10}$$

$$A_{2\ell} P_1 - A_{2\ell} P_2 - (\tau_{\ell2} S_{\ell2} + \tau_{i2} S_{i2}) \delta z + \rho_\ell g \left(h_1 A_{\ell1} - h_2 \frac{A_{\ell2}}{2} \right)$$
$$= P_\ell Q_\ell (u_{\ell2} - u_{\ell1}) + \Omega_\ell(\theta) - \Omega'(\theta) \tag{2.11}$$

where $\Omega_1(\theta)$ and $\Omega_g(\theta)$ represent the losses due to the turning of the indicated fluid and $\Omega'(\theta)$ represents the turning-momentum-exchange between the fluids.

If the friction, gravity and "turning" terms in equations 2.10 and 2.11 are neglected, addition of these two equations will yield the expression developed by Romie[14] (equation 1.3) indicating that the losses due to sudden expansion have been properly accounted for.

To choose the form of the Ω's, we again extrapolate from single phase procedures

$$\Omega_j(\theta) = \Omega_0 \rho_j A_j \frac{u_{j1}^2}{2} \sin^a \theta \tag{2.12}$$

The constant Ω_0 is assumed to be the same for single-phase flow and is given a value of 1.55. The angular dependence has been included to fit the extreme conditions. The exponent, a, will be experimentally determined.

The remaining term in the junction equations, $\Omega'(\theta)$, represents the momentum exchange between the two fluids at the junction. The expression for this exchange term is chosen to be similar in structure to that of equation 2.12. However, the properties of both fluids should be accounted for and so the following is used

$$\Omega'(\theta) = \Omega_0' \rho_{2\phi} A \frac{(u_g - u_\ell)^2}{2} \sin^a \theta \tag{2.13}$$

where

$$\rho_{2\phi} = (A_g \rho_g + A_\ell \rho_\ell) \frac{1}{A} \tag{2.14}$$

In the above equation Ω'_0 is an experimentally determined constant. With the evaluation of the Blasius parameters from the linear two
-phase configuration, the flow dividing system is described to within two constants Ω'_0 and a.

The initial conditions, "far away" from the junction, are determined by assuming that the fluids do not "see" the junction. In other words, the acceleration and gravity terms are equal to zero. Thus, the initial speeds may be determined for given flow rates by setting ψ (equation 2.5) equal to zero.

With the inclusion of the equations of continuity

$$u_i A_i = Q_i \tag{2.15}$$

and experimental determination of the Blasius constants used in evaluation of the friction terms, the above set of equations is closed.

2.2   Numerical Evaluation of the Momentum Equations

The momentum equations listed above are evaluated using the finite difference iterative scheme shown in the flow chart of figure 2. The finite difference forms used are shown below. Before and after the junction the momentum equations may be written

43

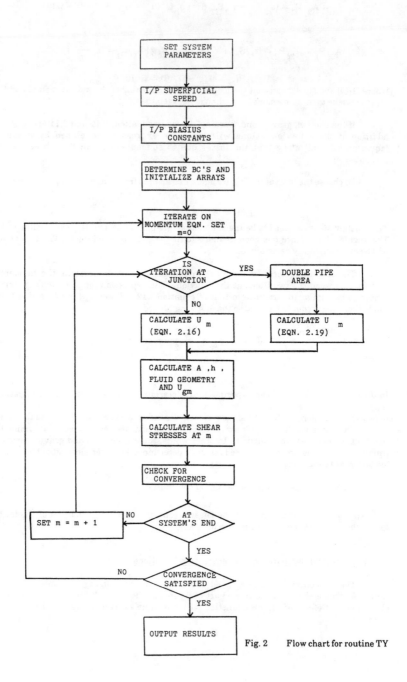

Fig. 2   Flow chart for routine TY

44

$$u_{\ell,m}^{n+1} = \left[\frac{\Delta z \Psi}{\rho_\ell} - g(h_m^n - h_{m-1}^{n+1}) + \frac{\rho_g}{\rho_\ell} u_{g,m}^n (u_{g,m}^n - u_{g,m-1}^{n+1}) \right] \frac{1}{u_{\ell,m}^n} + u_{\ell,m-1}^{n+1} \qquad (2.16)$$

$$\Psi = -\tau_{\ell,m}^n \frac{S_{\ell,m}^n}{A_{\ell,m}^n} + \tau_{g,m}^n \frac{S_{g,m}^n}{A_{g,m}^n} + \tau_{i,m}^n S_{i,m}^n \left(\frac{1}{A_{\ell,m}^n} + \frac{1}{A_{g,m}^n} \right) \qquad (2.17)$$

At the junction the momentum form used is

$$P_m^{n+1} = \left[A_{g,m}^n P_{m-1}^{n+1} - f_g \Delta z - \rho_g Q_g (u_{g,m}^n - u_{g,m-1}^{n+1}) - \Omega_g^n(\theta) - \Omega^{'n}(\theta) \right] \frac{1}{A_{g,m}^n} \qquad (2.18)$$

$$u_{\ell,m}^{n+1} = \left[A_{\ell,m}^n (P_{m-1}^{n+1} - P_m^n) - f_\ell \Delta z + \rho_\ell g \left(h_{m-1}^{n+1} A_{\ell,m-1}^{n+1} - \frac{h_m^n A_{\ell,m}^n}{2} \right) - \right.$$
$$\left. \Omega_\ell^n(\theta) + \Omega_\ell^{'n}(\theta) \right] \cdot \frac{1}{\rho_\ell Q_\ell} + u_{\ell,m-1}^{n+1} \qquad (2.19)$$

$$f_g = \tau_{g,m}^n S_{g,m}^n + \tau_{i,m}^n S_{i,m}^n \qquad (2.20)$$

$$f_\ell = \tau_{\ell,m}^n S_{\ell,m}^n - \tau_{i,m}^n S_{i,m}^n \qquad (2.21)$$

Here m represents the node and n represents the iteration value. The above equation are used in their respective locations along with the equations of continuity

$$A_{\ell,m}^{n+1} = \frac{Q_\ell}{A_{g,m}^{n+1}} \qquad (2.22)$$

$$u_{g,m}^{n+1} = \frac{Q_g}{u_{\ell,m}^{n+1}} \qquad (2.23)$$

The boundary conditions are determined as discussed in section 2.1. The variables are then initialized to their respective values "far" upstream and downstream of the junction. Equations 2.16 to 2.22 are then iterated on as shown in figure 2 until convergence on all three of u_1, u_g and h_1 is achieved. In the numerical evaluation the convergence criterion is set in the following form for the variable Λ.

$$\text{eps} \geq \left| 1 - \frac{\Lambda^n}{\Lambda^{n-1}} \right| \qquad (2.24)$$

Evaluation of these equations is performed by computer code TY. The results of this numerical evaluation are, of course, dependent on the Blasius coefficients used in equations 2.16, 2.19 and 2.20.

3. EXPERIMENTAL APPARATUS

A schematic of the top view of the experimental apparatus for the two-phase experiments is shown in figure 3. Water and compressed air are taken from the laboratory

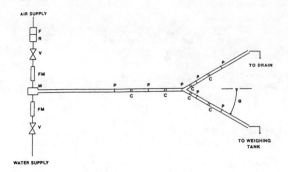

Fig. 3   Schematic of experiment apparatus for dividing two-phase flows.

C   Capacitance Transducer
F   Filter
FM   Flow Meter
M   Mixing Section
P   Pressure Tap
R   Pressure Regulator
V   Control Valve

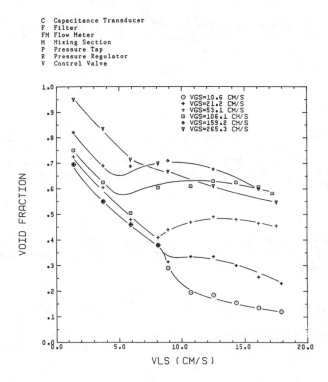

Fig. 4   Experimentally observed void fraction behaviour in the straight pipe two-phase system.

supply at approximately room temperature. The air is passed through a filter and a pressure regulator. Each fluid is passed through its own series of valves and flow meters which are used to monitor the flow. The two fluids then enter a horizontal pipe through a glass bead mixing section which is used to enhance the onset of fully developed two-phase flow.

The pipe itself is constructed of 2 cm diameter segmented glass rods connected together by means of brass joiners which also supply the pressure taps for the system. The Wyes are one-piece smooth-edged glass sections of the same inner diameter as the main pipe. The different Wyes are attached to the main pipe by means of the above mentioned brass joiners. The two ends of the flow dividing, the section are open to the atmosphere. A weighing tank is located at one of the Wye section ends for the purpose of flow quality measurement while the other branch is connected directly to a central drain.

Void fraction determination was through the use of four ring type capacitance transducers [16] positioned, as shown, on the outer surface of the tube. The ring electrodes are used in conjunction with a Booton 72-B capacitance meter. The void measurement system is calibrated in situ using stratified water conditions (since the capacitance ring configuration is nearly independent of flow regime [16] and the ultrasonic liquid level measurement system [17].

Differential pressure drop measurements were obtained using Validyne DP15 and DP105 pressure transducers, for the large and small pressure drops respectively, along with a Validyne CD-123 pressure demodulator. Differential pressure measurements are taken with respect to the pressure tap furthest upstream of the junction. The pressure sampling position is controlled through multiple input switching valves.

The void fraction and pressure waveforms were sampled "simultaneously" at a rate of 30 samples/second by a Nova III mini-computer. Time averaged values of these quantities were obtained through multiple averaging of waveforms of ten seconds duration. The important statistical parameters for these averages, standard deviation and maximum deviation, are obtained for each measurement. A sample waveform along with time averaged data for each measurement was recorded on magnetic tape for later access. The pressure and void fraction waveforms are also used for characterization of flow regimes [18].

4.   EXPERIMENTAL RESULTS

4.1   Straight Pipe Flow

Typical experimentally observed time averaged void fraction and pressure drop behaviour is shown in figures 4 and 5 respectively as a function of liquid superficial velocity for various gas superficial velocities. The flow regime transitions and behaviour are indicated clearly by the trends on these graphs as discussed below. To further illustrate the flow regime dependence of void fraction, figure 6 presents a contour map in which lines of constant void are shown on a graph whose axes are those used in the flow regime maps. The observed flow regime transition boundaries for the straight pipe system are superimposed on figure 6 and are indicated by the dotted lines. Comparison of this contour map with the flow regime map for the straight pipe system shows that the points of inflection of the contours correspond closely to the flow regime transition boundaries.

The general behaviour of void with flow structure may be seen by consideration of figure 4. The region $u_{\ell s} < 7.0$ cm/s constitutes the stratified flows. Here, the void fraction varies in a nearly linear fashion with liquid superficial velocity. For low gas superficial

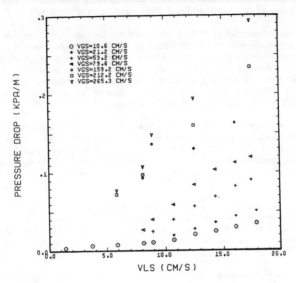

Fig. 5   Experimentally observed pressure drop behaviour in the straight pipe two-phase system.

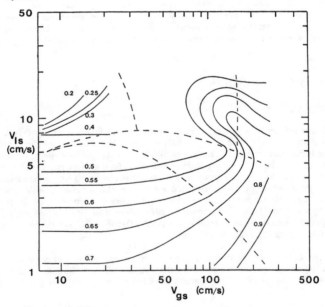

Fig. 6   Void fraction contour map for the straight pipe system.

48

velocities ($u_{gs} < 21.2$ cm/s) there is no void fraction dependence on gas flow rate. In other words, there is negligible interaction between the two phases. This corresponds to a stratified smooth flow.

With increasing gas flow rate, the void fraction also increases (owing to the larger pressure drop in the gas phase). At the same time the formation of ripple waves on the liquid surface was noted. The effect on the void fraction appears to be non-linear, with small gas flow rates having a negligible effect and the large gas flow rates have a comparatively large effect. This behaviour is contrary to the results obtained with the one-dimensional pressure equilibrium model where doubling the gas flow rate requires a doubling of the liquid flow rate to maintain a constant void fraction. This non-linear behaviour could be due to the appearance of waves and the different type of wave structure on the water surface. However, experimental and theoretical analysis in this report is insufficient to make a firm conclusion.

The transition from stratified to slug flow is clearly marked in both the pressure and void fraction curves by a sudden increase in the quantity of interest. This transition requires a gas superficial velocity of at least 50 cm/s.

Upon further increase of the gas flow rate, the time averaged pressure drop increases; however, its behaviour remains approximately linear with liquid flow rate. The large value of pressure drop observed from these flows is not due to the wall friction but results from the acceleration of the water, to a speed near that of the gas, at the time of slug formation.

Figures 4 and 5 show that the transition to intermittent flow occurs at lower liquid superficial velocities with increasing gas flow rate. It should also be noted that within the intermittent region, the void fraction hits a maximum (at $u_{1s} = 10$ cm/s, $u_{gs} = 159$ cm/s) and on further increase of the gas flow rate the void fraction decreases. This clearly marks the beginning of the transition to annular (or wavy-annular) flow. The mechanism which controls this behaviour is the aeration of the newly formed slug. With increasing gas flow rate the slug holds less water and more of the liquid phase remains in the water film which follows the slug. The final result of this slug aeration mechanism is a negligible effect on the void fraction at the point of transition from stratified to wavy annular flow, as shown in Figure 4 for $u_{gs} = 265$ cm/s. A plausible explanation for the observed behaviour lies in the nature of the flows. For transition between annular and stratified, the flow structure remains continuous rather than changing to an intermittent regime.

The transition to plug flow from stratified flow is characterized by a decrease in void fraction ($u_{gs} < 50$ cm/s) and a small increase in the slope of the pressure drop curve. The difference between the slug and plug flows is clearly dependent on the gas flow rate. The decrease in void arises in a plug flow due to the formation of a liquid bridge which necessarily fills part of the pipe. However, for large gas flow rates, the formation of a water bridge becomes a very effect method for "pushing" the water out of the pipe resulting in the observed increase in void.

From an analysis of experimental results, the constitutive laws arrived at, through application of the equilibrium flow model, are given in Table 1.

Since the results in Table 1 have been obtained for a single tube size no universality is suggested here. In fact, since there are four parameters in two-phase momentum equation which may be varied independently there exist many possible combinations which may yield the correct results for any one physical system. Hopefully, within each flow regime, there is only one set of these parameters which fully describes the general two-phase behaviour.

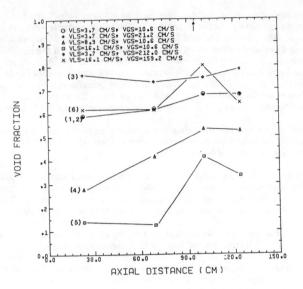

Fig. 7   Experimentally observed axial void fraction behaviour in a symmetric tee.

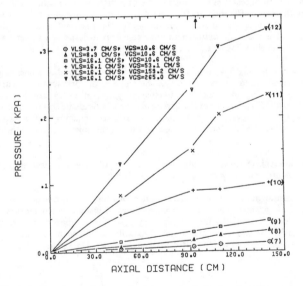

Fig. 8   Experimental observed axial pressure behaviour in a symmetric tee.

50

This section of this report was included to check the feasibility of applying a one-dimensional two fluid model to predict void fraction and pressure drop in straight pipe two-phase system. From the results obtained it is clear that the model has good potential for describing the flows away from the transition boundaries. Section 4.3 further tests the above constitutive laws through consideration of accelerating systems.

4.2   Symmetric Tee Flow Dividing System

Several experimentally observed results of void fraction ad pressure drop for various fluid superficial velocities (and hence flow regimes) are shown in figures 7 and 8 respectively for the symmetric Tee. Here, the junction location is indicated by an arrow in the upper portion of the graph. On these figures the lines connecting data points are not meant as an interpolation of the experimental results, but are included to help distinguish one data set from another. In figure 8 and in all of the following pressure curves, the absolute value of pressure with respect to the first pressure tap is shown hence, (0,0) is a data point.

The behaviour displayed in these graphs will be discussed below and then extended to a comparison between different flow dividing structures in the next section. In the following discussion references to flow regime indicate the structure well upstream of the junction unless otherwise specified.

For the low gas flow rate stratified flows (curves 1,2 and 7) there is an increase in void fraction as the fluids approach the junction. After the junction the void fraction is approximately constant. As observed before, for these low gas flow rates, variation in the gas flow rate has no effect on the void fraction of pressure drop. The increase in void fraction along the pipe upstream of the junction may be likened to an exit effect (due to the halving of the flow rates at the junction). This increase in voidage creates a corresponding increase in pressure drop due to the gravitational contribution.

For the stratified wavy flow (curve 3), there is an overall increase in voidage across the junction due to the halving of the flow rates. However, before the junction, there is a decrease in void. This is most probably due to an alteration in the gas velocity profile near the junction which allows the liquid level to rise.

Two instances of plug flow are shown in Fig. 7 each with $u_{gs} = 10.6$ cm/s and one with $u_{1s} = 8.9$ cm/s, the other with $u_{1s} = 16.1$ cm/s. In the former case (curve 4) the plug flow regime "collapses" into a stratified wavy structure, causing a large increase in voidage before the junction. However, after the junction, there are no flow regime alterations (other than dissipation of the surface waves) and the void fraction remains constant.

For the later case of $u_{1s} =$ cm/s (curve 5) the flow regime change occurs after the junction accounting, in part, for the large increase in void fraction observed there. Near, but just downstream of the junction, the void fraction is larger than "far" downstream of the junction because of the combined effect of fluid deceleration and the halving of the volume flow rates.

When the two fluids enter the junction they must decelerate and come to their new velocities. The low density gas phase decelerates quickly while the liquid moves through the junction with a velocity near that which it possessed just upstream of the flow division. This causes a large increase in voidage near the junction site. The above explanation for the flow behaviour near the junction is verified numerically in section 4.3. It should be noted that there is no observable pressure recovery at the junction for the above mentioned flows (curves 8 and 9), this effect being eliminated through the turbulent and gravitational losses.

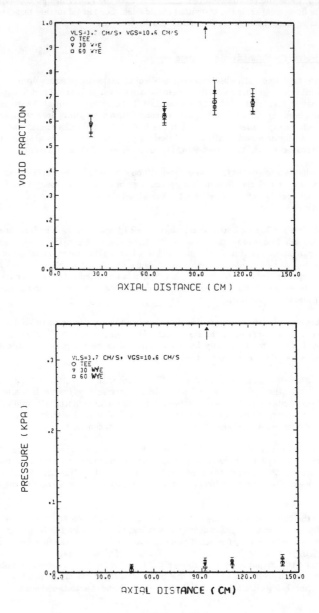

Fig. 9   Comparison between experimentally observed two-phase behaviour in various flow dividing junctions for a stratified flow.

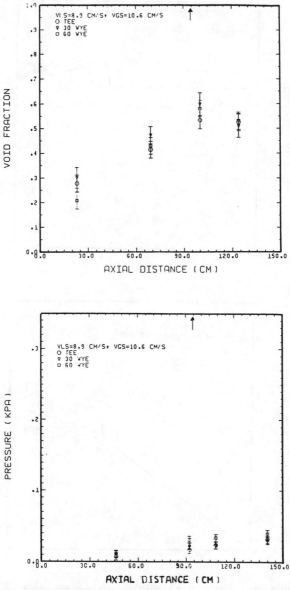

Fig. 10   Comparison between experimentally observed two-phase behaviour in various flow dividing junctions for a plug flow.

53

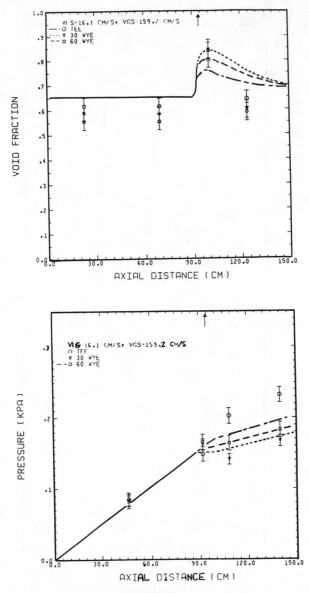

Fig. 11   Comparison between observed and predicted two-phase behaviour in various flow
dividing junctions for a slug flow.

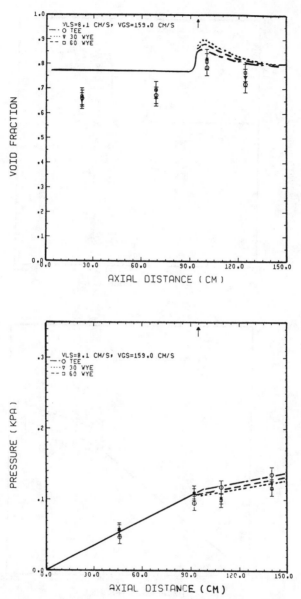

Fig. 12 Comparison between observed and predicted two-phase behaviour in various flow dividing junctions for a slug flow.

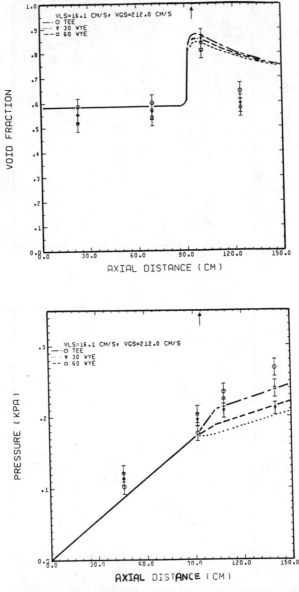

Fig. 13 Comparison between observed and predicted two-phase behaviour in various flow dividing junctions for a wavy annular flow.

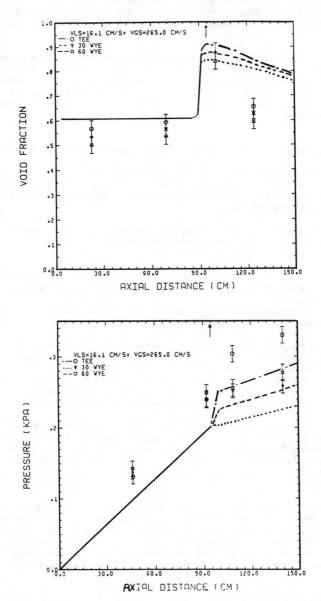

Fig. 14 Comparison between observed and predicted two-phase behaviour in various flow
dividing junctions for a wavy annular flow.

One case of slug flow is shown in Figure 7 (curve 6). Before the junction, there is no effect on void fraction. This corresponds to the limiting case where information (specifically about the flow division) is only transported in the direction of momentum flow and hence and the fluids do not "see" the junction until they arrive at the flow dividing site. A large increase followed by a decrease in void occurs along the system after the junction. The reasons for this behaviour, namely flow division and deceleration, are discussed above.

Axial pressure distribution is shown for two cases of slug flow (curves 10 and 11) and one case of wavy annular flow (curve 12). Before the junction the pressure drop is linear. At the junction site there is a large change in pressure, the pressure drop across the junction increasing with increasing gas flow rate. Note that for none of the conditions shown here is there any pressure recovery at the junction. Presumably this is due to the irreversible losses causes in the fluids by the Tee.

4.3   Comparison of Experiment with Theory for Various Flow Dividing Junctions

Figures 9 and 10 depict the time averaged two-phase behaviour for a stratified and a plug flow respectively. As can be seen from these figures, there is no junction angle dependence, to within error, on the quantities measured. Thus, the discussion of section 4.2 applies to these flow structures.

With respect to the above mentioned low momentum type flows no comparison with model predictions is possible as the numerical method used did not converge in these regions. Two reasons which may explain this problem are:

1)   Numerical method: A simple upwind finite difference scheme was used here whereas a higher order or more complicated scheme which lets information flow upstream may be necessary.

2)   One-dimensional time average model: The restrictions which this type of analysis imposes on the system may be severe enough to remove the model far from reality.

For high momentum flows (slug and wavy annular), a junction angle dependence is apparent as shown in figures 11 to   14. These figures compare the observed and predicted axial two-phase behaviour for different junction angles. In the model the two constants Ω' and a are given the value 0.03 and 5.0 respectively, and a convergence criterion of eps = 0.0001 is used. The large value of a indicates that the two-phase behaviour changes drastically with decreasing junction angle as shown by the pressure distribution curves.

For all of the above mentioned curves there is a significant increase in void fraction at the junction, the size of the increase varying with flow rates. For high flow rates the liquid speed upstream of the junction is large resulting in a large increase in void just downstream of the junction as discussed in section 4.2. The model predicts this general behaviour quite well. To within error no void fraction dependence on junction angle can be distinguished from the experimental data.

The general behaviour (both void and pressure) depicted in these figures shows that before and after the flow division the flows are nearly independent of the junction angle. Also note that in general the void fraction prediction downstream of the junction is consistently higher than that observed. The most likely explanation for this behaviour is that at these points the flows are in "transient" regimes and have a much higher momentum than they would possess in the linear system. Thus, the constitutive laws developed previously may not apply.

The pressure losses which occur at the junction can be quite large for the wavy annular type flows. In the regions strong junction angle dependence is apparent with significantly larger pressure drops occurring in the Tee junction. This greater pressure drop in the large angle junctions is principally due to the losses incurred by the fluid as it turns. Clearly the symmetric TEE, which requires a 90° turning of the fluid, will induce greater pressure losses in the system.

The two-phase system is capable of far greater losses at the junction than the equivalent single-phase system (by approximately one order of magnitude). This is due to the combined effects of periodically large liquid velocities, and change in void fraction and flow regime at the junction site. The inability of the model to consistently predict maximum pressure drop may be due to the lack of a flow regime energy loss term in the describing equations (that is a term which depicts the energy lost in forming a new flow regime). The structure of such a term cannot be extrapolated directly from single-phase studies as there is no single-phase equivalent. To determine the form of this loss factor a much greater number of experiments need to be performed.

While, for the data presented here, pressure recovery does not occur in the Tee, pressure recovery may occur in the smaller angle junctions. As is expected, greater pressure recovery is possible with decreasing junction angle. However, the model presented here never predicts pressure recovery regardless of the junction configuration. This inconsistency arises from the Romie type analysis which always predicts a pressure loss at the flow division.

5. CONCLUDING REMARKS

The experimental and theoretical investigations have been conducted to study the time averaged void fraction and pressure drop behaviour of horizontal air-water two-phase flow in dividing horizontal tubes. Following concluding remarks are obtained:

(1)   Two-phase behaviour (flow regime, void fraction and pressure drop) is strongly affected by the presence of a flow division in the system. These effects extend far upstream of the junction for low momentum flows and far downstream for high momentum flows.

(2)   A method for predicting time average pressure drop and void fraction in a two-phase flow, based on a two fluid separated flow model, has been proposed. The results and physical considerations show the method's validity.

(3)   Using a two fluid quasi-separated flow model the axial time averaged void fraction and pressure behaviour was predicted. Numerical and experimental results show that there occurs a large increase in void just downstream of the junction owing to the halving of the fluid volume rates and the liquid deceleration. In general, across the junction, there is an increase in voidage because of the flow division. Furthermore, the presence of junctions may produce either pressure recovery or significant pressure loss at the flow dividing site with pressure loss always occurring for the symmetric Tee.

NOMENCLATURE

A	fluid cross sectional area
D	pipe diameter
D_j	hydraulic diameter for fluid j
f	friction factor
g	acceleration due to gravity
h	fluid level
k	separation constant
P	pressure
Q	flow rate
R	pipe radius
Re	Reynolds number
S	perimeter of pipe "wetted" by fluid
u	fluid velocity
V	volume
W	mass flux
v	fluid velocity
X	Blasius exponent
(X,Y)	cartesian coordinates describing plane normal to the axis of the tube
(x,y)	non-dimensional cartesian coordinates describing plane normal to the axis of the tube
z	axial distance along the tube
ε	constant of integration
ε_g	void fraction (A_g/A)
θ	junction angle
λ	ratio of difference to summation of fluid viscosities
μ	viscosity
ρ	density
σ	surface tension
τ	shear stress
Ψ	equilibrium shear stress distribution
Ω	general single-phase loss coefficient
$\Omega(\theta)$	two-phase turning loss term
$\Omega'(\theta)$	two-phase turning-momentum-exchange term

Subscripts

d	deviation
g	gas phase
gs	gas superficial
h	height
i	interface
j	signifies fluid "j"
l	liquid phase
ls	liquid superficial
m	node value
s	slug
$\perp$	perpendicular
$\parallel$	parallel
$\circ$	indicates a constant
1	indicates upstream of flow perturbation or signifies fluid 1
2	indicates downstream of flow perturbation or signifies fluid 2
2ϕ	indicates a "mixed" two-phase quantity

Superscripts
*	indicates a "reduced" quantity
'	indicates a perturbed quantity
n	iteration value

Abbreviations
A	annular
ll	laminar liquid phase, laminar gas phase
tt	turbulent liquid phase, turbulent gas phase
lt	laminar liquid phase, turbulent gas phase
tl	turbulent liquid phase, laminar gas phase
PA	periodic annular
PACH	periodic annular churn
PL	plug
SL	slug
SS	stratified smooth
SW	stratified wavy

Acknowledgement

The authors express their thanks to S.I. Osamusali, R. Girard M. Shouki and S.T. Revankar for valuable discussions and comments. This work is supported by the Ministry of Universities and Colleges, the Ontario Government, Canada under, BILD grant and the Natural Sciences and Engineering Research Council of Canada.

Table 1.

Constants Used for Evaluation of Consitutive
Laws for Straight Pipe Two-Phase Flows

Flow Regime	C_ℓ	C_g	X_ℓ	X_g
ll Stratified	16.0	16.0	1.0	1.0
lt Stratified	16.0	0.046	1.0	0.2
tl Stratified	0.046	16.0	0.2	1.0
tt Stratified	0.046	0.046	0.2	0.2
Plug	0.06	20.0	0.2	0.8
Slug	0.06	100.0	0.2	0.8
Wavy Annular	0.10	60.0	0.2	0.8

where

$$f_\ell = C_\ell \left(\rho_\ell \frac{D_\ell u_\ell}{\mu_\ell} \right)^{-x_\ell}, \; f_g = C_g \left(\rho_g \frac{D_g u_g}{\mu_g} \right)^{-x_g}$$

61

REFERENCES

1. McNown, J.S., "Mechanics of Manifold Flow", Trans. ASCE, Vol. 119, 1954, pp. 1103-11143.

2. Modi, P.N., Ariel, P.D., Dondekar, M.M., "Conformal Mapping for Channel Junction Flow", Proc. ASCE, Vol. 107. 1981.

3. Pollard, A., Spalding, D.B., "The Prediction of the Three-Dimensional Turbulent Flow Field in a Flow-Splitting Tee-Junction", Comput. Meth. App. Mech. Eng., Vol. 13, 1978, pp. 294-306.

4. Pollard, A., Spalding, D.B., "On the Three-Dimensional Laminar Flow in a Tee-Junction", J. Heat Mass Transfer, Vol. 23, 1980, pp. 1605-1607.

5. Pollard, A., "Computer Modelling of Flow in Tee-Junctions", Physical Chemical Hydrodynamics, Vol. 2, 1981, pp. 203-227.

6. Tsuyama, M., Taga, M., "On the Flow of the Air-Water Mixture in the Branch Pipe", Bull. JSME, Vol. 2, 1959, pp. 151-156.

7. Fouda, A.E., Rhodes, E., "Two-Phase Annular Flow Stream Division in a Simple Tee", Trans. I. Chem. Engrs., Vol. 52, 1974, pp. 354-360.

8. Fouda, A.E., Rhodes, E., "Two-Phase Annular Flow Stream Division", Trans. I. Chem. Engrs., Vol. 50, 1972, pp. 353-363.

9. Henry, J.A.R., "Dividing Annular Flow in a Horizontal Tee", Int. J. Multiphase Flow, Vol. 7, 1981, pp. 343-355.

10. Honan, T.J., Lahey, R.T., "The Measurement of Phase Separation in Wyes and Tees", Nuc. Eng. Design, Vol. 64, 1981, pp. 93-102.

11. Azzopardi, B.J., Whalley, P.B., "The Effects of Flow Patterns on Two-Phase Flow in a Tee-Junction", Int. J. Multiphase Flow, Vol. 8, 1982, pp. 491-507.

12. McCreery, G.E., "A Correlation for Phase Separation in a Tee", 3rd Multiphase Symp., 1983, Miami.

13. Schutt, H.C., Trans. ASME, HYD 51, 1929, pp. 83-87.

14. Lottes, P.A., "Expansion Losses in Two-Phase Flow", Nuc. Sci. Eng., Vol. 9, 1961, pp. 26-31.

15. Lightstone, L., Osamusali, S.I. and Chang, J.S., "REGIME-3 Code for Prediction of Flow Regime Transition in a Two-Phase Manifold Flow", Proc. 13th Simulation Symposium, 1987.

16. Chang, J.S., Girard, R., Raman, R. and Tran, F.B.P., "Measurement of Void Fraction by Ring Type Capacitance Transducers" in Mass Flow Measurements - 1984", T.R. Hedrick and R.M. Reimer, Eds., ASME Press, New York, FED-Vol. 17, pp. 93-99, 1984.

17. Chang, J.A., Ichikawa, Y., Irons, G.A., "Flow Regime Characterization and Liquid Film Thickness Measurement in Horizontal Gas-Liquid Two-Phase Flow by an Ultrasonic Method", in "Measurements in Two-Phase Flow", T.R. Heidrick, Ed., ASME, 1982, pp. 7-12.

18. Lightstone, L. and Chang, J.S., "Flow Regime Observation in Horizontal Two-Phase Dividing Flow in Tees", and "Wyes", Proc. 33rd Cdn. Chem. Eng. Conf., Vol. 1, pp. 281-286. (1983).

19. Seeger, W., Reimann J. and Miller, V. "Two-phase flow in a T-junction with a Horizontal Inlet-Part I: Phase Separation" Int. J. Multi-Phase Flow, 12, 575-585, 1986.

20. Reimann, J. and Seeger, W., Two-Phase Flow in a T-junction with a Horizontal Inlet-Part II: Pressure Differences", Int. J. Multi-Phase Flow, 12. 587-608, 1986.

21. Ballyk, J.D., Shoukri, M., and Chan, A.M.C. "Steam-Water Annular Flow in a Horizontal Dividing T-Junction" Int. J. Multi-Phase Flow, 14. 265-285, 1988.

Two-Phase Flow in Curved Pipes

R. ULBRICH and S. WITCZAK
Heat Technique and Chemical Engineering Department
Opole Technical University
45-233 Opole uL. ZSP 5, Poland

Abstract

Based on the experimental investigations of hydrodynamics of two-phase gas-liquid flows in bends linking horizontal straight pipes, flow regimes,lengths of disturbance zone in linking pipes, and method of pressure drop calculations in curved pipes have been determined.

1. INTRODUCTION

The structure of compact tube apparatuses makes it necessary to use bends and elbows, located at different places, which affect the direction of the flow. However, it should be noted that the presence of the bends and elbows has a considerable, adverse effect on two-phase gas-liquid flow, which is not the case with one-phase flow, where the disturbance caused by them, as compared with tee junction or sudden contraction, is negligible. In the case of bends being placed between vertical pipes, we witness the change of flow direction from upward to downward, or the other way round which, thus, causes a change of two-phase flow regimes in the bend.

Similar effect occurs also in bends linking horizontal pipes. At the same time, bends and elbows, irrespective of their location, give rise to local disturbances affecting the hydrodynamics of two-phase gas-liquid flow both in the elements themselves and in the straight pipes connected to them.

When investigating two-phase flow in bends connecting vertical pipes, Oschinovo and Charles found differences in the observed flow structures of these elements , as compared with the structures in pipes, appearing as dry places on the inner or outer surface of the bend as well as local zones of liquid recirculation. Similar results were also obtained by Golan and Stenning [2], Chen and Yang [3], Usui, Aoki and Innoue [4] and Fuji et al. [5].

Much less information exist , on the other hand, on bends connecting horizontal pipes. A small number of papers [6,7] dealing with this problem, except for demonstrating similar effects, like those occuring in vertical bens, do not allow, however, quantitative determination of both the dimension of disturbance zones and the effect on the value of pressure drop.

In order to solve this problem, adequate investigations have been carried out by the authors of the present paper.

65

2.   EXPERIMENTAL TESTS

Description of work station

The diagram of the work station is shown in Fig.1. Test section 1 consisting of 4 horizontal channels (Fig.2) connected by means of bends is the main part of the installation. The whole constraction enabled the change of the angle of inclination of the flow plane to the level of -90° to $+90^{\circ}$. Air taken from the atmosphere was pumped by compressor 4 through cooler 6 to mixing chamber 5, to which water was also pumped by centrifugal pump 3. The measurement of the volume flux of liquid and water was effected by rotameters R_1- R_{5} respectively.
Both phases flowed from the mixing chamber to test section and then to tank 2 which was also used as a separator. To determine volume flux and physical properties of the air and water accurately, sensors testing tempetrature, pressure and humidity were installed.

Scope of investigations

The investigation comparised recognition of two-phase flow regimes and pressure drop values in a curved and straight pipes, the length of disturbance zone of flow in straight pipes before and behind the bend was determined. In order to estimate the effect of the bend location on the hydrodynamics of two-phase flow, investigations were carried out for five flow cases through the bend:
- vertical downward (-90°)   H -90
- vertical upward ($+90^{\circ}$)   H +90
- horizontal (0°)   H 0
- at the angle of 45° upwards   H +45
- at the angle of 45° downwards H -45.

In order to obtain in horizontal pipes all the typical flow regimes of two-phase flow (Fig.3) for each five positions of the test section, invstigations were carried out concerning mass flux of both phases : G_G = (0.001 - 4.38) 10^{-3} kg/s   and G_L =(8.3 - 278) 10^{-3} kg/s, which made it possible to obtain the following parameter values characteristic of two-phase flow :

$$x = 0 - 0.35$$
$$g_T = 15 - 575 \text{ kg/(m}^2 \text{ s)}$$
$$u_{Gs} = 0.017 - 7.45 \text{ m/s}$$
$$u_{Ls} = 0.017 - 0.566 \text{ m/s}$$

3.   INVESTIGATION RESULTS AND ANALYSIS

On comparing the obtained two-phase flow regimes with the flow patterns typical in horizontal pipe (Fig.3) on the modified Baker flow regime maps [9], it was found that curved pipe greatly affects the rise of various flow structures, which are usually different from those observed in a straight   pipe (Fig.4).
It is doubtless connected with the simultaneous effect of a

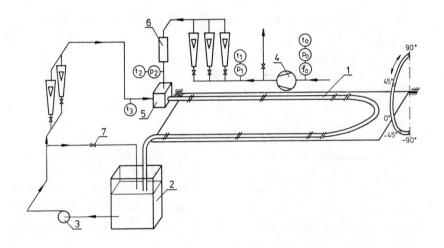

Fig. 1. Scheme of installation
1-test section, 2-tank, 3-pump
4-compressor, 5-mixing chamber
6-cooler, 7-bypass

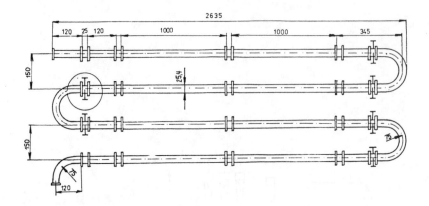

Fig. 2. Test section

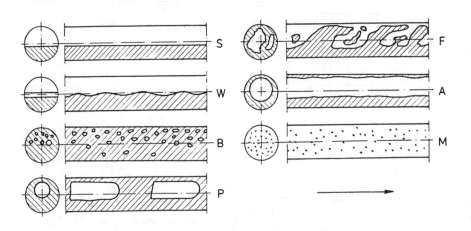

Fig.3. Flow regimes for two-phase gas-liquid flow
in horizontal pipes, by [8]:
B-bubble   P-plug   S-stratified
W-wave   F-froth   A-annular
M-dispersed-mist

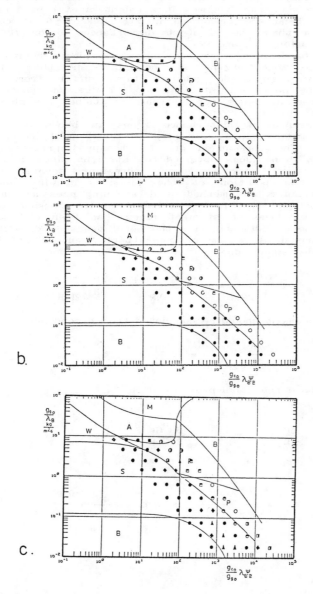

Fig.4. Comparison of observed flow regimes and flow regime maps

$\bullet$ - S $\blacklozenge$ - W $\circ$ - P $\odot$ - F $\square$ - B-P $\boxdot$ - P-F

$\blacksquare$ - F-A $\ominus$ - S-P $\boxslash$ - S-F $\blacktriangle$ - W-F ∇ - W-A

centrifugal force and the force of gravity on particular phases.
This gives rise, as shown on Fig.5 and Fig.6, to the change of
flow regime in straight pipe before the bend and its restructe-
ring in the straight pipe behind the bend. The measured relative
disturbance zones of flow regimes in straight pipes caused by the
effect of the bend are presented in Fig.7 and Fig.8. It can be
seen therefrom that the greatest disturbances occur for the
angles of inclination +90° and -90° .
On the other hand, in the case of the bend placed horizontally,
the disturbance zone lengths are negligible and the observed
two-phase flow regimes do not, in fact, differ from those
observed in the horizontal straight pipe. Typical two-phase flow
regimes obtained in bends for extreme positions are shown in
Fig.9 and Fig.10.
    The occurrence of disturbance zones in bends and their
effect on connecting pipes give rise to substantial increase in
pressure drop. The comparison of the measured pressure drop
values in bends with values obtained for straight pipes points to
the possibility of pressure drop increasing several times,
depending on the position of the pipes . Drawing on the
investigation results, the method of determining pressure drop in
two-phase, gas-liquid flow in bends, based on the Lockhart-Martinelli method [10], true of straight pipes, has been elaborated.
    In the case of adiabatic flow the total pressure drop in the
bend is determined as a sum of constituents caused by friction
and by elevated pressure drop

$$ \left(\frac{\Delta P}{\Delta L} \right)_{2F,c} = \left(\frac{\Delta P}{\Delta L} \right)_{2F,R,c} \pm \left(\frac{\Delta P}{\Delta H} \right)_{2F,H} \qquad (1) $$

The frictional element should be calculated from dependence

$$ \left(\frac{\Delta P}{\Delta L} \right)_{2F,R,c} = \left(\frac{\Delta P}{\Delta L} \right)_{G,c} \Phi_{G,c}^{2} \qquad (2) $$

in which one-phase frictional pressure drop

$$ \left(\frac{\Delta P}{\Delta L} \right)_{G,c} = \zeta_c \frac{\rho_G u_{Gs}^2}{2 \Pi R} \qquad (3) $$

$$ \zeta_c = \lambda_c \frac{\Pi R}{D} + 0.32 \left(\frac{R}{D} \right)^{-0.49} \qquad (4) $$

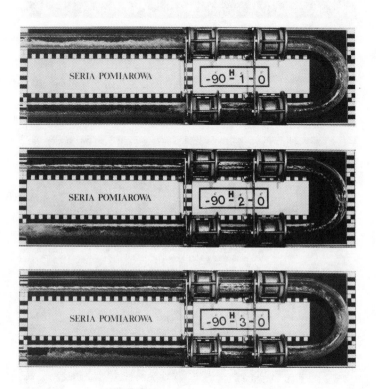

Fig.5. Flow regimes in test section for constant gas velocity
u_{Gs} = 6.58 m/s (bend H-90)

H-90-1-0	u_{Ls} = 0.548 m/s
H-90-2-0	u_{Ls} = 0.274 m/s
H-90-3-0	u_{Ls} = 0.137 m/s

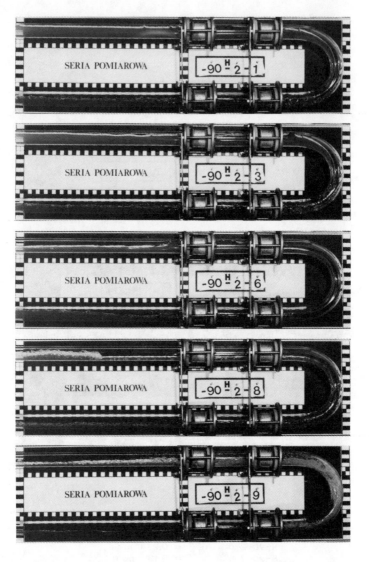

Fig. 6. Flow regimes in test section for constant liquid velocity
u_{LS} = 0.274 m/s (bend H-90)
H-90-2-1 u_{Gs} = 0.0164 m/s   H-90-2-3 u_{Gs} = 0.126 m/s
H-90-2-6 u_{Gs} = 0.548 m/s   H-90-2-8 u_{Gs} = 2.30 m/s
H-90-2-9 u_{Gs} = 3.82 m/s

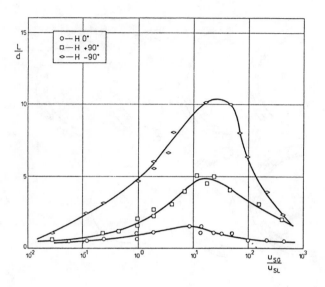

Fig.7. Length of disturbance zone before bend

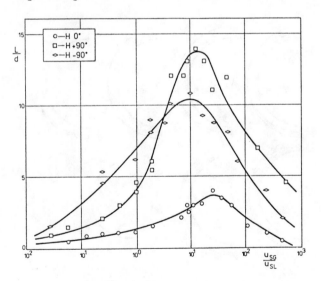

Fig.8. Length of disturbance zone behind bend

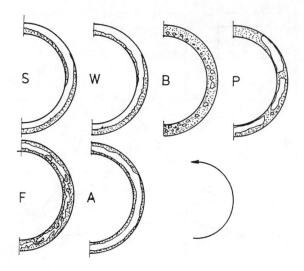

Fig.9. Flow regimes in horizontal upward bend H+90

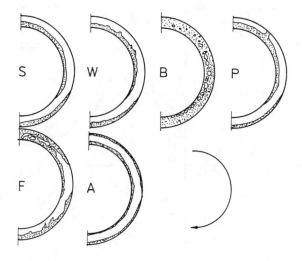

Fig.10. Flow regimes in horizontal downward bend H-90

74

$$\lambda = \begin{cases} 64 \ \text{Re}_G^{-1} & \text{for} \quad \text{Re}_G < 2100 \\ 0.3164 \ \text{Re}_G^{-0.25} & \text{for} \quad \text{Re}_G > 2100 \end{cases} \qquad (5)$$

$$\text{Re}_G = \frac{u_{Gs} \ D \ \rho_G}{\eta_G} \qquad (6)$$

Correction parameter

$$\Phi_{G,c} = C \ \Phi_G \qquad (7)$$

where

$$C = 1 + 1.25 \ \sin \left| \gamma \right| \qquad (8)$$

for upward flow, and

$$C = 1 + 0.25 \ \sin \left| \gamma \right| \qquad (9)$$

for downward flow.

The values of correction $\Phi_G = f(X_{LM})$ can be determined on the basis of the Lockhart and Martinelli plots [10] or dependence given in the paper [9].

It is advisable that the consistent of pressure drop brought about by the elevated pressure drop should be calculated from the dependence

$$\left(\frac{\Delta P}{\Delta H} \right)_{2F,H} = g \ \rho_{2F} \ \sin \phi \qquad (10)$$

where

$$\rho_{2F} = \alpha \ \rho_G \ (1 - \alpha) \ \rho_L \qquad (11)$$

Gas void fraction is calculated according to Stomma method [11]

$$\alpha = 1 - \frac{\varepsilon^2 - x^2}{2 \ [\ \ln (\frac{1 - x}{1 - \varepsilon}) - (\varepsilon - x)]} \qquad (12)$$

75

where

$$x = \cfrac{1}{1 + \cfrac{u_{Ls}}{u_{Gs}} \cfrac{\rho_L}{\rho_G}} \qquad\qquad (\,13\,)$$

$$\varepsilon = \cfrac{1}{1 + \cfrac{u_{Ls}}{u_{Gs}}} \qquad\qquad (\,14\,)$$

The accuracy of the worked out method which is $\overset{+}{-}30$ % may be compared with that of Lockhart-Martinelli for straight pipes.
The exemplary results of the calculation are shown in Fig.11 and Fig.12.

4.   CONCLUSION

On the basis of the investigations of a two-phase gas-liquid flow in horizontal bends it has been found that :
1. Placing a bend between horizontal pipes affects the flow regimes in straight-axis sections
2. Lengths of disturbance zones in straight-axis before and behind the bend has been determined
3. Method for calculating pressure drop in horizontal bends has been proposed.

5.   NOMENCLATURE

C	- constant	
D	- diameter	m
G	- mass flow rate	kg/s
g	- mass velocity	kg/(m² s)
g	- gravitional acceleration	m/s²
H	- height	m
L	- length	m
ΔP	- pressure drop	Pa
R	- radius	m
Re	- Reynolds number	
u	- velocity	m/s
X	- Lockhart-Martinelli parameter	
x	- gas mass ratio	
α	- gas void fraction	
γ	- band inclination	o
ε	- gas volume flux quality	
ζ	- local friction factor	
η	- viscosity	Pa s
λ	- friction factor	
λ_B	- Baker correction	
Φ_B	- correction	
Ψ_B	- Baker correction	

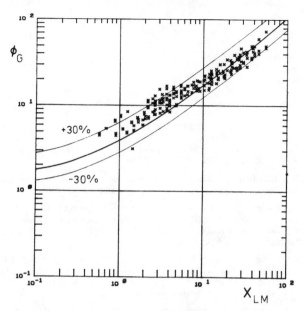

Fig.11. Correction $\Phi_{G.C}$ versus Lockhart-Martinelli parameter for horizontal bend H0

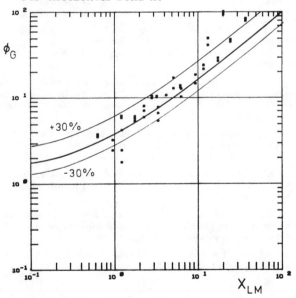

Fig.12. correction $\Phi_{G.C}$ versus Lockhart-Martinelli parameter for horizontal downward bend H-90

Subscripts

c   - curved
G   - gas
H   - elevated
L   - liquid
L-M - Lockhart-Martinelli
R   - frictional
s   - superficial
T   - total
2F   - two-phase

REFERENCES

1   Oschinovo T., Charles M.E., Vertical two-phase flow, Can. J. Chem. Engng, 1974, vol.52, no.1, pp.25-34.

2   Golan L.P., Stenning A.H., Two-phase vertical flow maps, Proc. Inst. Chem. Engrs, 1969-1970, vol.184, no.Pt3C.

3   Chen T.K., Yang Z.H., Flow pattern transitions for gas-liquid flow in the bend of vertical hairpin tubes,Multiphase Flow and Heat Transfer , ed. Veziroglu T.N., Bergeles A.E., Elsiver Publishers Corp., Amsterdam 1984.

4   Fuji T., Akagawa K., Kutsuna K., Yamada I., Kishimoto K., Flow structure of annular-mist two-phase flow in tube bends-distribution of entrained liquid flow rate, Mem. of Fac. of Engng, Kobe University, 1982, vol.29, pp.189-202.

5   Akagawa K., Fuji T., Kutsuna H., Matsui H., Yamada T., Flow structure of annular-mist two-phase flow in tube bends- gas velocity profile and pressure drop distribution, Mem. of Fac. of Engng, Kobe University, 1983, vol.30, pp.115-126.

6   Usui k., Aoki S., Inoue A., Flow behavior and pressure drop of two-phase through C-shaped bend in vertical plane- upward flow, J. of Nuclear Science and Technol., 1980, vol.17, no.2, pp.875-887.

7   Usui K., Aoki S., Inoue A., Flow behavior and pressure drop of two-phase through C-shaped bend in vertical plane-downward flow, J. of Nuclear Science and Technol., 1981, vol.18, no.3, pp.179-190.

8   Troniewski L., Ulbrich R., The analysis of flow regime maps of two-phase gas-liquid flow in pipes, Chem. Eng. Sci., 1984, vol.39, no.7/8, pp.1213-1224.

9   Ulbrich R., Hydraulics of two-phase gas-liquid flow in rectangular channels, Ph. D. Thesis, Technical University , Wroclaw 1981.

10 Lockhart R.W., Martinelli R.C., Proposed correlation of data for isothermal two-phase two-component flow in pipes, Chem. Engng. Progr., 1949, vol.45, no.1, pp.39-48.

11 Stomma Z., Two-phase flows- void fraction value determination, Report INR-1818, Warszawa-Świerk, 1979.

Relation between Gas Void Fraction in Upward and Downward Flow

R. ULBRICH
Heat Technique and Chemical Engineering Department
Opole Technical University
45-233 Opole uL. ZSP 5, Poland

Abstract

Relation between gas void fraction in upward and downward flow in vertical pipe has been presented. The Stomma method of gas void fraction calculation in upward flow has been suggested.
For downward flow in which the number of experimental data is not numerous, method of gas void fraction has not been worked out, but, making use of the drift flux model of Zuber and Findlay, a relation between gas void fraction in upward and downward flow in vertical pipe was determined.
A good accuracy between experimental data consisting of 347 data points and proposed methods of calculations has been established.

1. INTRODUCTION

In a large number of apparatuses [1,2] we can observe both upward and downward two-phase gas-liquid flows occuring simultaneously. The occurrence of this phenomenon can be noticed, for example, in such types of apparatus as heterogeneous chemical reactors, natural gas and oil transmission lines and transmission installations from geothermal sources.
However the information concerning the hydrodynamics of both flows is far out of proportion. Thus, there is a great number of experiments carried out regerding the upward flow and there exist relatively reliable methods of determining flow regimes and calculating pressure drop and void fraction of phases (e.g. the method Lockhart-Martinelli [3], Hughmark [4], Spedding and Chen[5], Stomma [6]), on the other hand for downward flow the information concerning investigations or calculation methods is barely sufficient. However, these two cases are very often treated jointly and as a result the same methods of calculations are proposed for both of them.
Friedel [7] suggests a method of pressure drop calculation for two-phase gas-liquid flow mixture, based on the equivalent properties, such as density

$$\rho_{2F} = \left(\frac{x}{\rho_G} + \frac{1-x}{\rho_L} \right)^{-1} \qquad (1)$$

makes as assume which gas void fraction to be equal to gas volume flux quality in two-phase flow mixture

79

$$\alpha = \varepsilon \qquad (2)$$

where

$$\varepsilon = \cfrac{1}{1 + \cfrac{u_{Ls}}{u_{Gs}}} \qquad (3)$$

In Fig.1 and Fig.2 are shown gas void fraction during simultanuous upward and downward flows of two-phase gas-liquid mixture in vertical pipe, in neightbouring branches of installations. Information on the experimental data [8-11] made use this paper is presented in Table 1.

It can seen that for upward flow (Fig.1)

$$\alpha_u < \varepsilon \qquad (4)$$

and for downward

$$\alpha_d > \varepsilon \qquad (5)$$

In the paper [1] the following formula is given without derivation

$$\alpha_u + \alpha_d = 2\varepsilon \qquad (6)$$

Likewise, Yamazaki and Yamaguchi [12] have found gas void fraction for downward flow to be always greater than gas void fraction for upward flow (Fig.3).

When investigating a deep shaft reactor, Clark and Flemmer [13] tested bubble flow coccurent upward and downward flow (Fig.4).

2.   DETAILS OF MODEL

In analysing upward and downward two-phase flow we should allow for the differences resulting from the change of the direction of gravity force against the force affecting the flow.

Zuber and Findlay [14] proposed to analyse two-phase gas-liquid flow on the basis of the so called drift flux model. According to this model the real velocity of gas structures (bubble, plug) is the superposition of the move (Fig.5) of :
-    two-phase continuous phase (liquid)
-    gas structure in stagnant continuous phase (liquid)

$$u_G = C u_T + u_{oo} \qquad (7)$$

where C is a distribution parameter.

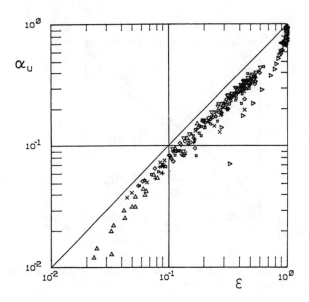

Fig. 1. Gas void fraction in upward flow in vertical pipe
(notation of data points in Table 1)

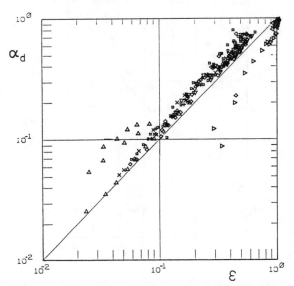

Fig. 2. Gas void fraction in downward flow in vertical pipe
(notation of data points in Table 1)

81

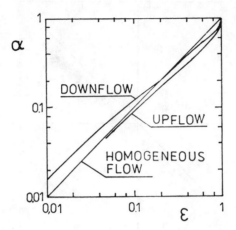

Fig. 3. Gas void fraction for upward and downward flow
calculated by Yamazaki and Yamaguchi [12]

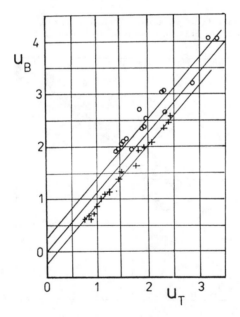

Fig. 4. Velocity of bubble for upward (o) and downward (+)
flow in vertical pipe measured by
Clark and Flemmer [13]

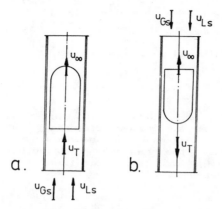

Fig. 5. Model of plug flow in vertical pipe
a) upward flow
b) downward flow

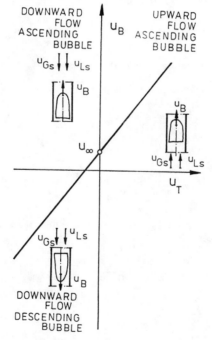

Fig. 6. Possibility of plug flow in vertical pipe

For bubble flow drift velocity of a bubble in the liquid is calculated by Harmathy formula [15]

$$u_{oo} = 1.53 \; [\; \frac{\sigma \; \hat{g} \; (\; \rho_L - \rho_G \;)}{\rho_L^2} \;]^{0.25} \qquad (8)$$

For plug flow, Griffith and Wallis [16] suggested

$$u_{oo} = 0.35 \; (\; \frac{\hat{g} \; D \; (\; \rho_L - \rho_G \;)}{\rho_L} \;)^{0.5} \qquad (9)$$

Investigators opinions [16,19] concerning the value of distribution parameter are rather divergent, and values ranging from 1.1 to 1.3 are proposed. In his investigations [20], Martin has pointed out that the values of distribution parameter for downward flow are lower than for upward flow.
Making use of the dependence

$$u_G = \frac{u_{Gs}}{\alpha} \qquad (10)$$

$$u_T = u_{Gs} + u_{Ls} \qquad (11)$$

and substituting it for (7) we obtained

$$\frac{u_{Gs}}{\alpha} = C \; (\; u_{Gs} + u_{Ls} \;) \; \overset{+}{\underset{-}{}} \; u_{oo} \qquad (12)$$

In Fig.6 are presented possible cases of the flow of a gas plug.
For upward flow (Fig.5a) we obtained

$$\frac{u_{Gs}}{\alpha_u} = C_u \; (\; u_{Gs} + u_{Ls} \;) + u_{oo} \qquad (13)$$

whereas for downward flow

$$\frac{u_{Gs}}{\alpha_d} = C_d \; (\; u_{Gs} + u_{Ls} \;) + u_{oo} \qquad (14)$$

After summing up of the formula (13) and (14) and transformation we obtained

$$\alpha_d = \frac{\varepsilon \; \alpha_u}{C_T \; \alpha_u + \varepsilon} \qquad (15)$$

where

$$C_T = C_u + C_d \qquad (16)$$

Thus, in order to determine gas void fraction for downward flow we sholud first calculate gas void fraction for upward flow and determine total distribution parameter.

During many years of research on two-phase flow [21-23], it has been found that the best accuracy for upward flow is secured by gas void fraction calculated by the Stomma method

$$\alpha = 1 - \frac{\varepsilon^2 - x^2}{2 \left[\ln \left(\frac{1-\varepsilon}{1-x} \right) - (\varepsilon - x) \right]} \qquad (17)$$

where

$$x = \frac{1}{1 + \frac{u_{Ls}}{u_{Gs}} \frac{\rho_L}{\rho_G}} \qquad (18)$$

3. EXPERIMENTAL INVESTIGATION

In order to determine distribution parameter, velocities of gas structures were investigated in cocurrent gas-liquid flow in a pipe 20 mm in diameter and 2 m long. In Fig.7 a double resistivity sensor with data processing system is shown.

Making use of crosscorrelation function of two signals (Fig.8) and then the velocity of plugs were found. Fig.9 presents the investigation results of plug velocity in liquids of different viscosity. Good agreement was found with the staight line having coefficients proposed by Nicklin and Davidson [17]

$$u_B = 1.2 \, u_T + 0.35 \left(\frac{\hat{g} \, D \, (\rho_L - \rho_G)}{\rho_L} \right)^{0.5} \qquad (19)$$

4. RESULTS AND DISCUSSION

In Fig.10 there is shown the total distribution parameter for the set of experimental data presented in Table 1 according to the formula

$$\overline{C}_T = -\frac{1}{N} \sum_{i=1}^{N} \left[\varepsilon_i \left(\frac{1}{\alpha_{u,i}} + \frac{1}{\alpha_{d,i}} \right) \right] \qquad (20)$$

After carrying out calculations we obtained mean value

$$\overline{C}_T = 2.32$$

Assuming $C_u = C_d$ we obtained

$$C_u = C_d = 1.16$$

i.e. the values proposed by many investigators.

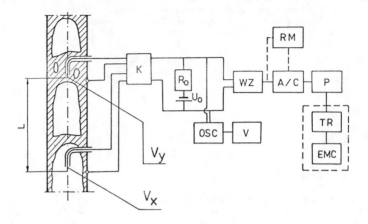

Fig. 7. Double resistivity probe and data processing system
K- commutator, R- resistor, U- power supply,
OSC- osciloscope, V- voltameter, WZ- amplifier,
A/C- analog/digital converter, P- puncher,
TR- tape reader, EMC- computer, RM- tape recorder

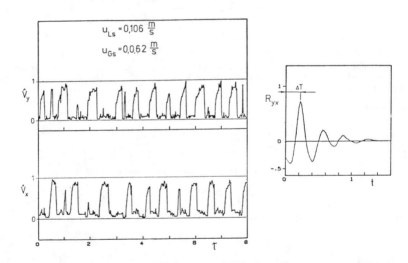

Fig. 8. Crosscorrelation function for upward plug flow
in vertical pipe

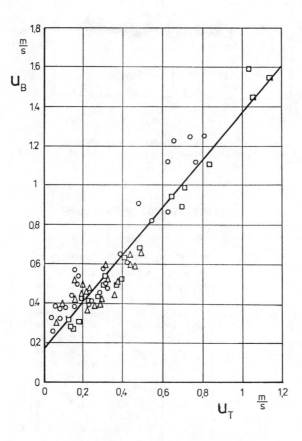

Fig. 9. Velocity of plug in upward flow
viscosity of liquid

    □ -   1 cP
    △ -   6 cP
    ○ -   30 cP

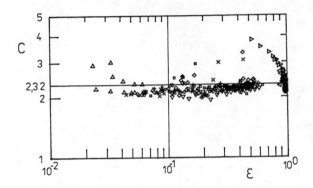

Fig. 10. Total distribution parameter
(notation of data points in Table 1)

TABLE 1. RANGE OF VARIABLES

No	Investigator	D mm	η_L cP	u_{6S} m/s	u_{LS} m/s	N	
1	Tisin [8]	32	0.75	0.1-0.4	1.0-2.13	8	+
2	Tisin [8]	32	1.18	0.1-0.8	0.62-2.15	8	×
3	Tisin [8]	32	3.49	0.1-0.8	1.05-1.96	12	×
4	Davydov [9]	50	1.0	0.12-1.73	0.6-2.0	81	□
5	Davydov [9]	50	2.15	0.12-1.56	0.8-2.0	23	◇
6	Davydov [9]	50	7.25	0.12-1.35	0.6-2.0	25	◇
7	Davydov [9]	50	20.5	0.12-1.33	0.6-1.8	19	◇
8	Davydov [9]	50	1.2	0.12-1.33	0.8-2.0	30	▽
9	Lorenzi-Sotgia [10]	44	1.0	0.01-0.31	0.37-1.43	19	△
10	Oschinovo-Charles [11]	25	1.13	0.22-10.9	0.02-0.45	33	▷
11	Oschinovo-Charles [11]	25	1.8	1.1-7.9	0.05-0.13	9	◁
12	Oschinovo-Charles [11]	25	3.3	0.66-7.5	0.01-0.13	11	◁
13	Oschinovo-Charles [11]	25	8.75	1.3-8.3	0.04-0.13	7	◁
14	Oschinovo-Charles [11]	25	11.3	1.1-9.4	0.01-0.13	11	◁
15	Oschinovo-Charles [11]	25	1.13	14-27	0.02-0.04	15	▷
16	Oschinovo-Charles [11]	25	1.8	15-26	0.01-0.13	12	◁
17	Oschinovo-Charles [11]	25	3.3	15-26	0.01-0.13	8	◁
18	Oschinovo-Charles [11]	25	8.75	15-26	0.04-0.13	6	◁
19	Oschinovo-Charles [11]	25	11.3	14-27	0.021	10	◁

If we, then, assume the constant $C_u = 1.29$ proposed in the paper [19] and $C_d = 1.05$ proposed by Martin [20], we shall, likewise, obtain the value of total parameter of distribution aproximating the one obtained in Fig.10.

Fig.11 presents a comparison of investigation results of gas void fraction for upward flow in a vertical pipe with those calculated by Stomma method (equation (17)); 80 % of data points can be placed on $\pm$15 % error area.

Fig.12 presents a comparison of investigation results of gas void fraction for downward flow in a vertical pipe with those calculated according to the formula

$$\alpha_d = \frac{\varepsilon - \alpha_u}{2.32 \, \alpha_u - \varepsilon} \qquad (21)$$

where gas void fraction for upward flow being calculated by Stomma method; 80 % data points can be placed on $\pm$ 20 % error area. Histograms of relative errors presented in Fig.13 show that the shape is similar to normal distributions.

5.   CONCLUSIONS

1.   Direction of two-phase flow mixture in vertical pipe affects the value of gas void fraction
2.   Based on the drift flux model of Zuber and Findlay, a relation between gas void fraction in upward and downward flow was determined
3.   Gas void fraction in upward flow can be calculated by Stomma method (equation (17)) with accuracy of $\pm$15 %
4.   Gas void fraction in downward flow can be calculated by equation (21) with accuracy of $\pm$ 20 % .

5.   NOMENCLATURE

C	- distribution parameter	
D	- diameter	m
g	- gravitional acceleration	m/s^2
L	- length	m
n	- fraction of data points in interval	
P	- probability	
R_{yx}	- crosscorrelation	
t	- time	s
u	- velocity	m/s
V	- voltage	V
x	- gas mass ratio	
α	- gas void fraction	
δ	- relative error	
ε	- gas volume flux quality	
η	- viscosity	Pa s
ρ	- density	kg/m^3
σ	- surface tension	N/m

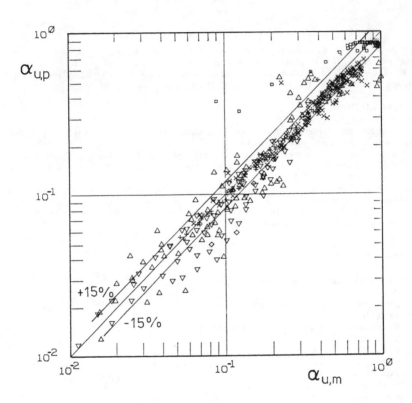

Fig. 11 Comparison of measured gas void fraction
in upward flow with those predicted
by Stomma method [6]
(notation of data points in Table 1)

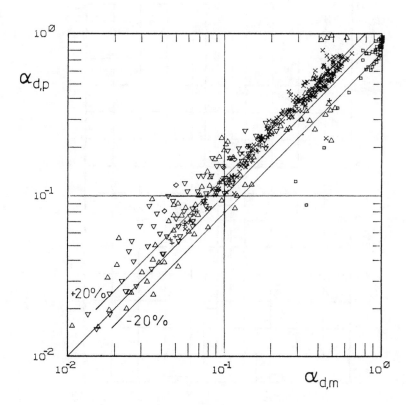

Fig. 12. Comparison of measured gas viod fractin
in downward flow with those predicted
by equation 〈 21 〉
〈 notation of data points in Table 1 〉

91

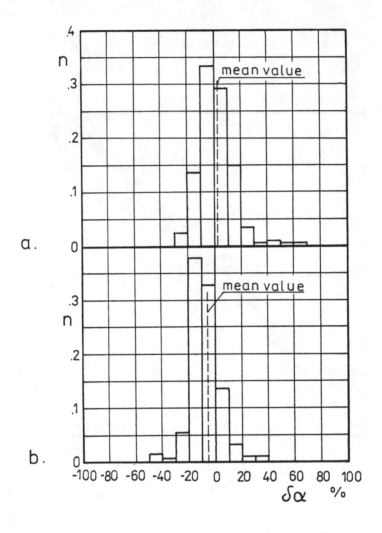

Fig. 13. Histograms of relative errors of gas void fraction
a) upward flow
b) downward flow

Subscripts

B   - bubble
d   - downward
G   - gas
L   - liquid
m   - measured
p   - predicted
s   - superficial
T   - total
u   - upward
oo   - stagnant liquid

REFERENCES

1   Sokolow W.N., Domanski I.W., Two-phase flows reactors, Machinebuilding, Leningrad 1976.

2   Charpentier J.C., A review of data on mass transfer parameters in most of gas-liquid reactors, in Two-Phase Flow and Heat Transfer , ed. Kakac S., Mayinger F., Hemisphere Publishing Corp., Washington 1976.

3   Lockhart R.W., Martinelli R.C., Proposed correlation of data for isothermal two-phase two-component flow in pipes, Chem. Engng. Prog., 1949, vol.45, no.1, pp.39-48.

4   Hughmark G., Holdup in gas-liquid flow, Chem. Engng Progr., 1962, vol.58, no.4, p.62.

5   Spedding P.L., Chen J.J.J., Holdup in two-phasae flow, Int. J. Multiphase Flow, 1984, vol.10, n0.3, pp.307-339.

6   Stomma Z., Two-phase flows- void fraction value determination, Report INR-1818, Warszawa-Swier 1979.

7   Friedel L., Improved friction pressure drop correlations for horizontal and vertical two-phase pipe flow , Europen Two-Phase Flow Group Meeting, Ispra 1979.

8   Tisin W.B., Ph. D. Thesis, Institute of Technology, Leningrad 1969.

9   Davydov I.W., Ph. D. Thesis, Institute of Technology, Leningrad 1969.

10   Lorenzi A., Sotgia G., Comparative investigation of some characteristics quantities of two-phase cocurrent upward and downward flow, Two-Phase Flow and Heat Transfer Symp.- Workshop, Miami 1976.

11   Oschinovo T., Charles M.E., Vertical two-phase flow, Can. J. of Chem. Engng, 1974, vol.52, no.5, p.438.

12   Yamazaki Y., Yamaguchi K., Characteristics of cocurrent two-phase downflow in tubes, J. of Nuclear Science and Technol., 1979, vol.16, no.4, pp.245-255.

13   Clark N.N., Flemmer R.L., Gas-liquid contacting in vertical two-phase flow, Ind. Eng. Chem. Process Dev., 1985, vol.24, pp.231-236.

14   Zuber N., Findlay J.A., Average volumetric concentration in two-phase flow systems, J. of Heat Transfer, 1965, vol.87, pp.453-468.

15   Harmathy T.Z., Velocity of large drops and bubbles in media of infinite or restricted extend, A I Ch E J., 1960, vol.6, no.4, p.281.

16   Griffith P., Wallis G.B., Two-phase slug flow, J. of Heat Transfer, 1961, vol.83, pp.307-319.

17   Nicklin D.J., Davidson J.F., The onset of instability in two-phase slug flow, Symposium on Two-Phase Fluid Flow, London 1962.

18   Taitel Y., Barnea D., Dukler A.E., Moddeling flow pattern transitions for steady upward gas-liquid flow in vertical tubes, A I Ch E J., 1980, vol.26, p.345.

19   Fernandes R.C., Semiat R., Dukler A.E., Hydrodynamic model for gas-liquid slug flow in vertical tubes, A I Ch E J., 1983, vol.29, no.6, pp.981-989.

20   Martin C.S., Vertically downward two-phase slug flow, Trans. ASME, J. of Fluids Engng, 1976, no.6,p.715.

21   Burian T., Troniewski L, Ulbrich R., Pressure drop in two-phase one-component flow in evaporation in tube, Papers of Opole Technical Univ., 1983,vol.95,p.25-40.

22   Babadagli T., Spisak W., Troniewski L., Ulbrich R., Pressure drop in two-phase gas-high viscosity liquid flow in pipes, Papers of Opole Technical Univ., 1987, vol.124, no.33, p.5.

23   Ulbrich R., National Conference on Fluid Mechanics, Białowieża 1988.

Downflow Two-Phase Pressure Drop for Multi-holed Plates

T. S. ANDREYCHEK, S. K. CHOW, L. E. HOCHREITER,
and M. F. McGUIRE
Westinghouse Electric Corporation
Nuclear and Advanced Technology Division
P.O. Box 355
Pittsburgh, Pennsylvania 15236, USA

Abstract

A series of steady-state single- and two-phase flow experiments were performed with perforated plates to study the effect of different plate thickness-to-hole diameter ratios (t/D) on measured pressure drop under downflow conditions. The test article consisted of a vertical cylindrical flow channel with three plates positioned in series. Two different plate t/D ratios were studied; a thin plate (t/D < 1.0), and a thick plate (t/D > 1.0). The working fluids used were water and gaseous nitrogen.

The two-phase downflow pressure drop data was reduced into its, gravitational, acceleration, frictional, and form loss components by using measured local void fraction and the differential pressure drops measured along the test section. A two-phase pressure drop multiplier for the plates was then reduced from the single- and two-phase form loss pressure drop components. The two-phase pressure drop multipliers obtained were in general agreement with those evaluated from the methodology found in the open literature.

1.0 INTRODUCTION

As recommended by Chisholm [1] and Collier [2], the methods used to calculate the two-phase pressure drop across a plate is to treat the plate as being either a thin, orifice-like plate, or a thick plate such as the junction of two pipes. For a thin plate, the flow contracts to the vena contracta and does not reattach to the plate; rather, it expands downstream of the plate. For a thick plate, the flow contracts in the same fashion, but, as it expands, it reattaches to the plate and then expands from the plate to the downstream flow area. Figure 1 shows a comparison of the postulated flow patterns through a thin and thick plate. Whether the flow reattaches within the plate is dependant on the plate thickness-to hole diameter (t/D) ratio. Thin plates are generally considered to be those having t/D ratios less than unity, with thick plates having t/D ratios greater than unity.

95

The current test program provided data from two plate designs. One design characterized thin plate effects (t/D < 1.0) and one design characterized thick plate effects (t/D > 1.0). These plates contained multiple holes, providing for the possibility of downstream interaction between jets that may be generated as the flow passed through those holes. There was also an interest in assessing plate-to-plate interaction of the two-phase flow. Thus, the test article was designed to accomodate three plates in series; however, during a given test run, the three plates installed in the test article were all of the same design (all thin or all thick plates). Experiments conducted with the test apparatus included single-phase downflow tests to obtain the single phase loss coefficient of the plates, and two-phase downflow experiments over a range of flows and qualities of interest.

The two-phase downflow pressure drop data was reduced into its, gravitational, accelerational, frictional, and form loss components by using measured void fraction and differential pressure drop measurements. Two-phase pressure drop multipliers were then calculated from the single-phase and two-phase form loss pressure drop components. These experimentally derived multipliers were then compared with those obtained using the methodology recommended by Chisholm [1] and Collier [2].

2.0 TEST HARDWARE DESCRIPTION

Described within this section is the test loop, test article, and test instrumentation.

2.1 Test Loop

The test facility consisted of an open loop nitrogen system and a closed loop water system that supplied flow to the test article. Mixing of the nitrogen and water flows was accomplished immediately upstream of the test article. After passing through the test article, the liquid and gas flows were separated in a lower plenum vessel. Nitrogen was discharged to the atmosphere, and the liquid was ducted to a head tank for recirculation through the test article. The test loop is shown schematically in Figure 2.

2.2 Test Article

The test article consisted of a vertically oriented cylindrical flow channel 154.4 inches in length. The test article was designed to accomodate three perforated plated in series. To accomodate the use of the void fraction measurement instrumentation, the flow channel was made of plastic, electrically insulating, 8-inch schedule 80 polyvinal cloride (PVC) piping. The elevation of the perforated plates is shown in Figure 3.

2.3 Perforated Plates

As noted earlier, two designs of plates were used; one thin plate (t/D < 1.0) and one thick plate (t/D > 1.0). The physical characteristics of the plates are summarized in Table I. The design of the thin plate is shown schematically in Figure 4, and the design of the thick plate is shown schematically in Figure 5.

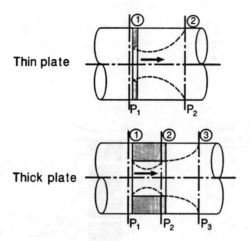

Fig. 1. - Vena Contracta for Thin and Thick Plates.

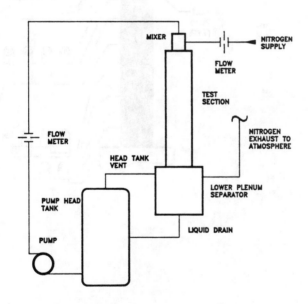

Fig. 2. - Test Loop Schematic Diagram.

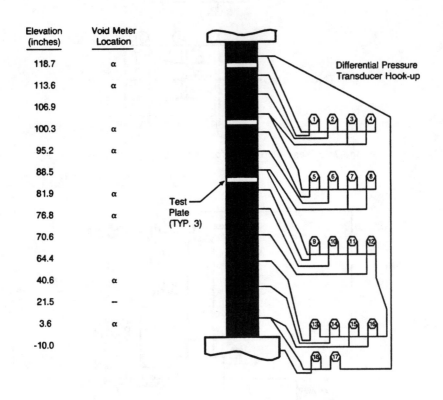

Elevation (inches)	Void Meter Location
118.7	α
113.6	α
106.9	
100.3	α
95.2	α
88.5	
81.9	α
76.8	α
70.6	
64.4	
40.6	α
21.5	–
3.6	α
-10.0	

Fig. 3. - Test Article Schematic Diagram.

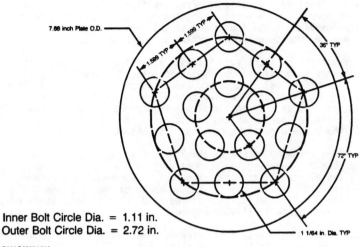

Inner Bolt Circle Dia. = 1.11 in.
Outer Bolt Circle Dia. = 2.72 in.

D026 D27861.006

Fig. 4. - Thin Plate Design.

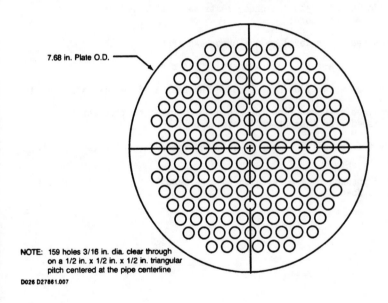

7.68 in. Plate O.D.

NOTE: 159 holes 3/16 in. dia. clear through
on a 1/2 in. x 1/2 in. x 1/2 in. triangular
pitch centered at the pipe centerline

D026 D27861.007

Fig. 5. - Thick Plate Design.

2.4 Test Instrumentation

Two types of measurements were made in the test article; differential pressure drops and local void fractions. Pressure taps were installed using standard industry practice. A total of eighteen differential pressure transducers were installed along the length of the test article. The differential pressure transducer hookup scheme is shown in Figure 3.

Direct measurements of local void fraction at various locations within the test article were made using Model 1081 void meters manufactured by Auburn International. The Model 1081 void meters use a two-phase rotating electric field so as to uniformly monitor the flow channel cross section. Thus, two pairs of probes are used at a given elevation; each probe traverses about one-fourth of the channel perimeter and is connected, through the meter electronics, to the probe directly across the flow channel.

The void meters actually determine the liquid fraction between a set of electrodes or probes by measuring the conductivity of the liquid medium enclosed within the electrodes and comparing that measurement to a reference conductivity. The void fraction (α) is then taken to be the compliment of the liquid fraction ($1-\alpha$). The location of the void meter probes within the test article are shown schematically in Figure 3.

Also, the liquid and gas mass flow rates were measured using standard orifice meter sections prior to the introduction of those flows into the mixing section just upstream of the test article.

3.0 TEST PROCEDURE

Both the liquid-solid and two-phase tests were conducted using essentially the same procedure;

1. The desired water level in the pump head tank was established.

2. Loop and test article instrumentation was checked.

3. The desired water flow rate through the test article was established.

4. The desired nitrogen flow rate through the test article was established (for two-phase flow tests only).

5. Steady flow conditions were established; data was collected for five minutes.

6. Steps 3 through 5 were repeated for the next test point, or the loop was shut down.

4.0 TEST DATA

Single- and two-phase downflow testing was conducted over a range of liquid and, for two-phase tests, vapor flow rates. The flow conditions used in conducting tests with both the thin and thick perforated plates are listed in Table II. Note that both thin and thick plate designs were tested at the same flow conditions.

The working fluids were driven through the test article by increasing the supply pressure just upstream of the test article until the desired liquid and, for two phase flow tests, vapor flow rates were achieved. To accomplish the test matrix given in Table II, this supply pressure varied from about 20 to 40 psia. During a test at a given flow condition, however, the supply pressure of the working fluids was maintained at a constant value. The temperature of the working fluids was maintained constant at about 80°F for all flow conditions tested.

A typical plot of the static pressure distribution along the test article for thin plates under two-phase flow conditions is given in Figure 6. As was expected, the total pressure drop across the test article increased as the flow quality increased; this was particularly true in the flow channel just downstream of a perforated plate. However, the static pressure gradient in the lower, constant flow area region of the test article downstream of the perforated plates was noted to always be nearly zero. This implies that the gravity head offset the frictional losses in this region such that the net static pressure remained nearly constant.

A plot of measured void fraction distribution along the test article that corresponds to the static pressure distribution of Figure 5 is given in Figure 7. The variation in measured void fraction along the test article indicates a two-phase recovery effect in the flow field as it proceeds downstream of a perforated plate. The large magnitude of the measured void fraction is typical of a downflow annular flow regime, and is consistant with the flow regime map shown in Figure 8.

5.0 DATA ANALYSIS

This section describes the method used to reduce the data collected from the tests identified in Table II.

5.1 Single-Phase Tests

The single-phase form loss coefficient, C_D, for the perforated plates was calculated from the data using a one dimensional momentum balance:

$$\Delta P_{meas} = \frac{C_D \, \rho \, U_P^2}{2g_c} + \frac{f \, \Delta Z \, \rho U^2}{D_e \, 2g_c} \tag{1}$$

1678v:1D/121588

TABLE I: PHYSICAL CHARACTERISTICS OF TEST PLATES

Plate Type	Plate Thickness (inches)	Hole Diameter (inches)	Number of Holes	Total Flow Area (square inches)	$\frac{t}{D}$
Thin	0.561	1.015625	15	12.152	0.552
Thick	0.561	0.3125	159	12.195	1.795

TABLE II: DOWNFLOW TEST MATRIX THIN AND THICK PERFORATED PLATES

Liquid Flow (GPM)	QUALITY (Percent)								
	0.00	0.25	0.50	1.00	2.00	3.00	5.00	7.50	10.00
25	2S 3T								
50	2S 2T	1S 1T	1S 1T	2S 1T	1S 1T	1S 1T	3S 2T	1S 1T	2S 2T
75	2S 2T								
100	2S 2T	2S 1T	2S 1T	2S 1T	2S 1T	1S 1T	2S 2T	1S 1T	3S 3T
150	2S 2T								
200	2S 3T	1S 1T	1S 1T	1S 1T	1S 1T	2S 2T	2S 2T		
250	2S 3T								
300	2S 3T	1S 1T	1S 1T	1S 2T	1S 2T				
375	2S 2T								

Note: Leading digit denotes number of tests performed, trailing character denotes type of plate tested; thin (S) or thick (T)

Pressure (psi)

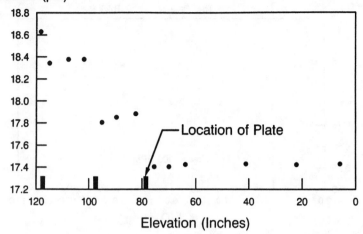

Fig. 6. - Typical Axial Static Pressure Distribution Along Test Article at 100 gpm Flow Rate 2 Per Cent Flow Quality.

Void Fraction

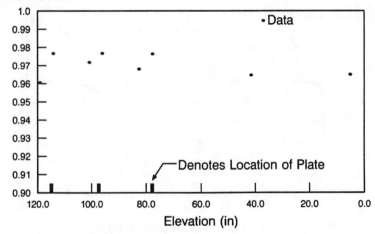

Fig. 7. - Typical Void Fraction Distribution Along Test Article at 100 gpm Flow Rate and 2 Per Cent Flow Quality.

103

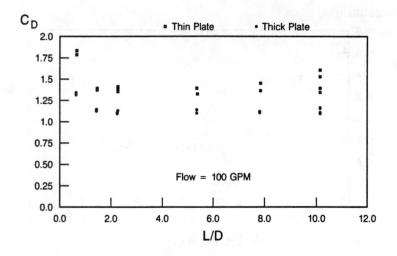

Fig. 8. - Downflow Two-Phase Flow Regime Map.

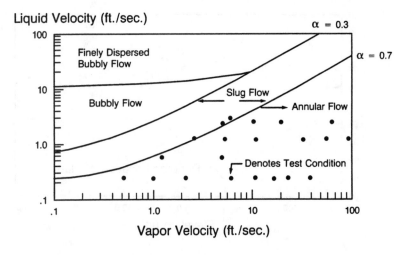

Fig. 9. - Single Phase Hydraulic Loss Coefficients, Thin and Thick Plates.

Where the friction factor, f, was defined as:

$$f = 0.005 + \frac{0.55}{Re^{1/3}} \tag{2}$$

The only unknown parameter in Equation (1) is the single phase form loss, C_D. All other parameters are known either directly from the data (pressure drop), are calculated from test data (fluid velocities, fluid properties, friction factor), or are known dimensions (hydraulic diameter, elevation change). There was no pressure drop due to acceleration of the flow as the test article flow area was a constant. The differential pressure cells had a water solid reference leg such that the elevation head was zeroed out.

To allow flow redevelopment effects to be studied, the test article had a long, constant diameter, instrumented flow channel downstream of the last perforated plate. A typical plot of calculated C_D as a function of the ratio of flow channel length to flow channel diameter (L/D) for both thin and thick plates is given in Figure 9. From this data, it is noted that pressure drop data collected over a span in excess of L/D > 2 will provide for a developed flow single phase loss coefficient.

Some plate-to-plate variation in calculated single phase loss coefficients at given test conditions was observed. For thin plates, the plate-to-plate variation in the values of calculated loss coefficients was beyond the data scatter observed in the experiments. For the thick plates, calculated loss coefficients from the second and last plates in the flow field were similar, with only the first or leading perforated plate's, performance being noticably different. In the case of the thin plates, individual single phase loss coefficient correlations were developed for each plate, depending upon its location in the flow field. For the thick plates, a unique loss coefficient correlation was developed for the leading plate, with the same correlation being applied to the second and last plates.

5.2 Void Fraction And Pressure Drop Analysis

The separated flow momentum equation that includes all components of momentum exchange between the two phases may be written as:

$$- \left[\frac{\partial}{\partial Z} \left(\alpha \, \rho_g \, U_g^2 + (1-\alpha) \, \rho_f U_f^2 \right) \right] + \frac{dP}{dZ} - g \left(\alpha \, \rho_g + (1-\alpha) \, \rho_f \right) = $$

$$\frac{1}{2 \, De} \left[f_f \rho_f U_f^2 + f_g \rho_g U_g^2 \right] + \frac{C_D}{2 \Delta Z} \left[\alpha \, \rho_g \, U_g^2 + (1-\alpha) \, \rho_f U_f^2 \right] \tag{3}$$

Void fraction, total pressure drop, and flows are provided from the test data. Therefore, given friction factors for each phase, the acceleration, gravitational, and frictional pressure drop components of Equation (3) can be calculated for each measured pressure drop location for which there is a void meter reading. It was found that using a conventional single phase friction factor relationship with the measured void fraction over-prediction expected frictional losses in the test article.

105

As shown in Figure 8, the two-phase flow conditions tested were considered to provide for a predominately annular flow regime within the test article. This flow regime is characterized by a liquid film on the flow channel wall, with vapor flow and possibly some entrained droplets in the channel's core region. It is postulated that, due to the perforated design of the plates, some of the liquid flow would be entrained as it passed through one plate, de-entrain on the top surface of the next plate in the flow channel, then re-entrain as it passed through that plate and continued its fall through the test article.

The Model 1081 void meters require a liquid-continuous medium to accurately measure void fraction; they cannot detect entrained liquid flow. Once the flow regime was observed to be annular, the 1081 meter was specially calibrated for high void fraction annular flows. It was judged that there was entrained liquid flow that the void meters were not detecting; the void meters were therefore providing an excessively large void fraction, and corresponding excessively small liquid fraction, for the calculation of the pressure drop components of Equation (3). The impact of these measurements is to distort the phase velocities, making the liquid velocity larger than it is in the test, and causing the frictional pressure drop to be over estimated.

The approach then taken was to evaluate two-phase friction factors from test data, using measured void fractions and pressure drops. This approach preserves both the pressure drop data and the relationship between the fictional pressure drop and gravitational pressure drop terms by adjusting the friction factor to compensate for the larger than expected void fraction. For this approach, the difference between differential pressure transducer readings located along the span between perforated plates is used to calculate the frictional pressure drop along the test article between the adjacent plates. The measured void fractions taken at the elevation of those differential pressure transducer readings were also used to evaluate the acceleration component of the pressure drop.

A one-dimensional mixture momentum equation for the span between two differential pressure transducers downstream of a plate can be written as;

$$DP_{meas} = DP_{fric} + DP_{accel} - DP_{gravity} \tag{4}$$

Expanding the gravitational and acceleration terms and solving for the frictional term between the pressure taps gives

$$DP_{fric} = DP_{meas} - [(\alpha \, \rho_g \, U_g^2 + (1-\alpha) \, \rho_f \, U_f^2)_2$$

$$- (\alpha \, \rho_g \, U_g^2 + (1-\alpha) \, \rho_f \, U_f^2)]_1 \, / \, g_c \tag{5}$$

$$+ \frac{\Delta Z g}{g_c} (\bar{\alpha} \, \rho_g + (1-\bar{\alpha}) \, \rho_f)$$

Where:

$$\bar{\alpha} = (\alpha_1 + \alpha_2)/ \, 2 \tag{6}$$

DP_{meas} = Difference between two differential pressure transducer sharing a common tap upstream of a perforated plate, but extending different lengths downstream of the plate

The frictional pressure drop term is expanded as:

$$DP_{fric} = \Delta Z \left(f_f\, \rho_f\, U_f^2 + f_g\, \rho_g\, U_g^2 \right) / (2D_e\, g_c) \tag{7}$$

From hand calculations, it is evaluated that $\rho_f\, U_f^2 \gg \rho_g\, U_g^2$ such that the vapor momentum flux may be ignored in Equation (7). Substituting the expanded expression for frictional loss into Equation (5) and solving for the two-phase liquid friction factor:

$$f_f = \frac{2_{gc}\, D_e}{\Delta Z \rho_f\, U_f^2} \cdot DP_{meas} - [(\alpha\, \rho_g\, U_g^2 + (1-\alpha)\, \rho_f\, U_f^2)_2$$

$$- (\alpha\, \rho_g\, U_g^2 + (1-\alpha)\, \rho_f\, U_f^2)_1]/g_c \tag{8}$$

$$+ \frac{\Delta Z\, g}{g_c} (\bar{\alpha}\, \rho_g + (1-\bar{\alpha})\, \rho_f)$$

The two-phase liquid fraction factor may be correlated in the form:

$$f_f = A + \frac{B}{Re_f^C} \tag{9}$$

Where A, B, and C are constants evaluated from the data for the two-phase liquid friction factor. A similar expression may be assumed for the two-phase vapor friction factor.

The pressure drop data between the plates was reduced in this fashion and is shown on Figure 9. A best fit expression for this data was calculated as:

$$f_f = 0.008 - \frac{0.002}{140,000}\ (Re - 10,000) \tag{10}$$

Where the phase velocity in the Reynolds number is evaluated based on the measured void fraction:

$$R_{e_f} = \frac{(1-\alpha)\, U_f\, \rho_f\, De}{\mu_f} \tag{11}$$

or

$$Re_g = \frac{\alpha\, \rho_g\, U_g\, De}{\mu_g} \tag{12}$$

107

The total two-phase pressure drop was then solved for the form loss associated with the perforated plates using the pressure drop measurements across the plates as:

$$DP_{form_{2\emptyset}} = DP_{meas} - DP_{accel} + DP_{gravity} - DP_{fric_{2\emptyset}}$$

(13)

The two-phase form loss component of the pressure drop was then normalized to the single phase form loss pressure drop component, assuming the total flow is liquid, to obtain a two-phase multiplier as:

$$\emptyset_{fo}^2 = \frac{DP_{form_{2\emptyset}}}{DP_{form_{1\emptyset}}}$$

(14)

The experimentally determined two-phase multipliers were then compared to those predicted from the method recommended by Collier and Chisholm.

Another approach to solving Equations (5) and (7) would have been to solve for the void fraction using the assumed single-phase friction factors given by the separated flow equation, Equation (3). However, since the uncertainty of applying a single-phase friction factor to a two phase flow would only increase the uncertainty of the calculated void fraction, it was judged that adjusting the friction factor would provide for the most accurate two-phase hydraulic form loss coefficients.

5.3 Comparison With Predictive Methods of Collier and Chisholm

Chisholm [1] and Collier [2] recommend a methodology for calculating two-phase pressure drop based on collected data and numerous experiments performed to characterize two-phase flow pressure drop and void fraction behavior. This methodology is an extension of the Lockhart-Marinelli [3] approach, in which a pressure drop multiplier is developed from two-phase data using single-phase relationships. This approach calculates the Martinelli parameter:

$$\underline{X} = \left[\left(\frac{dP}{dZ} \right)_f / \left(\frac{dP}{dZ} \right)_g \right]^{1/2}$$

(15)

Where $\left(\frac{dP}{dZ} \right)_f$ and $\left(\frac{dP}{dZ} \right)_g$ are the

single-phase pressure drops calculated for each phase flowing separately in the channel. Expanding these terms:

$$\left(\frac{dP}{dZ} \right)_f = \frac{f_f \, \rho_f \, \dot{M}_f^2}{D_e \, \rho_f^2 \, A_T^2 \, 2g_c} = \frac{f_f \, \nu_f \, (1-x)^2 \, \dot{M}_T^2}{De \, A_T^2 \, 2g_c}$$

(16)

and similarly,

$$\left(\frac{dP}{dZ}\right)_g = \frac{f_g \, \nu_g \, x^2 \, \dot{M}_T^2}{D_e \, A_T^2 \, 2g_c} \qquad (17)$$

Substituting Equations (16) and (17) into Equation (15) and simplifying:

$$\underline{X} = \left[\frac{f_f \, \nu_f \, (1-x)^2}{f_g \, \nu_g \, x^2}\right]^{0.5} \qquad (18)$$

The two-phase pressure drop multiplier for the liquid phase flowing in a two phase mixture is defined as:

$$\emptyset_f^2 = 1 + \frac{C}{\underline{X}} + \frac{1}{\underline{X}^2} \qquad (19)$$

Where, Collier gives the parameter C as:

$$C = \left[\lambda + (C_2 - \lambda) \, \frac{\nu_{fg}}{\nu_g}\right]^{0.5} \left[\left(\frac{\nu_g}{\nu_f}\right)^{0.5} + \left(\frac{\nu_f}{\nu_g}\right)^{0.5}\right] \qquad (20)$$

For cases where $\nu_g \gg \nu_f$, Equation (20) can be simplified to:

$$C = \left[C_2\right]^{.5} \left[\left(\frac{\nu_g}{\nu_f}\right)^{0.5} + \left(\frac{\nu_f}{\nu_g}\right)^{0.5}\right] \qquad (21)$$

The values of λ and C_2 are dependent upon the plate thickness-to-diameter ratio, t/D. Chisholm [1] recommends values of $\lambda = 1.0$ and $C_2 = 0.5$ for thin orifice plates for which there is no reattachment of the flow (t/D < 1.0). For thick plates where the vena contracts reattaches to the inside diameter of the hole (t/D > 1.0), Chisholm [1] recommends values of $\lambda = 1.0$ and $C_2 = 1.5$. It should be noted that there exists a considerably larger data base for thin plates as compared to that for thick plates; only one or two thick plate experiments were noted. Using the values of λ and C_2 recommended by Chisolm, values of $\emptyset_f^2$ were calculated for the test conditions.

The two-phase pressure drop multiplier that is generally used, denoted as $\emptyset_{fo}^2$, and assumes the total flow is liquid. The value of $\emptyset_{fo}^2$ is related to $\emptyset_f^2$ as follows:

$$\emptyset_{fo}^2 = (1-x)^2 \, \frac{f_f}{f_{fo}} \, \emptyset_f^2 \qquad (22)$$

109

Therefore, the total two phase pressure drop of a two phase mixture flowing through a plate is expressed as:

$$\frac{dP}{dZ} = \frac{f_{fo}\ G_T^2\ \nu_f}{2\ g_c\ D_e}\ \phi_{fo}^2 \tag{23}$$

in which the total flow is taken to be liquid and the friction factor, f_{fo}, is the single phase friction factor which is also evaluated assuming the total flow is liquid.

Two phase pressure drop multipliers were calculated for the thin ($t/D < 1.0$) and thick ($t/D > 1.0$) perforated plates for each flow condition identified in the test matrix given in Table II. Figure 10 provides a comparison of the two phase flow multipliers predicted by the methodology recommended by Chisholm [1] and Collier [2] and those reduced from test data for a thin plate at a liquid flow rate of 50 gallons per minute. Figures 11 and 12 provide similar comparisons between predictions and data for a thin plate at liquid flow rate of 100 and 200 gallons per minute flow rate, respectively. Similarly, Figures 13, 14, and 15 provide comparisons of between predicted two phase pressure drop multipliers and those reduced from test data for a thick plate at a liquid flow rate of 50, 100, and 200 gallons per minute, respectively.

A review of the thin and thick plate experimental data given in Figures 10 through 15 suggests the two phase pressure drop multipliers reduced from test data are similar. Generally, the two-phase pressure drop multipliers predicted using the separated flow pressure drop prediction methodology recommended by Chisholm [1] and Collier [2] for a thin plate agrees with the multipliers calculated from the experimental data.

All of the two-phase pressure drop multipliers, ϕ_{fo}^2, reduced from test data were divided by the corresponding two-phase pressure drop multiplier predicted from the thin plate methodology given by Collier and Chisholm. This ratio of measured-to-predicted (M/P) two phase pressure drop multipliers was then plotted as a function of quality, as shown in Figure 16. The average value of all the ratio's plotted in Figure 16 is

$$\left(\frac{\bar{M}}{P}\right) = 0.9423 \tag{24}$$

The scatter of the data shown in Figure 16 suggests the two phase pressure drop multipliers reduced from the data agree well with predicted thin plate methodology. It is noted that for the thick plate data taken at a liquid flow rate of 100 gallons per minute, the two phase pressure drop multipliers reduced from the data at higher qualities do lie somewhat above those multipliers predicted for a thin plate, but are far less the multipliers predicted for a thick plate.

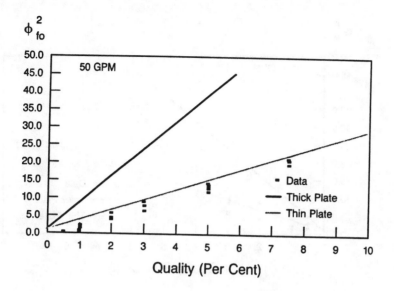

Fig. 10. - Two-Phase Pressure Drop Multipliers,
Thin Plate, 50 GPM Liquid Flow.

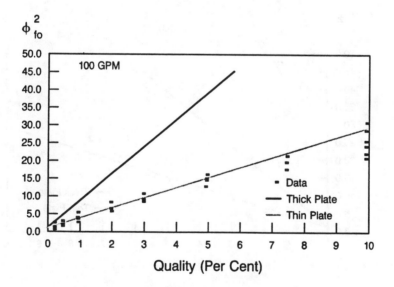

Fig. 11. - Two-Phase Pressure Drop Multipliers,
Thin Plate, 100 GPM Liquid Flow.

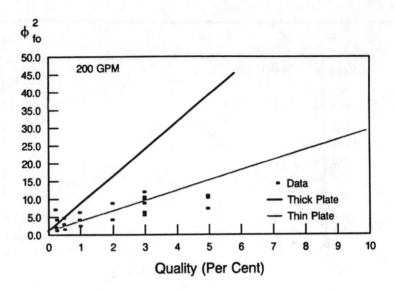

Fig. 12. - Two-Phase Pressure Drop Multipliers
Thin Plate, 200 GPM Liquid Flow.

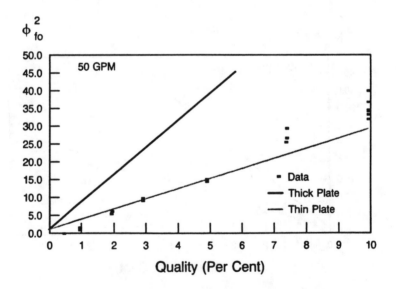

Fig. 13. - Two-Phase Pressure Drop Multipliers,
Thick Plate, 50 GPM Liquid Flow.

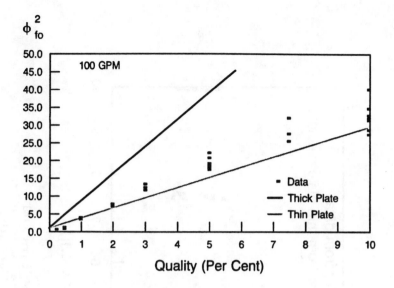

Fig. 14. - Two-Phase Pressure Drop Multipliers,
Thick Plate, 100 GPM Liquid Flow.

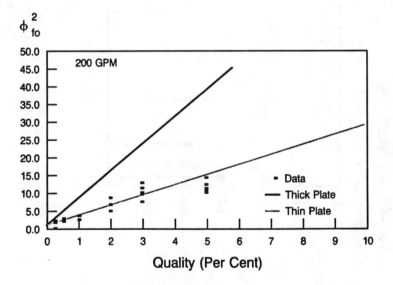

Fig. 15. - Two-Phase Pressure Drop Multipliers,
Thick Plates, 200 GPM Liquid Flow.

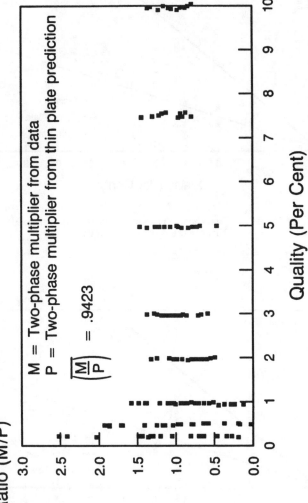

Fig. 16. – Ratio of Measured-to-Predicted
Two Phase Pressure Drop Multipliers
as a Function of Flow Quality.

114

6.0 FUTURE WORK

In reviewing the downflow two-phase pressure drop literature, it was noted that there is little experimental data for thick plates. The results of the current test program provides a limited addition to that body of information, as the thick plate tested had a t/D ratio of only 1.795. Indeed, the two phase pressure drop multiplier predicted for a thin plate agree well with those multipliers reduced from test data collected using the thick plate. Thus, it is suggested that additional experimental work with thick plate is required to adequately understand their two phase pressure drop behavior.

7.0 SUMMARY

A two-phase downflow test program was conducted with two multi-hole plate designs for the purpose of assessing the effects of thin (t/D < 1.0) and thick (t/D > 1.0) plates on measured two phase pressure drop multiplier. No significant differences were noted in the multipliers reduced from the experimental data collected using the two phate designs. The two phase pressure drop multipliers reduced from test data for both thin and thick plates agreed well with those multipliers predicted by the separates flow pressure drop prediction methodology recommended by Chisholm [1] and Collier [2] for thin plates.

ACKNOWLEDGEMENTS

The test program described in this paper was conducted as part of the Advanced Pressurized Water Reactor (WPWR) Development Program which was funded jointly by Westinghouse Electric Corporation, Mitsubishi Heavy Industries, LTD., and the Kansai Electric, Kyushu Electric, Shiksku Electric, Hokkaido Electric, and Japan Atomic Power Companies. Testing was conducted at Westinghouse's Forest Hills Test Facility. The efforts and patience of D. W. Sklarsky and D. B. Henderson are also acknowledged and gratefully appreciated.

8.0 NOMENCLATURE

A = Flow area; ft^2

C_D = Hydraulic loss coefficient

D = Diameter of test article; ft

D_e = Hydraulic diameter; ft

f = Friction factor

G = Mass flux; $lb_m/(sec - ft^2)$

g = Gravitational constant; 32.2 ft/sec^2

g_c = Conversion constant; 32.2 $(lb_m\text{-}ft)/(lb_f\text{-}sec^2)$

$\dot{M}$ = Mass flow rate; lb_m/sec

P = Pressure; lb_f/ft^2

ΔP, dP = Pressure drop; lb_f/ft^2

Re = Regnolds number

U = Velocity, ft/sec

dZ, ΔZ = Change in elevation along the test article, ft

α = void fraction

ρ = density; lb/ft^3

ν = Specific volume; ft^3/lb_m

x = quality

Subscripts

f = liquid phase

g = vapor phase

f_g = difference between liquid and vapor phase

1, 2 = denotes location of differential pressure measurements between two adjacent plates

T = Referring to the total for the flow; as in total mass flow rate or total flow area.

REFERENCES

1. Chisholm, D., Two-Phase Flow in Pipelines and Heat Exchangers, George Godison Publisher, (1983).

2. Collier, J. G., Convective Boiling and Condensation, McGraw-Hill Book Company, (1972).

3. Lockhart, R. W. and R. C. Martinelli, Proposed Correlation of Data for Isothermal Two-Phase Two-Component Flow in Pipes," Chemical Engineering Progress, Volume 45, pg. 39, (1949).

Comparison of Methods
for the Determination of the Head Loss
for Mixture Flow in Horizontal Pipes,
Heterogeneous Regime

V. FRANCO
Engineering Institute
National University of México
Apartado Postal 70-472
Coyoacán 04510, México, DF

Abstract

An important concept of solid-liquid slurry transport in pipeline is related to the so called head loss developed along the pipeline. That has been investigated generally from an empirical point of view for different slurries, e.g. water with sands, iron ore or coal particles. This paper is concerned with a critical analysis of the several methods available for the computation of the head loss in heterogeneous regime compared against experimentals results. The discrepancy between the calculated and measures values is discussed.

1. INTRODUCTION

Nowadays in the iron and steel industry, for instance, or in the dredging of coasts and rivers, particle transport by means of pipelines in a common practice. Most of the facilities built in this way in different countries are a sign of their efficiency, and also, in some cases, of their economical advances when compared to other forms of transport.

Let's consider an horizontal pipeline containing a mixture of solid particles and water. If there is no velocity in the duct or if it is too small, those particles deposit on the bed producing the stationary plane bed condition. But when velocity increases the layer surface changes its form while its width diminishes in such a way that at a certain moment all the particles remain in suspension.

Related to this there is a head loss i_m, which is a function of velocity and of the way particles are, that is, whether they are deposited of they travel in suspension, as seen in fig 1. Its analysis shows that for the stationary bed condition, the head loss is greater than for the heterogeneous one, because the area becomes smaller for the first condition thus increasing the velocity which, at the same time, is reflected in the head loss value.

This paper shows the different criteria used to estimate the

117

head loss for heterogeneous regime in horizontal pipelines and presents a comparison between the obtained values and the experimental data.

2. TYPES OF FLOW REGIME

Based upon the distribution of the particles in the vertical (concentration distribution) Graf and Acaroglu (1967) have defined different flow regimes of the solid-liquid transport through horizontal pipelines. Acoording to this the following regimes may be distinguished: homogeneous, heterogeneous, with moving bed (some particles may be transported in saltation) and stationary bed. Due to the fact that from the practical point of view the last two are not pertinent, only the first two will be defined here.

- Homogeneous regime

In this case all the particles are uniformly distributed in the liquid medium used as transport. This condition generally implies fine sediment and great discharge velocities.

- Heterogeneous regime

Is that where all the solid particles are not uniformly distributed in the transversal section of the conduction, but in spite of this the whole material moves in suspension and there is not bed load transport. Some sediments may move in saltation.

3. HEAD LOSS FOR THE HETEROGENEOUS REGIME

This regime is the most frequently encountered in the practical problems and the most generally dealt with by the specialized literature. The heterogeneous flow is limited by two velocities: a transition one V_H and the other, which is called critical, V_C (fig 1). They are defined by

$$V_H = \sqrt[3]{1800 \ \omega \ g \ D} \qquad\qquad (3.1)$$

$$V_C = F_L \ \sqrt{2 \ g \ D(S_S - 1)} \qquad\qquad (3.2a)$$

$$S_S = \gamma_S / \gamma \qquad\qquad (3.2b)$$

where V_H, and V_C correspond, respectively, to the transition and critical velocities, in m/s. The first was defined by Newitt et al (1955) and the other by Durand (1953); ω is the fall velocity of particle, in m/s; g gravity acceleration, in m/s^2; D pipeline diameter, in m; F_L is know as Froude modified number, it represents an adimensionless parameter whose value is a funtion of the solid concentration and of the particles mean diameter, which can be calculated using the criteria proposed by Durand (1953), Durand and Condolios (1956) or McElvain and Cave (1972); S_S is the specific gravity of solids, adimensionally;

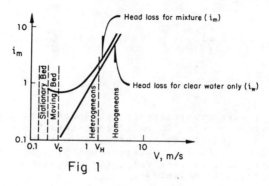

Fig 1

γ_s and γ are, respectively, the specific weight of the particles and of water, in kgf/m^3.

V_H is the flow maximum velocity for the heterogeneous regime, because if this point is not fulfilled, the regime will be homogeneous; for V_C the opposite occurs, since for smaller values particle deposit takes place. Now will be presented the different criteria used to estimate the head loss in the heterogeneous regime, which, as experimentally proved, is bigger than that of water.

One of the first researches on the subject was done by Durand and Condolios (1952), who made their tests with mixtures of sand and water and gravel and water. The results they obtained are summarized in the following equation

$$i_m = i_w \left[1 + 176\, C_v \left(\frac{\sqrt{gD}}{V} \right)^3 \left(\frac{1}{\sqrt{C_D}} \right)^{1.5} \right] \qquad (3.3a)$$

$$C_D = \frac{4}{3} \frac{gd(S_s - 1)}{\omega^2} \qquad (3.3b)$$

$$i_m = \left(\frac{\Delta h}{\Delta L} \right)_m = f_m \frac{V^2}{2g\,D} \qquad (3.3c)$$

$$i_m = \left(\frac{\Delta h}{\Delta L} \right)_w = f_w \frac{V^2}{2g\,D} \qquad (3.3d)$$

where i_m, i_w are, respectively, the head loss due to the mixture and the water, in m/s; C_v solid concentration expressed in volume, in decimals; V discharge velocity, in m/s; C_D drag coefficient for spherical particles, adimensionally; d particle diameter, in m; f_m and f_w are, respectively, the friction factor of the mixture and of water, adimensionless.

These researchers point out that the effect of particle size in the computus of i_m has no meaning when the diameter is more than 1.5 to 2 mm; the S_s they used to solve eq. 3.3a was of 2.65.

At the same time as Durand and Condolios, Worster (1952) also studied big particles, mainly coal, and found an equation similar to 3.3a, which he expressed as

$$i_m = i_w \left[1 + 120\, C_v \left(\frac{\sqrt{gD}}{V} \right) \sqrt{S_s - 1} \,\right)^3 \right] \qquad (3.3e)$$

The difference between both equations lies in the fact that 3.3a takes into account the characteristics of the particle and that this does not happen in 3.3e.

The eq. 3.3e modified for particles other than coal is written as

$$i_m = i_w \left[1 + 81 \; C_V \left(\frac{gD(S_S - 1)}{v^2 \sqrt{C_D}} \right)^{1.5} \right] \tag{3.3f}$$

Later, Durand (1953), after analyzing the values obtained by SOGREAH and using the Dimensional Analysis theory, proposed the following relation

$$i_m = i_w \left[1 + C_V \; K_D \; f_1(S_S - 1) \; f_2 \left(\frac{v^2}{gD} \right) f_3 \left(\frac{\omega^2}{gd} \right) \right] \tag{3.4}$$

where K_D, f_1, f_2 and f_3 are values that have to be obtained in the laboratory.

Following with their studies and using eq. 3.4 Durand and Condolios (1956) proposed the following equations

$$i_m = i_w \left[1 + C_V \; K'_D \left(\frac{v^2}{g \; D(S_S - 1)} \sqrt{C_D} \right) \right]^{-3/2} \tag{3.5}$$

$$i_m = i_w \left[1 + C_V \; K_D \left(\frac{v^2}{g \; D(S_S - 1)} \sqrt{\frac{g \; d \; (S_S - 1)}{\omega^2}} \right) \right]^{-3/2} \tag{3.6}$$

According to Newitt et al (1955), $K_D = 121$ and Bonnington (1959) considers that $K'_D = 150$.

The range of values for which the deduction of eq. 3.4 was done was with D between 40 and 580 mm; d of uniform size between 0.2 and 25 mm, S_S between 1.6 and 3.95 while volume concentrations varied between 2 and 22.5%.

Smith (1955) points out that for concentrations up to 33%, sands with ω between 0.03 and 0.12 m/s and D of 50.8 and 76.2 mm, eq. 3.6 yields good results when $K_D = 121$.

Due to the fact that eqs. 3.4, 3.5 and 3.6 were developed for material with uniform size, Condolios et al (1963) suggested that they can also be applied for mixtures of different diameters calculating C_D with the following equation

$$C_D = p_1 \sqrt{C_{D_1}} + p_2 \sqrt{C_{D_2}} + \ldots + p_i \sqrt{C_{D_i}} \tag{3.7}$$

where p_1, p_2,...., p_i are the weight percentages of the material with their corresponding drag coefficients C_{D_1}, C_{D_2},...., C_{D_i}

Bonnington (1959) proposes to use a C_D calculate with weighted average, which means that

$$C_D = \frac{\Sigma C_{D_i} \, P_i}{\Sigma P_i} \tag{3.8}$$

From tests done by different authors it is important to emphasize the following with reference to coefficient K_D and to the exponent in eq. 3.6. For Koch (1962), K_D is equal to 81; Ellis et al (1963) consider that $K_D = 85$; for suspensions of nickel particles and water Ellis and Round (1963) find that $K_D = 385$ (with $S_S = 8.9$ and $d = 0.106$ mm); Babcock (1964) agrees with the value of K_D proposed by Koch; Hayden and Stelson (1971) propose that $K_D = 83.3$ and the exponent equal to 1.3.

However, Garde and Ranga Raju (1985) point out that big mistakes can be done when using eq. 3.6 mainly if the term

$$\frac{V^2}{g \, D(S_S - 1)} \sqrt{\frac{gd(S_S - 1)}{\omega^2}}$$ is big, even if they do not say when it

is big. Toda et al (1969) agree that eq. 3.6 yields good results, but marks some differences in the data.

Newitt et al (1955) divide the heterogeneous regime into two, one corresponding to the condition according to which particles move in suspension (even if their vertical distribution is asymmetric) and the other to the existence of sediment deposit over the pipeline bed (he calls this condition as flow with a moving bed). For the first case this author proposes the use of the following equation

$$i_m = i_w \left[1 + 1100 \, C_V \, (S_S - 1) \, \frac{\omega \, g \, D}{V^3} \right] \tag{3.9}$$

This equation was calibrated for sand, gravel and coal particles, with ω between 0.01 and 0.25 m/s, S_S between 1.18 and 4.60, and concentrations between 0 and 37%, but the tests were limited to only pipeline with a 25.4 mm diameter. For Graf (1971) eqs. 3.6 and 3.9 present differences even if small; however, Hayden and Stelson (1971) say that they found a low correlation between the measured and calculated values with eq. 3.9.

For different diameters the representative ω should be calculated from

$$d_m = \Sigma d_i \, P_i \tag{3.10}$$

Kriegel et al (1966) suggest the use of a relation between the particles fall velocity and the turbulence of the flow. Upon these bases they propose the following relations

$$i_m = i_w + i_s \tag{3.11a}$$

$$i_s = f_K \frac{1}{D} \frac{V}{2g} \tag{3.11b}$$

$$f_K = 0.282 \ C_{\mathcal{V}} \ (S_S - 1) \left[\frac{\omega^3}{g\nu} \right]^{1/3} \left[\frac{gD}{V^2} \right]^{4/3} \tag{3.11c}$$

where i_s is the head loss due to the solids, in m/m; f_K the friction factor, adimensionless; ν the kinematic viscosity of the transporting fluid, in m^2/s. The tests to obtain this equation were carried out in two pipelines with D equal to 26.2 and 53.5 mm, ω varied between 0.037 and 1.07 m/s, S_S between 1.38 and 4.62 and the concentrations used were smaller than 22%. Besides, this researcher used the data by Durand (1953) and Führböter (1961).

Zandi and Govatos (1967) proposed the use of a adimensional number called N_I, known as the Zandi number or the flow regime index, which is given by

$$N_I = \frac{V^2 \sqrt{C_D}}{C_{\mathcal{V}} \ D_g \ (S_S-1)} \tag{3.12a}$$

For heterogoneous regime N_I should be bigger than 40, if not, it will be the case of sediment transport with deposit over the pipeline bed. However, its analysis showed that it was a conservative criterion and Babcock (1971) suggested the use of 10 instead of 40. Considering all this the equation recommended is

$$i_m = i_w \left[1 + 280 \ C_{\mathcal{V}} \ \left(\frac{V^2 \sqrt{C_D}}{g \ D(S_S - 1)} \right) \right]^{-1.93} \tag{3.12b}$$

These researchers used the data (2549) by different authors and point out that with the i_m value calculated with eq. 3.12b, a $\pm$ 40% approximation is obtained, which they consider acceptable.

Charles (1970) proposes a relation which is the result of combining eq. 3.5 and the one used for the homogeneous regime; it is given by

$$i_m = i_w \left[1 + 120 \ C_{\mathcal{V}} \left(\frac{V^2 \sqrt{C_D}}{g \ D(S_S - 1)} \right)^{-1.5} + (S_S - 1)C_{\mathcal{V}} \right] \tag{3.13}$$

Eq. 3.13 was calibrated for mixtures of sand and water and nickel with water, with S_S = 2.65 and 8.9, respectively . His author points out that it has to be used with reserve if $C_{\mathcal{V}} > 0.25$.

Babcock (1971) made a detailed analysis of the different variables that take part in the previously mentioned equations. From his observations, among others, the following stand out: for small particles and low concentrations the exponent to be used in eq. 3.5 is equal to -0.25; for coarse quartz particles or for heavy minerals, using a pipeline with D = 25.4 mm, he proposes the following equation

$$i_m = i_w \left[1 + 60.6 \ C_{\Psi} \left(\frac{v^2}{gD} \right)^{-1} \right]$$ (3.14)

However, the use of this equation is restricted to velocities smaller than 2.85 m/s and the particles drag over the pipeline bed.

Turian and Yuan (1977) used 2848 experimental data to establish a correlation to estimate the friction factor f_m in heterogeneous regime, which is given by

$$f_m = f_w + 0.5513 \left(C_{\Psi}^{\ 0.8687} \right) (f_w)^{1.2} (C_D^{\ -0.1677}) \left[\frac{v^2}{Dg(S_S-1)} \right]^{-0.6938}$$ (3.15)

The value of i_m is obtained with eq. 3.3c.

Wasp et al (1977) suggested the use of a technique to analize this kind of regime, based upon the studies done in a pipeline with D = 304.8 mm, which consists in calculating two losses per friction: one due to the "vehicle" (corresponding to the condition of homogeneous regime) and the other corresponding to the particles that move near the bed (which is calculated, for instance, with eq. 3.6); their sum is the total loss. The disadvantage of this method lies in the fact that it is iterative.

Lazarus and Neilson (1978) say that the methods used to predict the head loss seem to be inadequate, because there are great variations in the intervening values of the variables, like, for instance, pipe diameter, density and specific weight of the fluid, size, specific weight and concentration of the particles. Using the experimental results from different laboratories they propose a relation and compare the values it yields with the equations generally used. Finally, they found that their expression is the one that better adjusts to the data; it is given by

$$f_m = f_w \ 1.45 \ (\lambda^{-0.4})$$ (3.16a)

$$\lambda = \frac{v^2}{gD} \sqrt{\frac{\nu}{V \ D \ M_*}} \frac{1}{S_S} \left[100 \ (d/D)^{(0.44 \ \log(d/D)+1.31)} \ \tan h(1+M_*) \right]$$ (3.16b)

$$M_* = S_S \left(\frac{C_{\Psi}}{1-C_{\Psi}} \right)$$ (3.16c)

Eq. 3.16a for heterogeneous regime may be applied when $0.8 < \lambda < 2.53$. In the same way i_m is obtained with eq. 3.3c.

4. ANALYSIS OF RESULTS

A comparison was performed between the head loss values measured by Babcock (1971) and those obtained when applying each one of the methods described in chapter 3, see figs 2 and 3. This was done for two values of C_V, one equal to 0.2 and the other for 0.3.

With eqs. 3.1 and 3.2 was first obtained the range of values between which fall velocity should be to correspond to the homogeneous regime. The results were $V_H = 3.3$ m/s, $V_C = 1.16$ m/s ($C_V = 0.2$) and $V_C = 1.19$ m/s ($C_V = 0.3$).

It is important to point out that the values reported by Babcock correspond to sand particles with a mean diameter of 0.72 mm, $S_S = 2.651$ and $\omega = 0.08$ m/s. He performed the tests in an horizontal pipeline of transparent plastic with D = 25.4 mm.

From the analysis of figs. 2 and 3 the following points stand out:

- All the criteria yield different values, except those by Durand and Worster which even if equal do not adjust to the experimental ones.

- For velocities bigger than V_H the methods by Durand, Worster, Bonnington, Kriegel et al and Zandi tend to be almost equal to the loss of water alone; this takes place only when $C_V = 0.2$.

- The equation proposed by Newitt et al yields and envelope curve with respect to the measured values, since for the two C_V values, a part passes over the experimental data and another part below them. This is valid when the constant that appears in eq. 3.9 is 1800, but if its 1100 it passes below.

- The equation suggested by Charles for velocities bigger than 2.6 m/s tends to adjust to the experimental results.

- The criteria by Turian and Yuan and by Lazarus and Neilson give values almost equal to those of the head loss of water alone, and that is why they were not drawn in figs. 2 and 3.

- Figs. 2 and 3 emphasize the fact that the minimum value of velocity V_C, in general, fulfills in almost all the cases, except in the criteria by Newitt et al and Zandi and Govatos.

- Due to the fact that the eq. 3.9 proposed by Newitt et al is the one that better adjusts to the data, a comparison was performed of the results that would be obtained if the value of the constant 1100 that appears in that equation were changed, figs. 4 and 5 show this. From the analysis of these

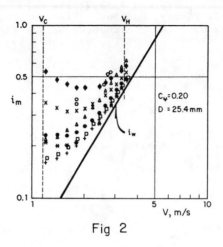

Fig 2

+ Durand-Condolios (1952) and Worster (1952)
x Newitt et al (1955),1100
♦ Newitt et al (1955),1800
● Bonnington (1959)
□ Kriegel et al (1966)
* Zandi Govatos (1967)
△ Charles (1970)
○ Measures values for Babcock (1971)

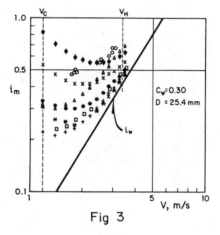

Fig 3

126

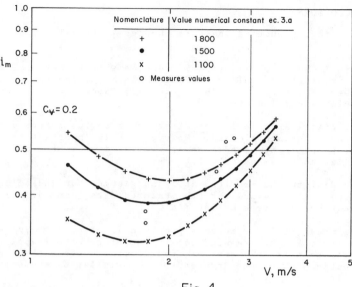

Fig 4

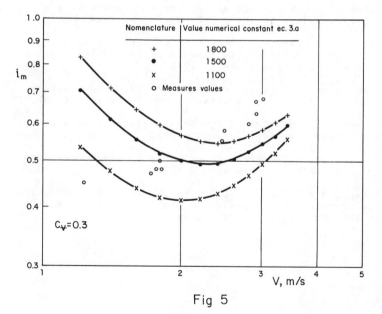

Fig 5

figures it can be concluded that a change in the value of the
constant also produces a change in the adjusting curve.   In
this particular case the best results are obtained when using
1800 instead of 1100.

5. CONCLUSIONS

According to what was said in chapter 4 the following can be
concluded

- There is not one criterion to define with enough precision
  the transport of solids and water in an horizontal pipeline,
  rather, for every case in particular, laboratory studies
  should be carried out.

- The criterion by Newitt et al seems to be the better adjusted
  to the experimental data. However, it is important to
  remember that the conditions for which it was obtained are
  similar to those used by Babcock.

- According to the results, it is suggested that and adjusting
  equation is obtained for the measured data, considering equal
  conditions and type of material.

- In general, all the methods proposed in chapter 3 give values
  smaller than the measured ones. Therefore, their application
  should be extremely careful, considering whether the
  conditions they were deduced for are similar to those of the
  problem under study.

6. REFERENCES

1. Babcock, H, The state of the art of transporting solids in
    pipelines, Chem Eng Rev, vol 60, No 48, 1964

2. Babcock, H, Heterogeneous flow of heterogeneous solids, in
    Advances in Solid-Liquid Flow in Pipes and its Application,
    ed I Zandi, Pergamon Press, 1971

3. Bonnington, S, Experiments on the hydraulic transport of
    mixed-sized solids, British Hydrom Res Assoc, RR637, 1959

4. Condolios, E and Chapus,E, Transporting solid materials in
    pipelines, Chem Eng, june-july 1963

5. Charles, M, Transport of solids by pipeline, Paper A3,
    Hydrotransport 1, Coventry, England, 1970

6. Durand, R and Condolios, E, Experimental investigation on
    the transport of solids in pipes, Le Journels d'Hydraulique,
    Societe Hydrotecnique de France, june 1952

7. Durand, R, Basic relationships of the transportation of solids in pipes-experimental research, Proc Intern Assoc Hydr Res, 5th Congrs, Minneapolis, vol 5, pp 89-103, 1953

8. Durand, R and Condolios, E, Donneés tecniques sur le refoulement hydraulique des matériaux solides en conduite, Rev l'Industrie Minerals, Special number 1F, june 1956.

9. Ellis, H, *et al*, Slurries: Basic principles and power requirements, Ind Eng Chem, vol 55, No 8, aug 1963

10. Ellis, H and Round, G, Laboratory studies on the flow of nickel--water suspensions, Canadian Mining and Metallurgical Bulletin,No 56, oct 1963

11. Führböter, A, Uber die forbeninu von sand-wasser-gemischen in rohrleitungen, Mitteil Franzius-Inst, Techn Hochschule Hannover, Heft 19, 1961

12. Garde, R and Ranga Raju, K, Mechanics of Sediment Transportation and Alluvial Stream Problems (second edition), Halsted Press Book, jul 1985

13. Graf, H and Acaroglu,E, Homogeneous suspensions in circular conduits, Journal of Pipeline Division, vol 93, No PL2, 1967

14. Graf, H, Hydraulics of Sediment Transport, McGraw-Hill Book Co., 1971

15. Hayden, J and Stelson,T, Hydraulic convenyance of solids in pipes, in Advances in Solid-Liquid Flow in Pipes and its Application, ed I Zandi, Pergamon Press, 1971

16. Lazarus, J and Neilson, I, A generalized correlation for friction head losses of settling mixtures in horizontal smooth pipelines, Paper B1, Hydrotransport 5, Hannover, 1978

17. Koch, L, Solids in pipes, International Science and Technology No 26, feb 1962

18. Kriegel, E, *et al*, Hydraulischer transport körniger feststoffe durch waagerechte rohrleitungen, VDI-Forshungsh, No 515, 1966

19. McElvain, R and Cave, I, Transportation of tailings, World Mining Tailings Symposium, 1972

20. Newitt, D, *et al*, Hydraulic conveying of solids in horizontal pipes, Trans Inst Chem Engrs, vol 33, No 2, 1955

21. Smith, R, Experiments on the flow of sand-water slurries in horizontal pipes,Trans Inst Chem Engrs, vol 33, No 2, 1955

22. Toda, M, *et al*, Hydraulic conveying of solids through horizontal and vertical pipes, Chem Eng Japan, vol 33, No 1, 1969

23. Turian, R and Yuan, T, Flow of slurries in pipelines, Journal AICHE, vol 3, No 3, may 1977

24. Wasp, E, *et al*, Solid-Liquid Flow Slurry Pipeline Transportation, Trans Tech Publications, 1977

25. Worster, R , The hydraulic transport of solids, Proceedings Colloquim on the Hydraulic Transport of Coal, London, nov 1952

26. Zandi, I and Govatos, G, Heterogeneous flow of solids in pipelines, Journal of the Hydraulics Division, ASCE, vol 93, No HY3, Proc Paper 5244, may 1967

A Two-Phase Pressure Drop Calculation Code Based on a New Method with a Correction Factor Obtained from an Assessment of Existing Correlations

MOON-HYUN CHUN and JAE-GUEN OH
Department of Nuclear Engineering
Korea Advanced Institute of Science and Technology
P.O. Box 150, Cheongryang, Seoul, Korea

ABSTRACT

Ten methods of the total two-phase pressure drop prediction based on five existing models and correlations have been examined for their accuracy and applicability to pressurized water reactor conditions. These methods were tested against 209 experimental local and bulk boiling conditions : Each correlations were evaluated for different ranges of pressure, mass velocity and quality, and best performing models were identified for each data subsets. A computer code entitled 'K-TWOPD' has been developed to calculate the total two-phase pressure drop using the best performing existing correlations for a specific property range and a correction factor to compensate for the predicted error of the selected correlations. Assessment of this code shows that the present method fits all the available data within $\pm 11\%$ at a 95% confidence level compared with $\pm 25\%$ for the existing correlations.

I. INTRODUCTION

Although the average exit enthalpy of a current PWR core under normal operation does not exceed the saturation value, bulk boiling can occur in a hot channel. Moreover, in the event of a reactor transient, such as a small or a large break LOCA, bulk boiling can occur throughout the core. For both steady and transient conditions it is necessary to predict the pressure loss as accurately as possible for a two-phase mixture since pressure drop not only influences flow distribution and pumping power requirements but also the hot channel is limiting from DNB considerations.

Various approaches for two-phase flow pressure drop predictions have been proposed during the past two decades. The existing two-phase pressure drop models can be classified into : (a) homogeneous models [1-4], (b) separated flow models [5-6], and (c) mass flow correction models [7-8]. However, these existing models still have the following inherent drawbacks for application : (1) limited accuracy (e.g., the accuracy of the most widely used Martinelli-Nelson method has been quoted as +35%), (2) limited applicable range of properties (such as pressure, mass velocity, and quality), and (3) often complicated and cumbersome numerical procedures (or the use of tables and figures) is required to obtain correction factors.

Idsinger et al. [9] compared a large number of existing correlations against measured data for conditions representing transient as well as steady-state BWR operation, and showed that the correlation which had the least RMS error overall for the normal BWR operation conditions is the Armand-Treschev Correlation [10]. Also, they reported that the correlations which have shown the best overall performance were (a) homogeneous theory, (b) the Thom [6], and (c) the Baroczy model [7].

The purpose of this paper is : first, to present the results of an assessment of existing correlations for their applicability to PWR conditions ; and secondly, to present the outlines of a computer code entitled 'K-TWOPD' for two-phase pressure drop calculation based on

131

the best performing existing models for specific property ranges and a correction factor to compensate for the predicted errors of selected correlations [1-8].

2. SUMMARY OF EXISTING TWO-PHASE PRESSURE DROP CORRELATIONS EXAMINED

The existing pressure drop correlations may be classified into 3 broad categories : (1) the homogeneous model, (2) the separated flow model, and (3) the mass flow correction methods based on the homogeneous or separated flow models. A brief description of those models, that are selected for assessments of their applicability to PWR conditions and their accuracy of pressure drop predictions, is given here to aid in understanding and for convenience in discussion.

The Homogeneous Model

The total static pressure gradient evaluated from the homogeneous model can be represented by the following expression [11] :

$$(\frac{dp}{dz}) = (\frac{dp}{dz}F) + (\frac{dp}{dz}a) + (\frac{dp}{dz}z) \tag{1}$$

where

$$-(\frac{dp}{dz}F) = \frac{2f_{TP}G^2V_f}{D_e}[1+x(\frac{V_{fg}}{V_f})] \tag{2}$$

$$-(\frac{dp}{dz}a) = G^2[V_{fg}\frac{dx}{dz} + x\frac{dV_g}{dp}(\frac{dp}{dz})] \tag{3}$$

$$-(\frac{dp}{dz}z) = \frac{g \sin\theta}{V_f [1+x(\frac{V_{fg}}{V_f})]} \tag{4}$$

All the terms in Eqs. (1)-(4) are defineable except the two-phase friction factor f_{TP}. To use the homogeneous model it is necessary to apply a suitably defined single-phase friction factor to two-phase flow. A number of different approaches have been made to the definition of this two-phase friction factor : (1) Owens [1] has assumed that the friction factor f_{TP} is equal to that which would have occurred had total flow been assumed to be all liquid. (2) Another approach is to evaluate the friction factor f_{TP} using a mean two-phase viscosity $\bar{\mu}$ in the normal friction factor relationships. The three different forms of the relationship between $\bar{\mu}$ and the quality x proposed by McAdams et al. [2], Cicchitti et al. [3], and Dukler et al. [4] are

$$\frac{1}{\bar{\mu}} = \frac{x}{\mu_g} + \frac{(1-x)}{\mu_f} \qquad \text{(McAdams et al.)} \tag{5}$$

$$\bar{\mu} = x\mu_g + (1-x)\mu_f \qquad \text{(Cicchitti et al.)} \tag{6}$$

$$\bar{\mu} = \bar{\rho}[xV_g\mu_g + (1-x)V_f\mu_f] \qquad \text{(Dukler et al.)} \tag{7}$$

Thus, the methods proposed by Owens [1], McAdams et al. [2] and Dukler et al. [4] fall into the category of the homogeneous model.

The Separated Flow Model

The total static pressure gradient as evaluated from the separated flow model can be represented by substitution of the following equations into Eq. (1) :

$$-(\frac{dp}{dz}F) = [\frac{2f_{fo}G^2V_f}{D_e}]\phi^2_{fo} \tag{8}$$

$$-(\frac{dp}{dz}a) = G^2\frac{d}{dz}[\frac{x^2V_g}{\alpha} + \frac{(1-x)^2V_f}{(1-\alpha)}] \tag{9}$$

$$-(\frac{dp}{dz}z) = g\ sin\theta\ [\alpha\rho_g + (1-\alpha)\rho_f] \tag{10}$$

In order to apply Eqs.(8)-(10) it is necessary to develop expressions for the two-phase multiplier (ϕ^2_{fo}) and the void fraction (α) in terms of the independent flow variables. As in the case of the homogeneous model a number of different approaches have been made to obtain ϕ^2_{fo} and α : (1) Martinelli-Nelson [5] presented values of ϕ^2_{fo} as a function of mass quality and pressure (e.g., Fig. 2.4 and Table 2.2 in Ref. 11) and also values of α as a function of mass quality x with pressure as parameter (e.g., Fig.2.6 in Ref. 11). (2) Thom [6] proposed an alternative set of consistent values for the two-phase frictional multiplier (ϕ^2_{fo}) as a function of pressure and mass quality (e.g., Table 2.3 in Ref.11). Also Thom [6] proposed to fit mass quality-void fraction curves of the type.

$$\alpha = \frac{\gamma x}{1+x(\gamma-1)} \tag{11}$$

to their new data (Table 2.3 in Ref. 11) in which the slip factor γ is a constant at any given pressure.

Mass Flow Correction Methods for Use with the Homogeneous or Separated Flow Models

Attempts to correct existing models for the influence of mass velocity on the friction multiplier ϕ^2_{fo} have been published by Baroczy [7] and by Chisholm [8] : (1) The method of calculation proposed by Baroczy [7] employs two separate sets of curves. The first of these is a plot of the two-phase frictional multiplier ϕ^2_{fo} as a function of a physical property index $[(\mu_f/\mu_g)^{0.2}(V_f/V_g)]$ with mass quality x as parameter for a reference mass velocity of 1356 Kg/m^2-s (e.g., Fig.2.12 and Table 2.5 in Ref.11). The second is a plot of a correction factor Ω expressed as a function of the same physical property index for mass velocities of 339, 678, 2812, and 4068 Kg/m^2-s with mass quality as parameter (e.g., Fig.2.13 in Ref. 11). Thus,

$$(\frac{dp}{dz}F)_{TP} = \frac{2f_{fo}G^2V_f}{D_e}\phi^2_{fo}\ _{(G=1356)}\Omega \tag{12}$$

(2) Chisholm [8] has shown that the Lockhart-Martinelli [12] two-phase multiplier can be

transformed with sufficient accuracy for engineering purposes to

$$\phi^2_{fo} = 1+(\Gamma^2-1) \ [Bx^{(2-n)/2}(1-x)^{(2-n)/2}+x^{2-n}] \tag{13}$$

where

$$\Gamma = [\frac{(\frac{dp}{dz}F)_{go}}{(\frac{dp}{dz}F)_{fo}}]^{0.5} \tag{14}$$

n = exponent in Blasius' relation for friction factor,

and recommend values of B as a function of Γ and G (Tables 2 in Ref. 8).

Summary of Correlations Examined

A summary of the correlations selected for assessments of their applicability to PWR conditions and their accuracy of pressure drop predictions is shown in Table 1. This table also shows that a different combination of frictional terms and void fraction models were used in the 5 fundamental methods examined. This approach is taken becuase the use of different friction factor and void fraction models can obviously affect the pressure drop predictions made by the same separated flow models, in particular.

3. ASSESSMENT OF EXISTING CORRELATIONS AND DISSCUSTIONS

Methods Used to Evaluate Existing Correlations

For an objective evaluation and judgement of the relative superiority between the existing correlations it is necessary to define a quantitative parameter that can either measure or reveal the accuracy of each correlation. The most important parameters used to judge the relative superiority of a model in the present work is the fractional error, which is defined as the ratio of the error of the quantity to the true value of the quantity

$$\epsilon = \frac{(\Delta P)_{corr}-(\Delta P)_{exp}}{(\Delta P)_{exp}} \tag{15}$$

where ϵ is the fractional error of the predicted value of the correlation, $(\Delta P)_{corr}$ is the total amount of two-phase pressure drop calculated from the correlation, and $(\Delta P)_{exp}$ is the total amount of two-phase pressure drop obtained from the experimental data which is assumed to be the true value.

In order to compare the accuracy of the predicted values $(\Delta P)_{corr}$ of the correlations for groups of data, the mean error $\bar{\epsilon}$, root-mean-square (RMS) error ϵ_{RMS}, and standard deviation (SD) of the error from the mean σ_ϵ were also computed using the following equations. :

$$\bar{\epsilon} = \sum_{i=1}^{N} \frac{\epsilon_i}{N} = \frac{\epsilon_1+\epsilon_2+ \ ----- \ +\epsilon_{N-1}+\epsilon_N}{N} \tag{16}$$

$$\epsilon_{RMS} = [\sum_{i=1}^{N} \frac{\epsilon^2_i}{N}]^{1/2} \tag{17}$$

134

$$\sigma_\epsilon = [\frac{1}{N} \sum_{i=1}^{N}(\epsilon_i - \bar{\epsilon})^2]^{1/2} = [(\epsilon_{RMS})^2 - \bar{\epsilon}^2]^{1/2} \qquad (18)$$

Roughly speaking, the correlation that produces a smaller absolute values of $\bar{\epsilon}$, ϵ_{RMS}, and σ_ϵ of the predicted values $(\Delta P)_{corr}$ may be considered as a better correlation.

Experimental Data Used to Assess Existing Correlations

The experimental data represented several geometries (e.g., tube, annulus, rectangular channels, and rod array) and had the following property ranges :

Pressure (P)	:	$8.5\text{-}16.5 \ MN/m^2$
Mass Velocity (G)	:	$406.8\text{-}8000 \ Kg/m^2\text{-s}$
Exit Quality (x)	:	subcooled to 60%
Equivalent Diameter	:	4.83-38.10 mm

All the data were subdivided into 14 subsets based on the property and flow conditions to study how each correlations behave for different ranges of pressure, mass velocity, and quality. The property/flow condition groupings combined data of similar pressure ranges, quality ranges, and mass velocity ranges. Table II shows the property and flow condition ranges from which 14 data subsets were formed.

Calculation of Two-phase Pressure Drop $(\Delta P)_{corr}$

The total amount of two phase pressure drop from each correlation, $(\Delta P)_{corr}$ in Eq.(15), is obtained by stepwise integration of Eqs. (1)-(4) and Eqs. (8)-(10). To reduce the uncertainty associated with the two-phase pressure drop correlations the local and bulk boiling regions were subdivided into a finite number of smaller differential control volumes and thermodynamic properties for each control volumes were found as a function of the temperature computed from a heat balance equation for the given region and system pressure. The whole process of computation for $(\Delta P)_{corr}$ has been performed by the computer code entitled 'K-TWOPD'. The flow diagram of this code is given in Fig.1. Input data required for the 'K-TWOPD' to calculate $(\Delta P)_{corr}$ and the sample result are shown in Table III.

Comparison of Existing Correlations with Data

Table IV gives the results of the overall evaluation of existing correlations using 209 experimental data points : This table gives the mean $(\bar{\epsilon})$, the RMS (ϵ_{RMS}), the standard deviation (σ_ϵ) of the error, and the confidence limits of correlations at a 95% probability level. The correlations are identified by the numbers indicated in Table I. The correlation errors appearing in Table IV refer to the uncertainty in the two-phase pressure drop based on data and the discrepancy between data and correlations. The errors are obtained by Eqs. (15)-(18).

Table V, on the other hand, shows the results of the evaluations of each correlations for 14 data subsets shown in Table II. Also, in Fig.2 the RMS error (ϵ_{RMS}) ranges of existing correlations examined for each data subsets are shown in order of increasing RMS error along with the list of 4 best performing correlations for each data groups. From the Table V and Fig.2 it can be observed that the RMS error ranges of the correlation for low quality region $(x \le 0.1)$ are considerably smaller than those for high quality region $(x \ge 0.1)$.

As can be seen from Figs. 3-5, comparisons of experimental and predicted total pressure drop using the Owens model [1], Martinelli-Nelson model [5], and Thom model [6] show that the experimental data fall within the -28% and +23% spread of the correlation at a 95% probability level.

4. OUTLINES OF A NEW METHOD AND ITS ASSESSMENT

Outlines of the New Method Proposed

The basic concept of the new method proposed here is to calculate the total two-phase pressure drop using the best performing correlation for a specific data range and then apply a correction factor to compensate for the expected error of the selected correlation.

To summarize, the step-by-step calculation performed by the computer code developed here (i.e., the 'K-TWOPD' code) is as follows :

(1) Select the best performing correlation that has the smallest RMS error out of the 10 methods shown in Table I for a given specific property and flow conditions with the aid of Table V and Fig.2.

(2) Calculate the total two-phase pressure drop $(\Delta P)_{opt}$ using the selected correlation.

(3) Obtain a correction factor ψ to compensate the expected error of the selected correlation by

$$\psi = B_1 x + B_2(\frac{q''}{G}) + B_3 L^2 + B_4 \log_{10}(\frac{q''}{G}) + B_5 \log_{10}(\frac{L}{D_e}) + Co \qquad (19)$$

where

$$B_1 = 0.26739, \qquad\qquad B_2 = -0.36968,$$
$$B_3 = 0.00024, \qquad\qquad B_4 = -0.07543,$$
$$B_5 = -0.13777, \qquad\qquad Co = 1.35224,$$

x is the exit quality, q" is the heat flux in Btu/hr-ft^2, G is the mass velocity in lbm/hr-ft^2, L is the pipe length in ft, and De is the equivalent diameter in ft.

(4) Finally, the predicted value $(\Delta P)_{new}$ is obtained by substituting $(\Delta P)_{opt}$ and ψ into the the following expression :

$$(\Delta P)_{new} = \frac{(\Delta P)_{opt}}{\psi} \qquad (20)$$

Procedures Used to Obtain the Correction Factor ψ

If some input variables to evaluate $(\Delta P)_{corr}$ are very influential on the error ϵ, those input variables or their combinations require a close study in composing a correction factor that can reduce the error vectors. The stepwise regression technique (SRT) [13] is used to select these sensitive parameters, thereby to build a regression equation composed of not undue

136

member of input parameters while the constructed regression equation reveals the input-output relationship. This procedure enables to select or remove the most important variables sequentially. At each step, to decide the adequacy of the constructed regression model composed of the selected input parameters, the analysis of variance (ANOVA) calculation has been performed.

The procedure used to derive the correction factor ψ is as follows :

(1) Select the best performing correlation for each data subset with the aid of Table V and Fig.2, and collect the selected correlations.

(2) Calculate the error of those correlations collected in step 1.

(3) Select the most important parameters sensitive to produce errors obtained in step 2 by the stepwise regression technique.

(4) Introduce new variables by the combination of selected parameters, taking its logarithm or powers.

(5) The stepwise regression technique is used again to select or remove the introduced new variables in step 4.

(6) Build up the error response, $\hat{\epsilon}$, which is a function of the selected variables.

(7) Finally, from the definition of the error, Eq.(15), the correction factor ψ to compensate the expected error $\hat{\epsilon}$ in the $(\Delta P)_{opt}$ obtained by the best performing correlation is derived as

$$\psi = 1 + \hat{\epsilon} \tag{21}$$

where $\hat{\epsilon}$ is the predicted error regressed with the important parameters as shown in Eq. (19).

Evaluation of the New Method

As described in the previous section, once the best performing correlation that has the least RMS error is selected for a specific range of pressure, mass velocity, and exit quality, the two-phase pressure drop calculation with a correction factor ψ to compensate the expected error of the selected best performing correlation can be carried out by the computer code K-TWOPD : This code calculates the exit quality internally by the Bowring model [14]. To examine the goodness of fit and the accuracy of this new procedure the following statistical analysis has been performed.

(1) Stepwise Regression Results : the importance of variables and regression of the correction factor ψ, Eq. (19), are investigated by stepwise regression procedure [13]. In addition, the analysis of variance (ANOVA) has been performed to see the difference between ϵ and $\hat{\epsilon}$. The analysis of variance table obtained from the final stage of the stepwise regression is shown in Table VI. There are basically three factors that describe how well the model actually fits the observed data [13], (1) the mean square due to residual variation (MSE), (2) the coefficient of determination R^2, and (3) the calculated F value. The smaller the MSE value, the better is the regression. Notice that the MSE value is 0.00388 as shown in Table VI. Also one should be pleased if the sum of squares (SS) due to regression is much greater than the SS due to the residuals, or what amount

to the same thing if the ratio R^2 is not too far from unity. The F value serves to test how well the regression model fits the data. If the calculated F value exceeds the 'critical-F' value obtained from a statistics table for $F(k, n-k-1 ; \alpha)$ the regression is significant at a level of $100 (1-\alpha)\%$. It can be observed that all the values shown in Table VI pass the above tests for the precision of the estimated regression and the F-test for significance of regression.

(2) Histograms of Errors : to examine the frequency distribution of errors from the new method two histograms of errors are compared as shown in Fig.6 : top figure (Fig.6a) shows the distribution of errors obtained by using the best performing correlations for each 14 subsets (shown in Table II) with the aid of Fig.2, whereas lower figure (Fig.6b) shows the results obtained from the new procedure with a correction factor ψ. From these figures two facts can be observed : first, the distribution of error is nearly symmetrical for both cases, with the highest frequency occuring in the middle where the error is zero. Secondly, the magnitude of all errors, such as $\overline{\epsilon}$, ϵ_{RMS}, and σ_{ϵ} of the new procedure is appreciably smaller than those of the best performing correlations.

(3) Comparison of Experimental and Predicted Total Pressure Drop Using a New Procedure : Fig.7 shows the comparison between experimental data and predicted total two-phase pressure drop using the present new procedure with a correction factor. As can be seen from this figure, all the experimental data fall within the $+11\%$ spread of the new procedure at a 95% confidence level, whereas the same data has approximately $\pm25\%$ spread for the existing correlations examined in Figs. 3-5.

Applicability of the New Method to PWR Analysis

In previous sections 10 different approaches of existing correlations have been compared against measured data for conditions representing transient as well as steady state PWR operation. The data subsets investigated that are pertinent to the normal operation of the PWR, in particular, are those representing the following properties :

Pressure	: 15.1-15.8 MN/m^2 (2200-2300 psia) ;
Mass Velocity	: 3254.4-3525.6 Kg/m^2-s (2.4-2.6 Mlb/ft^2-hr)
Quality	: 0-1.

As can be seen in Fig.2, the correlation which had the least overall RMS error for the above conditions (equivalent to the subsets No.14 in Table II) is the Martinelli-Nelson correlation (ID No.5 in Table I).

In the event of a nuclear reactor transient such as the small and large LOCAs, the quality can be higher, whereas the mass velocity and the pressure in the core can be lower. Under these circumstances, conditions above may be changed and the best performing existing correlation for the changed conditions can be found from Fig.2. However, the new procedure outlined in previous sections is recommended for the more accurate analysis of PWR pressure drop at steady and transient core conditions. For the above property ranges representing the normal operation of the PWR, the RMS error of the new approach (for the subset No.14) is $\epsilon_{RMS} = 0.0569$ whereas that of the best performing correlation (ID No.5) is $\epsilon_{RMS} = 0.1015$. This indicates that the new approach is superior to the best performing correlation for the subset No.14. In general, this is true for all the subsets as can be deduced from Fig.6 and the derivation of the correction factor ψ.

138

5. CONCLUSION

The performance of the ten two-phase pressure drop prediction methods were evaluated for their accuracy and applicability to PWR conditions. Specifically, the best performing correlations were identified for each data subsets representing specific ranges of pressure, mass velocity, and quality and data sets representing PWR conditions. The root-mean-square error between model prediction and data was used as the criteria to evaluate model performance.

(1) Considering the total data bank, the existing models exhibiting minimum error were (a) Chisholm model (ID No.9 and 10), (b) Baroczy model (ID No.7 and 8), and (c) the homogeneous model with the two-phase viscosity term based on all-liquid flow (ID No.1).

(2) The best performing models or methods for each data range can be found with the aid of Fig.2 or Table V.

(3) The new method proposed here is to calculate the total two-phase pressure drop using the best performing correlation for a specific data range and apply a correction factor ψ, defined by Eq. (19), to compensate the expected error of the selected correlation : that is, use Eq. (20). This new approach fits all the pressure drop data collected in the present work within $\pm 11\%$ at a 95% confidence level compared with $\pm 25\%$ for the existing correlations.

TABLE I : A SUMMARY OF
TWO-PHASE PRESSURE DROP CORRELATIONS EXAMINED

Correlation Category (Authors) :	ID No.	Ref. No.	Frictional Terms (f_{TP} or o^2_{fo})	Void Fraction Models
Homogeneous (Owens)	1	1	$f_{TP} = f_{fo}$	Homogeneous
Homogeneous (McAdams et al.)	2	2	Eq.5	Homogeneous
Homogeneous (Cicchitti et al.)	3	3	Eq.6	Homogeneous
Homogeneous (Dukler et al.)	4	4	Eq.7	Homogeneous
Separated Flow (Martinelli-Nelson)	5	5	Fig.2.4 in Ref.11	Fig.2.6 in Ref.11 Table 2.3 in Ref.11
Separated Flow (Thom)	6	6	Table 2.3. in Ref.11	Eq.11 Table 1 in Ref.6
Mass Flow Correction (Baroczy)	7	7	Fig.2.12 in Ref.11 Fig.2.13 in Ref.11	Fig.2.6 in Ref.11 Table 2.3 in Ref.11
Mass Flow Correction (Baroczy)	8	7	Fig.2.12 in Ref.11 Fig.2.13 in Ref.11	Eq.11 Table 1 in Ref.6
Mass Flow Correction (Chisholm)	9	8	Eq. 13 (or Eq.26 in Ref. 8)	Fig.2.6 in Ref.11 Table 2.3 in Ref.11
Mass Flow Correction (Chisholm)	10	8	Eq. 13 (or Eq.26 in Ref. 8)	Eq.11 Table 1 in Ref.6

TABLE II : THE RANGES OF PHYSICAL PROPERTIES AND
FLOW CONDITIONS USED TO FORM DATA SUBSETS

Pressure MN/m^2	Mass Flux Kg/m^2-s	Quality by mass	Number of Data	ID No. of Subsets
P≤13.79 (P≤2000 psia)	G<126	x<0.03	14	1
		0.03≤x<0.10	13	2
		x≥0.10	32	3
	126≤G<252	x<0.03	28	4
		0.03≤x<0.10	10	5
		x≥0.10	16	6
	G≥252	x<0.03	22	7
		x≥0.03	12	8
P≥13.79 (P≥2000 psia)	G<126	x<0.03	19	9
		0.03<x<0.10	25	10
		x≥0.10	21	11
	126<G<252	x<0.03	30	12
		x≥0.03	14	13
	G≥252	All Region	22	14

TABLE III : INPUT DATA FOR K-TWOPD CODE AND SAMPLE RESULTS

Input Data : Model ID no.   1
             Pipe Length   (m)   0.146
             Heat Flux   (W/m^2)   1.81x10^6
             De   (m)   0.0045
             Inlet Temperature   (degree C)   310.11
             Mass Flux   (Kg/m^2-sec)   5072.00
             System Pressure   (Pa)   1.013x10^8
             e/De   0.0002
             Inclination Angle   (degree)   90.0
Output :   Void Departure Position   0.0396 m
             Saturation Position   0.162 m
             Bulk Boiling Position   0.247 m

Beginning Point (m)	Interval (m)	Quality	dPf (Pa)	dPa (Pa)	dPg (Pa)
0.00	0.0396	0.00	26268.8	2895.8	2620.0
0.0396	0.0427	0.22	29647.2	16271.5	2688.9
0.0792	0.0427	0.66	30957.2	16340.4	2551.0
0.122	0.0427	1.09	32336.1	16340.4	2482.1
0.165	0.0427	1.53	33646.1	16340.4	2413.1
0.204	0.0427	1.97	34887.2	16340.4	2275.3
0.247	0.0335	2.61	29923.0	32060.4	1792.6
0.280	0.0335	3.47	31922.5	32060.4	1654.7
0.314	0.0335	4.32	33990.9	32060.4	1585.8
0.347	0.0335	5.18	35990.3	32198.2	1516.8
0.384	0.0335	6.03	37989.8	32198.2	1378.9
			357559.1	245175.5	23028.3

Total Pressure Drop = 6.26 x 10^5 (Pa)

TABLE IV : RESULTS OF OVERALL EVALUATION OF EXISTING CORRELATIONS

Model ID No.	Mean Error	RMS Error	S.D. of Error	Confidence Limits(%)	
				Upper	Lower
1	-0.07445	0.12507	0.10049	9.08616	-23.97585
2	-0.08043	0.13257	0.10538	9.29221	-25.37895
3	-0.08508	0.13367	0.10309	8.45094	-25.46660
4	-0.09164	0.13903	0.10455	8.03537	-26.36278
5	-0.01396	0.14710	0.14644	22.69316	-25.48523
6	-0.09435	0.14458	0.10955	8.58529	-27.45583
7	-0.06085	0.10657	0.08749	8.30697	-20.47734
8	-0.05257	0.10995	0.09657	10.62806	-21.14264
9	-0.05155	0.09934	0.08493	8.81569	-19.12494
10	-0.04379	0.10469	0.09509	11.26289	-20.02135

TABLE V : MEAN ERRORS, RMS ERRORS, AND STANDARD DEVIATIONS OF THE ERROR OF EXISTING CORRELATIONS FOR DATA SUBSETS

subset \ model		1	2	3	4	5	6	7	8	9	10
1	$\bar{\varepsilon}$	-0.0437	-0.0439	-0.0467	-0.0450	-0.0374	-0.0458	-0.0410	-0.0313	-0.0425	-0.0328
	ε_{RMS}	0.0573	0.0573	0.0589	0.0578	0.0558	0.0671	0.0575	0.0619	0.0575	0.0629
	σ_{ε}	0.0370	0.0369	0.0359	0.0362	0.0415	0.0490	0.0404	0.0534	0.0387	0.0536
2	$\bar{\varepsilon}$	-0.1169	-0.1179	-0.1202	-0.1238	-0.0740	-0.0976	-0.0741	-0.0385	-0.0734	-0.0378
	ε_{RMS}	0.1354	0.1364	0.1388	0.1423	0.1074	0.1466	0.0994	0.1015	0.1031	0.1073
	σ_{ε}	0.0682	0.0687	0.0695	0.0700	0.0778	0.1095	0.0661	0.0939	0.0724	0.1005
3	$\bar{\varepsilon}$	-0.2212	-0.2311	-0.2460	-0.2594	-0.1054	-0.2456	-0.1199	-0.0981	-0.0685	-0.0470
	ε_{RMS}	0.2362	0.2452	0.2588	0.2717	0.1778	0.2646	0.1470	0.1353	0.1228	0.1183
	σ_{ε}	0.0830	0.0818	0.0806	0.0808	0.1432	0.0985	0.0851	0.0931	0.1019	0.1086
4	$\bar{\varepsilon}$	-0.0858	-0.0859	-0.0864	-0.0870	-0.0756	-0.0987	-0.0853	-0.0898	-0.0858	-0.0903
	ε_{RMS}	0.1092	0.1093	0.1094	0.1096	0.1070	0.1181	0.1087	0.1142	0.1084	0.1140
	σ_{ε}	0.0676	0.0675	0.0672	0.0666	0.0757	0.0648	0.0673	0.0705	0.0662	0.0695
5	$\bar{\varepsilon}$	-0.0069	-0.0135	-0.0167	-0.0291	0.0620	-0.0588	0.0187	0.0107	0.0026	-0.0053
	ε_{RMS}	0.0459	0.0497	0.0498	0.0560	0.0840	0.0866	0.0543	0.0646	0.0481	0.0653
	σ_{ε}	0.0453	0.0479	0.0469	0.0479	0.0568	0.0635	0.0510	0.0637	0.0480	0.0651
6	$\bar{\varepsilon}$	-0.0559	-0.0742	-0.1044	-0.1334	0.1583	-0.1432	-0.1046	-0.0901	-0.0995	-0.0850
	ε_{RMS}	0.0789	0.0912	0.1160	0.1421	0.2161	0.1546	0.1208	0.1095	0.1228	0.1131
	σ_{ε}	0.0557	0.0531	0.0506	0.0488	0.1471	0.0583	0.0604	0.0623	0.0720	0.0746
7	$\bar{\varepsilon}$	-0.0873	-0.0873	-0.0875	-0.0900	-0.0730	-0.0945	-0.0886	-0.0984	-0.0831	-0.0929
	ε_{RMS}	0.1156	0.1156	0.1156	0.1167	0.1215	0.1168	0.1157	0.1181	0.1160	0.1164
	σ_{ε}	0.0758	0.0757	0.0755	0.0743	0.0971	0.0686	0.0744	0.0654	0.0809	0.0700
8	$\bar{\varepsilon}$	0.0399	0.0330	0.0157	-0.0054	0.2000	-0.0928	-0.0990	-0.1360	-0.0618	-0.0988
	ε_{RMS}	0.1276	0.1215	0.1088	0.1004	0.2971	0.1411	0.1334	0.1564	0.1138	0.1328
	σ_{ε}	0.1212	0.1169	0.1077	0.1002	0.2197	0.1063	0.0893	0.0771	0.0955	0.0887
9	$\bar{\varepsilon}$	-0.0302	-0.0303	-0.0307	-0.0312	-0.0283	-0.0354	-0.0295	-0.0221	-0.0318	-0.0245
	ε_{RMS}	0.0506	0.0506	0.0507	0.0509	0.0495	0.0588	0.0496	0.0539	0.0499	0.0543
	σ_{ε}	0.0406	0.0405	0.0404	0.0402	0.0406	0.0470	0.0399	0.0492	0.0384	0.0484
10	$\bar{\varepsilon}$	-0.0375	-0.0381	-0.0393	-0.0407	-0.0216	-0.0271	-0.0140	0.0128	-0.0199	0.0068
	ε_{RMS}	0.0863	0.0869	0.0880	0.0893	0.0792	0.0994	0.0744	0.0851	0.0754	0.0845
	σ_{ε}	0.0778	0.0781	0.0787	0.0795	0.0763	0.0956	0.0731	0.0841	0.0727	0.0842
11	$\bar{\varepsilon}$	-0.1134	-0.1155	-0.1192	-0.1227	-0.0705	-0.1054	-0.0415	-0.0170	-0.0449	-0.0205
	ε_{RMS}	0.1781	0.1799	0.1831	0.1860	0.1785	0.2021	0.1593	0.1595	0.1553	0.1551
	σ_{ε}	0.1373	0.1379	0.1389	0.1397	0.1639	0.1725	0.1538	0.1586	0.1487	0.1537
12	$\bar{\varepsilon}$	-0.0836	-0.0837	-0.0841	-0.0846	-0.0724	-0.0984	-0.0784	-0.0913	-0.0788	-0.0917
	ε_{RMS}	0.1092	0.1093	0.1094	0.1096	0.1059	0.1189	0.1073	0.1146	0.1070	0.1144
	σ_{ε}	0.0702	0.0702	0.0699	0.0696	0.0772	0.0669	0.0733	0.0693	0.0724	0.0685
13	$\bar{\varepsilon}$	-0.0638	-0.0677	-0.0718	-0.0706	-0.0308	-0.1240	-0.0629	-0.0768	-0.0647	-0.0876
	ε_{RMS}	0.1010	0.1059	0.1100	0.1031	0.0972	0.1385	0.1173	0.1128	0.1036	0.1144
	σ_{ε}	0.0783	0.0815	0.0833	0.0752	0.0922	0.0618	0.0990	0.0827	0.0809	0.0741
14	$\bar{\varepsilon}$	-0.0897	-0.0899	-0.0902	-0.0928	-0.0796	-0.1026	-0.0983	-0.1083	-0.0908	-0.1008
	ε_{RMS}	0.1095	0.1097	0.1102	0.1120	0.1015	0.1260	0.1243	0.1351	0.1140	0.1240
	σ_{ε}	0.0629	0.0630	0.0633	0.0626	0.0631	0.0732	0.0761	0.0808	0.0690	0.0722

144

TABLE VI : THE ANALYSIS OF VARIANCE TABLE FOR FINAL STAGE OF STEPWISE REGRESSION OF ERROR

Degrees of Freedom	208
Variable Entering	5
Multiple R	.72977
R Square	.53256
Adjusted R Square	.52105
Standard Error	.06229

ANOVA

Source	D.F	S.S	M.S	F	Critical-F
Regression	5	.89725	.17945	46.25625	2.25
Residual	203	.78753	.00388		
Total	208	1.68478			

D.F	:	Degrees of Freedom
S.S	:	Sum of Squares
M.S	:	Mean Square
F	:	M.S. Due to Regression/M.S due to Residual
Critical-F	:	Critical F Value in 5% significance level

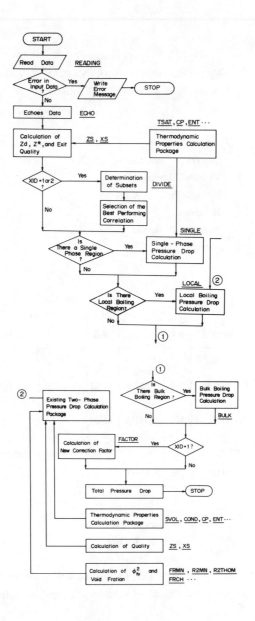

Fig.1   Flowchart of the Method for Evaluating a Single and
Two-Phase Pressure Drop by the K-TWOPD Code.

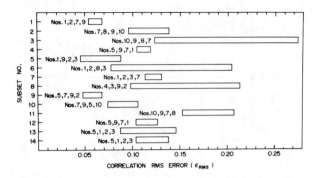

Fig.2   RMS Errors of All Correlations for Each Data Subsets.

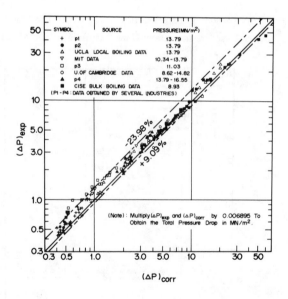

Fig.3   Comparison of Experimental and Predicted Local and Bulk
Boiling Total Pressure Drop Using Owens Homogeneous Model.

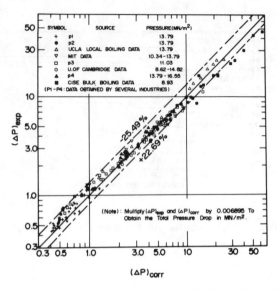

Fig.4   Comparison of Experimental and Predicted Local and Bulk
Boiling Total Pressure Drop Using Martinelli-Nelson Model.

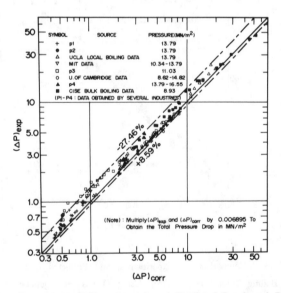

Fig.5   Comparison of Experimental and Predicted Local and Bulk
Boiling Total Pressure Drop Using Thom Model.

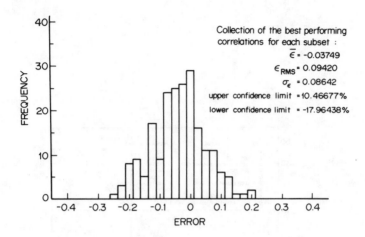

(a) Distribution of errors of the best performing correlations.

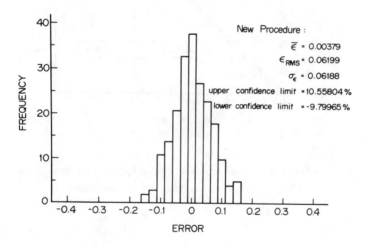

(b) Distribution of errors of the new procedure.

Fig.6   Comparison of Two Histograms of Errors : The Best
Performing Correlation VS. the New Procedure.

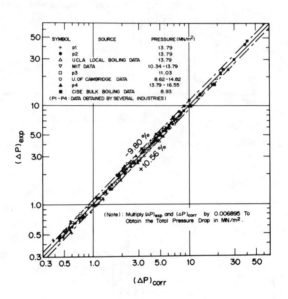

Fig.7   Comparison of Experimental and Predicted Total Two-Phase Pressure
Drop Using a New Procedure with a Correction Factor ψ.

NOMENCLATURE

B	Coefficient in Eq. (13)
De	Hydraulic equivalent diameter
f_{fo}	Friction factor based on total flow assumed liquid
$f_{fo,iso}$	Isothermal friction factor based on total flow assumed liquid
f_{TP}	Two-phase friction factor
G	Mass velocity
g	Acceleration due to gravity
L	Length of channel
n	Exponent in Blasius' relation for friction factor
P	Static pressure
q"	Surface heat flux
V_f	Specific volume of liquid
V_g	Specific volume of vapor or gas
V_{fg}	Difference in specific volumes of saturated liquid and vapor
x	Mass vapor quality
z	Axial coordinate
α	Void fraction
γ	Dimensionless slip factor used in Eq. (11)
Γ	Physical property coefficient defined by Eq. (14)
ϵ	Fractional error defined by Eq. (15)
ϵ	Pipe roughness
$\bar{\epsilon}$	Mean error
ϵ_{RMS}	Root-mean-square error
$\hat{\epsilon}$	Predicted error used in Eq. (21)
θ	Angle to horizontal plane
μ_f	Liquid viscosity
μ_g	Vapor viscosity
$\bar{\mu}$	Mean viscosity of homogeneous fluid
ρ_f	Liquid density
ρ_g	Vapor density
$\bar{\rho}$	Average density of homogeneous fluid
σ_ϵ	Standard deviation of the error from the mean

ϕ^2_{fo} Two-phase frictional multiplier based on pressure gradient for total flow assumed liquid

ψ Correction factor defined by Eq. (19)

$(\Delta P)_{corr}$ Total two-phase pressure drop calculated by correlations

$(\Delta P)_{exp}$ Total two-phase pressure drop obtained by experiment

$(\Delta P)_{new}$ Total two-phase pressure drop calculated by the new method proposed, Eq. (20)

$(\Delta P)_{opt}$ Total two-phase pressure drop calculated by the best performing correlation

$(\dfrac{dp}{dz})$ Total pressure gradient

$(\dfrac{dp}{dz}F)$ Pressure gradient due to friction

$(\dfrac{dp}{dz}a)$ Pressure gradient due to acceleration

$(\dfrac{dp}{dz}z)$ Pressure gradient due to static head

SUBSCRIPTS

fo   Assuming total flow to be liquid

f   Liquid

g   Gas or vapor

REFERENCES

1. W.L. Owens, Jr., "Two-Phase Pressure Gradient," Part II, International Developments in Heat Transfer, ASME (1961) 363-368.

2. W.H. McAdams et al., "Vaporization Inside Horizontal Tubes-II-Benzene-Oil Mixtures," Trans., ASME, 64 (1942) 193.

3. A. Cicchitti et al., "Two-Phase Cooling Experiments - Pressure Drop, Heat Transfer and Burnout Measurements," Energia Nucleare, 7(6) (1960) 407-425.

4. A.E. Dukler, Moye Wicks, III, and R.G. Cleveland, "Frictional Pressure Drop in Two-Phase Flow : Part A - A Comparison of Existing Correlations for Pressure Loss and Holdup," A.I.Ch.E. Journal 10(1) (1964) 38-51.

5. R.C. Martinelli and D.B. Nelson, "Prediction of Pressure Drop During Forced Circulation Boiling of Water," Trans. ASME, 70 (1948) 695.

6. J.R.S. Thom, "Prediction of Pressure Drop During Forced Circulation Boiling of Water," Int. J. Heat Mass Transfer, 7 (1964) 709-724.

7. C.J. Baroczy, "A Systematic Correlation for Two-Phase Pressure Drop," AIChE Reprint 37 Presented at 8th National Heat Transfer Conference, Los Angeles (1965).

8. D. Chisholm, "Pressure Gradients Due to Friction During the Flow of Evaporating Two-Phase Mixtures in Smooth Tubes and Channels," Int. J. Heat Mass Transfer, 16 (1972) 347-358.

9. W. Idsinga, N. Todreas, and R. Bowring, "An Assessment of Two-Phase Pressure Drop Correlations for Steam-Water Systems," Int. J. Multiphase Flow, 3 (1977) 401-413.

10. A.A. Armand and G.G. Treschev, "Investigation of the Resistance During the Movement of Steam-Water Mixtures in a Heater Boiler Pipe at High Pressures," AERE-Lib./Trans (1959) 816.

11. J.G. Collier, Convective Boiling and Condensation (McGraw-Hill New York, 1981) pp.30-55.

12. R.W. Lickhart and R.C. Martinelli, "Proposed Correlation of Data for Isothermal Two-Phase Two-Component Flow in Pipes," Chem. Eng. Prog., 45 (1949) 39.

13. N.R. Draper and H. Smith, Applied Regression Analysis (John Wiley and Sons, 1981).

14. R.W. Bowring, "Physical Model Based on Bubble Detachment and Calculation of Steam Voidage in the Subcooled Region of a Heated Channel," OECD Halden Reactor Project Report HPR-10 (1962).

MULTIPHASE TRANSIENTS

A Numerical Study of Transient Two-Phase Annular Flow across Convergent-Divergent Nozzles Using Separate Flow Model

K. M. AKYUZLU and M. J. GUILLOT
Mechanical Engineering Department
University of New Orleans
New Orleans, Louisiana 70148, USA

Abstract

A numerical simulation of transient two-phase annular flow across variable area sections is sought. The variable area sections studied are converging-diverging nozzles with several different constriction diameters. A two-fluid, separated flow model is used to describe the two-phase flow. The governing equations for this model are solved numerically with the pressure, density, and enthalpy of each phase held constant at the inlet, and the pressure held constant at the exit. The explicit finite-difference approximations adopted incorporate a dual velocity concept and a staggered spatial mesh with densities and enthalpies defined at the center of the cell and velocities defined at the cell boundaries. As a result, the steady-state pressure drop across the variable area section is determined, and pressure and velocity transients with superimposed acoustic oscillations due to perturbations upstream of the variable area section are predicted for each phase.

I. INTRODUCTION

Two-phase flow is extensively found in many practical engineering applications and has been studied experimentally and theoretically since the 1950's. However, the current literature indicates that little effort has been made in determining the unsteady characteristics of two-phase flows through variable area sections such as nozzles. Furthermore, the previous studies have been restricted to homogeneous models in which conservation equations are written only for the mixture and do not account for interactions between the phases which in some cases can significantly affect the characteristics of the flow.

Therefore, this study adopts a separated flow model for studying unsteady two-phase flow through variable area sections. Separate conservation equations of mass, momentum, and energy are written for each phase. Thermal non-equilibrium is allowed to exist between the phases, and each phase is allowed to move at a different velocity. However, the pressure is assumed to be the

157

same for each phase. Furthermore, the flow is assumed to be one-dimensional.

The numerical technique adopted to solve the governing equations is a fully explicit finite difference scheme. The two basic features of the scheme are the dual velocity [1] and the staggered spatial mesh concepts. The dual velocity concept replaces the single velocity at the cell boundary with two velocities, one is defined just upstream and the other just downstream of the cell boundary. This concept is adopted to overcome the numerical difficulties that arise in integrating the governing equations across abrupt area changes. The staggered spatial mesh used in this study defines the densities and enthalpies at the center of the cells and the velocities at the cell boundaries. This is necessary to overcome the numerical diffusion that eventually degenerates the pressure waves created by the perturbations at the inlet of the test section. Both concepts were previously used by Akyuzlu and Espat [2] to study transient, one-dimensional, two-phase flow across converging-diverging nozzles and orifices where the liquid and vapor phases were assumed to be a homogeneous mixture.

The study of transient two-phase flow through variable area sections is necessary in order to learn more about the two-phase flow instabilities induced by perturbations upstream of the flow. These flow instabilities, if sustained could cause mechanical vibrations of the system, and ultimately loss of flow control. The instabilities observed by numerical experimentation in this study are dynamic instabilities of the kind called pressure (or acoustic oscillations. These are high frequency oscillations and their frequency is of the same order of magnitude as the time required for a pressure wave to travel through the system.

II. MATHEMATICAL MODEL

This study adopts a one-dimensional "separated flow model" which considers each phase separately, i.e., thermal non-equilibrium is allowed to exist between the liquid and vapor phases and each phase is allowed to have a different velocity. However, the liquid and vapor pressures at any cross section is assumed to be the same and the flow is assumed to be adiabatic so that heat transfer with the walls of the pipe is not considered. This model is also known as the UVUT (Unequal Velocity Unequal Temperature) model.

II.a Conservation equations

The unsteady conservation equations of mass, momentum, and energy for the liquid and vapor phases for one-dimensional two-phase flow are, respectively:

For the liquid phase:

$$\frac{\partial}{\partial t}(\rho'_l) + \frac{\partial}{\partial z}(\rho'_l u_l) = -\Gamma_l^g \tag{1}$$

$$\frac{\partial}{\partial t}(\rho'_l u_l) + \frac{\partial}{\partial z}(\rho'_l u_l^2) + \Theta_l \frac{\partial P}{\partial z} = \tau'''_{in} - \tau'''_{wl} - \rho'_l g - \Gamma_l^g u_l \tag{2}$$

$$\frac{\partial}{\partial t}(\rho'_l h_l) + \frac{\partial}{\partial z}(\rho'_l u_l h_l) = -\Gamma_l^g h_{l,sat} \tag{3}$$

For the vapor phase:

$$\frac{\partial}{\partial t}(\rho'_g) + \frac{\partial}{\partial z}(\rho'_g u_g) = \Gamma_l^g \tag{4}$$

$$\frac{\partial}{\partial t}(\rho'_g u_g) + \frac{\partial}{\partial z}(\rho'_g u_g^2) + \Theta_g \frac{\partial P}{\partial z} = -\tau'''_{in} - \rho'_g g + \Gamma_l^g u_g \tag{5}$$

$$\frac{\partial}{\partial t}(\rho'_g h_g) + \frac{\partial}{\partial z}(\rho'_g u_g h_g) = \Gamma_l^g h_{g,sat} \tag{6}$$

The smeared densities in the above equations are related to the actual densities by the following relation (for the k th phase):

$$\rho'_k = \rho_k \Theta_k \tag{7}$$

where

$$\Theta_k = \frac{A_k}{A_{ref}} \tag{8}$$

The quantities are related by;

$$\Theta_l + \Theta_g = \Theta_T \tag{9}$$

II.b Constitutive Relations:

The above set of equations is closed by thermodynamic equation of state for each phase, and constitutive relations which specify the interfacial mass, momentum, and heat transfer, as well as the liquid-wall, and interfacial drag forces in terms of the independent variables of the above formulation.

The equations of state for Freon-II, which is used as working fluid in this and previous experimental and theoretical studies done by one of the authors, are presented by seventh order polynomials [3]. The other constitutive relations as well as the interfacial area and interfacial heat transfer coefficients are dependent on the flow regime and may be selected on the basis of a simple flow regime map. In this study, however, the flow is assumed to be annular. Therefore, only the constitutive relations for the annular two-phase flow is presented below.

The mass evaporation term is determined from energy balance at the interface:

$$\Gamma_l^g = \frac{-q_{in_g} - q_{in_l}}{h_{fg}}$$ (10)

where heat transferred to the th phase is given by,

$$q_{in_k} = h_{in_k} A_{in}'''(T_s - T_k)$$ (11)

and the interfacial area per unit volume is given for annular flow as

$$A_{in}''' = \frac{(4\pi A_g)^{1/2}}{A_{ref}}$$ (12)

The liquid wall drag force is given by the relation,

$$\tau_{wl}''' = K_l(u_l)$$ (13)

where

$$K_l = \left(\frac{f_l}{2D_h}\right)\left(\frac{A_l}{A}\right)^2 \phi^2 \rho_l |u_l|$$ (14)

The friction factor is based on the Blassius formula and is for the liquid only:

$$f_l = C\left(\frac{D|u_l|\rho_l}{\mu_l}\right)^{-1/4}$$ (15)

where C is a calibration parameter to be determined by numerical experimentation.

The interfacial drag force is given by the relation

$$\tau_{in}''' = K_{in}(u_r)$$ (16)

where the relative velocity is defined as,

$$u_r = u_g - u_l$$ (17)

The value of K in equation 16 is given by,

160

$$K_{in} = \frac{f_{in}}{2} \rho_y \ u_r! \ A_{in}^{'''} \tag{18}$$

where the interfacial friction factor is given by the following relation [4]:

$$f_{in} = .005 \left(1 + 75 \left(1 - A_g/A\right)\right) \tag{19}$$

III. NUMERICAL SOLUTION

A fully explicit finite-difference technique is adopted to solve the governing equations given above.

III.a. Discretization

The shape of the variable area section, the nozzle in this study, is approximated by a trigonometric function (see Figure 1). The parameters in the trigonometric function can be changed to approximate nozzles with different d/D ratios. This function is then used to determine the radius of each computational cell, and thus the cross sectional area [2]. The variable area section can be divided into as many cells as one wishes. In addition to the cells within the test section half-cells are placed upstream of the inlet and downstream of the exit as shown in Figure 2.

The staggered spatial mesh used in the numerical scheme is shown in Figure 3. In this scheme, the fluid properties such as density and enthalpy are defined at the center of the computational cells, while the liquid and vapor velocities are defined at the cell boundaries. Furthermore, a dual velocity concept is used at the cell boundaries [1] which defines a velocity just upstream and another one just downstream of the boundary. The relation between these two velocities k th phase is given by

$$\frac{v_{k_i}'}{v_{k_i}''} = \frac{A_{k_i}}{A_{k_{i-1}}} \tag{20}$$

and is based on the steady state mass conservation across the interface.

III.b. Finite-Difference Approximations

The basic concepts used in developing the finite-difference approximations of the governing equations is summarized below. The reader should refer to reference [5] for complete formulation of the numerical approximations. In developing the finite-difference approximations of the conservation equations, the following principles are used:

161

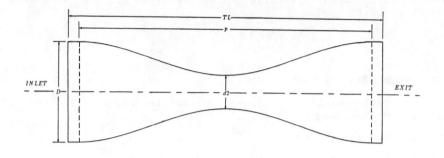

$$R = A \cdot COS(B \cdot \Theta) + C \qquad\qquad C = A + \frac{d_2}{2} \qquad\qquad P_{(nozzle)} = TL - \Delta z$$

$$\Theta = p - \left((IN - I)\Delta z + \frac{\Delta z}{2} \right) \qquad\qquad \Delta z = \frac{TL}{IN + 1}$$

Figure 1 - Convergent - Divergent Nozzle Approximated by the Area Generating Function

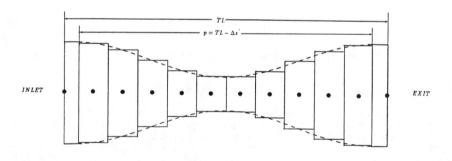

Figure 2 - Discretization of the Convergent - Divergent Nozzle

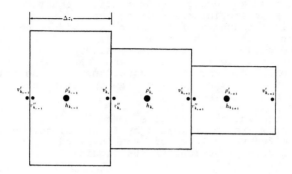

Figure 3 - Staggered Spatial Mesh Used for Finite-Difference
Approximation

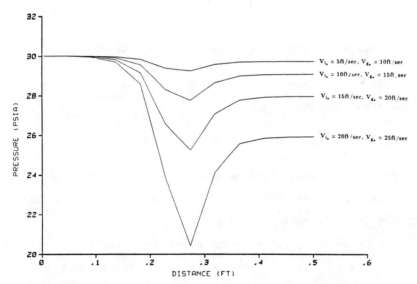

Figure 4 - Numerical Simulation of Steady State Pressure Drop
Through the Nozzle for various Liquid and Vapor Inlet
Velocities

(i) The mass and energy equations are approximated by forward differencing in time and space, while the momentum equations are approximated by forward differencing in time and central differencing in space.

(ii) The mass and energy conservation equations for both liquid and vapor are integrated from the right boundary of each cell to the right boundary of the next cell with the enthalpies and densities held constant over the length of each cell. The velocities are assumed to vary lineary within each cell between the value at the left boundary and the value at the right boundary. Momentum conservation equations, on the other hand, are integrated from the center of the computational cell to the center of the next cell.

Applying the above principles, the continuity equations, Equation 1 and 4 are approximated by finite-differences and then solved for the future values of the smeared densities which yield the following equation for the liquid:

$$\rho_{l_i}^{j+1} = \frac{\Delta t}{\Delta z}\left(\rho_{l_{i-1}}''v_{l_i}'' - \rho_{l_i}''v_{l_{i+1}}''\right) - (\Gamma_l^g)_i^j\Delta t + \rho_{l_i}'' \tag{21}$$

where the mass evaporation term in finite-difference form is given by

$$(\Gamma_l^g)_i^j = \frac{h_{in_g}A_{in}'''(T_g - T_s)_i^j + h_{in_l}A_{in}'''(T_l - T_s)_i^j}{(h_{fg})_i^j} \tag{22}$$

The equation for the vapor density is developed in a similar fashion [5].

The energy equation, Equation 2, is approximated by finite-differences and then solved for the future value of the enthalpies. For the liquid:

$$h_{l_i}^{j+1} = \frac{\Delta t}{\Delta z\rho_{l_i}^{n+1}}\left(\rho_{l_{i-1}}''v_{l_i}''h_{l_{i-1}}^j - \rho_{l_i}''v_{l_{i+1}}''h_{l_i}^j\right) + \left(\frac{\rho_{l_i}''}{\rho_{l_i}^{n+1}}\right)h_{l_i}^j - \frac{\Delta t(\Gamma_l^g h_{l,sat})_i^j}{\rho_{l_i}^{n+1}} \tag{23}$$

The enthalpy equation for the vapor is given in Reference [5].

The approximations for liquid and vapor continuity and energy equations as given above are applicable to all cells including the half cell located at the exit.

In integrating the momentum equations, the inertia term for both liquid and vapor is approximated by forward differencing in time. This requires some averaging. For example, the mass fluxes across the two cells are averaged as follows:

$$\rho'_k u_k = \frac{(\rho_{k_{i-1}} v'_{k_i}) + (\rho_{k_i} v''_{k_i})}{2} \tag{24}$$

Then the averaged inertia term is approximated for the liquid as,

$$\frac{\partial}{\partial t}(\rho'_l u_l) = \left[\frac{\left[\frac{\rho'_{l_{i-1}} v'_{l_i} + \rho_{l_i} v''_{l_i}}{2} \right]^{j+1} + \left[\frac{\rho'_{l_{i-1}} v'_{l_i} + \rho'_{l_i} v''_{l_i}}{2} \right]^{j}}{\Delta t} \right] \tag{25}$$

The convective term for both liquid and vapor is approximated by central differencing in space. For the liquid,

$$\frac{\partial}{\partial z}(\rho'_l u_l^2) = \frac{(\rho'_{l_i} u_{l_i}^2)^j - (\rho'_{l_{i-1}} u_{l_{i-1}}^2)^j}{\Delta z} \tag{26}$$

where the velocities are redefined at the center of the cells in terms of left and right boundary velocities as,

$$u_{k_i} = \frac{v''_{k_i} + v'_{k_{i+1}}}{2} \tag{27}$$

and

$$u_{k_{i-1}} = \frac{v''_{k_i-1} + v'_{k_i}}{2} \tag{28}$$

The pressure drop term is approximated by central differencing in space. For example, for the liquid:

$$\Theta_l \frac{\partial P}{\partial z} = \Theta_{l_{av}} \left[\frac{P_i^j - P_{i-1}^j}{\Delta z} \right] \tag{29}$$

where

$$\Theta_{k_{av}} = \frac{\Theta_{k_{i-1}} + \Theta_{k_i}}{2} \tag{30}$$

The other terms in the momentum equations are evaluated at the present j th time are averaged across the two cells in question. For example, the averaged interfacial drag force is given by

$$\tau^j_{in_{av}} = \left[\frac{\left(\frac{f_{in}}{2} \rho_g |u_r| u_r A'''_{in} \right)^j_{i-1} + \left(\frac{f_{in}}{2} \rho_g |u_r| u_r A'''_{in} \right)^j_i}{2} \right] \tag{31}$$

where the interfacial friction factor is given for the ith cell as

$$f^j_{in_i} = .005\,(1 + 75\,(1 - A_{g_i}/A_i))$$ (32)

Combining all the approximations for the terms in the momentum equation for the liquid and solving for the future value of the liquid velocity yields,

$$v^{\prime\prime+1}_{l_i} = \left[\frac{A^{j+1}_{l_i}}{\rho^{n+1}_{l_{i-1}} A^{j+1}_{l_i} + \rho^{n+1}_{l_i} A^{j+1}_{l_{i-1}}} \right] \bullet \left[\left(\rho^{\prime\prime}_{l_{i-1}} v^{\prime\prime}_{l_i} + \rho^{\prime\prime}_{l_i} v^{\prime\prime\prime}_{l_i} \right) \right.$$

$$+ 2\frac{\Delta t}{\Delta z} \left[\left(\rho^{\prime\prime}_{l_{i-1}} u^{2^j}_{l_{i-1}} - \rho^{\prime\prime}_{l_i} u^{2^j}_{l_i} \right) + \Theta^j_{l_{av}} \left(P^j_{i-1} - P^j_i \right) \right]$$

$$\left. + \Delta t \left(2\tau^{\prime\prime\prime j}_{in_{av}} - \left(\rho^{\prime\prime}_{l_{i-1}} + \rho^{\prime\prime}_{l_i} \right) g - 2\tau^{\prime\prime\prime j}_{wl_{av}} - 2\,(\Gamma^g_l u_l)^j_{av} \right) \right]$$ (33)

An equation for the vapor phase can be developed in a similar fashion [5].

The differencing techniques used in the momentum finite-difference equations require that the information is received from cells located upstream and downstream of the point in question; thus, restricting it to the central cells. An alternate method must be used for the boundary cells.

For the computational cell at the inlet, backward differencing in space is used. For the liquid, the inlet velocity is approximated by:

$$v^{\prime\prime+1}_{l_o} \cong v^{\prime\prime}_{l_o} + \frac{1}{\rho^{\prime\prime}_{l_o}} \left(\left(\frac{\Delta t}{\Delta z} \right) \left(\rho^{\prime\prime}_{l_1} v^{\prime\prime 2^j}_{l_1} - \rho^{\prime\prime}_{l_o} v^{\prime\prime 2^j}_{l_o} + \Theta_{l_{av}} \left(P^j_1 - P^j_o \right) \right) \right.$$

$$\left. + \left(\Delta t/2 \right) \left(\tau^{\prime\prime\prime}_{in_{av}} - \tau^{\prime\prime\prime}_{wl_{av}} - (\rho'_l g)_{av} - (\Gamma^g_l u_l)^j_{av} \right) \right)$$ (34)

Forward differencing is used to approximate the momentum equations for the exit cell. For example, the exit velocity for the liquid is given by

$$v^{\prime\prime+1}_{l_e} \cong \left(\frac{\rho^{\prime\prime+1}_{l_e}}{\rho^{\prime\prime}_{l_e}} \right) v^{\prime\prime}_{l_e} + \frac{1}{\rho^{\prime\prime+1}_{l_e}} \left(\left(\frac{\Delta t}{\Delta z} \right) \left(\rho^{\prime\prime}_{l_{IN}} v^{\prime 2^j}_{l_{in}} - \rho^{\prime\prime}_{l_e} v^{\prime 2^j}_{l_e} + \Theta_{l_{av}} \left(P^j_{IN} - P^j_e \right) \right) \right.$$

$$\left. + \left(\Delta t/2 \right) \left(\tau^{\prime\prime\prime}_{in_{av}} - \tau^{\prime\prime\prime}_{wl_{av}} - (\rho'_l g)_{av} - (\Gamma^g_l u_l)^j_{av} \right) \right)$$ (35)

III.c. Initial and Boundary Conditions:

The initial conditions are determined from the steady-state

solution. Mathematical formulation and the numerical techniques used to solve the steady-state governing equations are given in detail by Guillot [5]. The basic feature of the model is that a thermodynamic equilibrium is assumed to exist between the phases. However, each phase is allowed to move at a different velocity (Unequal Velocity Equal Temperature, UVET, model). The finite-difference techniques used to approximate the governing equations is similar to the ones used for the unsteady state case and can be found in Reference [5]. The algorithm developed to solve the resulting equations uses an implicit technique which iterates the unknown values until a correct solution is obtained.

The governing equations are solved for the following boundary conditions: (i) constant inlet pressure, (ii) constant exit pressure, (iii) constant inlet liquid and vapor densities, and (iv) constant inlet liquid and vapor temperatures (or enthalpies).

III.d. Numerical Stability Criteria:

Courant Stability Criteria is used to determine the computational time step and is given by,

$$\Delta t \leq \frac{\Delta z}{a + |u|} \tag{36}$$

where Δz is the width of the computational cell, a is the maximum acoustic speed, and u is the maximum fluid velocity within the flow domain.

III.e. Computational Procedure

The algorithm developed to solve the unsteady state finite-difference equations for the unknown variables uses an explicit technique and is given in Reference [5].

IV. RESULTS

A computer program was developed in Fortran 77 to implement the algorithm which solves the finite-difference equations. The program performs two tasks: first it solves for the steady-state solution, then it predicts the flow transients or oscillations due to perturbations upstream of the test section. The numerical study was limited to adiabatic, annular flow through a convergent-divergent nozzle with d/D ratio of 1/3, total length of 6 inches, and an inlet diameter of 3/8 inches. These dimensions match the one for the test section used in studying the homogeneous model [2].

The steady state results are given in Figures 4 and 5. In Figure 4, the inlet liquid and vapor velocities are varied as other parameters are held constant. The results show that for low velocities the pressure drop through the throat is fairly low, and there is a considerable pressure recovery after the throat. It can also be seen, as expected, that as the inlet velocities increase, the pressure drop through the throat increases substantially, and the pressure recovery after the throat is much less. Increasing the friction coefficient produces a larger pressure difference between the inlet and the exit, and this is shown in Figure 5. However, increasing the friction coefficient does not affect the pressure drop across the throat as drastically as it does in the case where the inlet velocities are increased.

The results of the unsteady-state case studies are shown in Figures 6, 7, 8 and 10. The program was run for different operating conditions for the convergent-divergent nozzle for a time increment of 10^{-6} seconds with the test section approximated by 10 computational cells. Operating conditions for the numerical results shown in Figures 6, 7, and 8 are: inlet pressure of 30 psia, inlet liquid velocity of 5 ft/sec, and inlet vapor velocity of 25 ft/sec. The system is perturbed by increasing the inlet pressure by 1 psia. As seen from the figures, this perturbation generates transients acoustic oscillations which gradually die out, and the system achieves a new steady-state. Figure 6 shows the pressure oscillations, and Figures 7 and 8 show the liquid and vapor velocity oscillations at the inlet, respectively. The period of the oscillations is around 0.0013 seconds and a new steady-state is reached in 0.025 seconds. The amplitude of the oscillations for the liquid velocity is small compared to the one for the vapor velocity. Figure 9 shows the transient pressure oscillations in the throat of the nozzle for operating conditions of 30 psia inlet pressure, 5 ft/sec inlet liquid velocity, 15 ft/sec inlet vapor velocity. Inlet mass flow rate is 0.4221 lbm/sec. The flow is perturbed at the inlet by increasing the pressure by 1 psia.

Comparison of these oscillations to the ones obtained by the homogeneous model [2] for the same operating conditions shows that the models predict similar transient conditions for the cases studied here. The period of oscillations is about 0.0015 seconds for the separated flow model and 0.0017 seconds for the homogeneous model. In both cases the system achieves a new steady state in about 0.5 seconds; however, separated flow model predicts a lower steady-state pressure than the one predicted by the homogeneous model.

V. CONCLUSIONS

A one-dimensional, separated flow model has been successfully adopted to study steady and unsteady annular two-phase flow

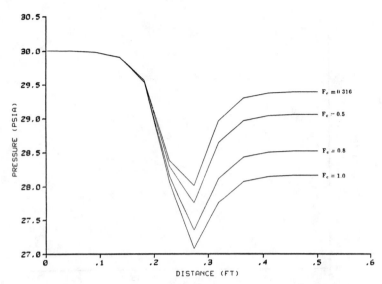

Figure 5 - Numerical Simulation of Steady State Pressure Drop
Through Nozzle for Various Friction Coefficients

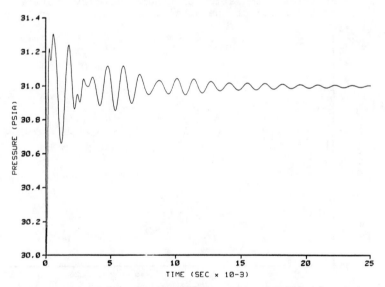

Figure 6 - Numerical Simulation of Pressure Oscillations at
the Inlet of the Nozzle

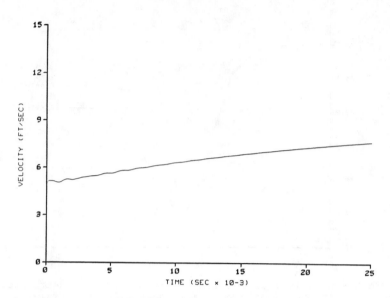

Figure 7 - Numerical Simulation of Liquid Velocity Oscillations at the Inlet of the Nozzle.

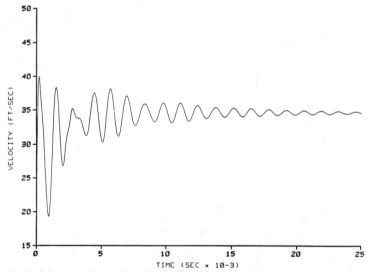

Figure 8 - Numerical Simulation of Vapor Velocity Oscillations at the Inlet of the Nozzle.

170

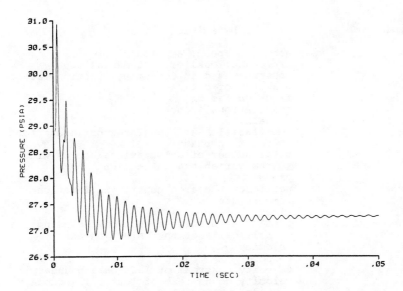

Figure 9 -Numerical Simulation of Pressure Oscillations at the
Throat of the Nozzle for same Operating Conditions Used
for the Homogeneous Model Simulations.

171

through variable area sections: a converging-diverging nozzle in this case. The explicit finite-difference approximations of the governing equations which incorporate a dual velocity concept and a staggered spatial mesh was necessary to eliminate the numerical instabilities experienced with other schemes. Furthermore, the momentum equations for the inlet and exit computational cells had to be approximated by backward and forward differences, respectively, since extrapolation schemes (which were successfully used in the homogeneous model) failed. The model was also successful in simulating the trends and typical characteristics of acoustic oscillations due to perturbations at the inlet of the test section.

NOMENCLATURE

Symbol	Definition
a	acoustic speed; amplitude
A	cross sectional or interfacial area
C	constant used in computing friction factor
D	diameter of cell
f	friction factor
g	gravitational constant
h	enthalpy
h_{in_k}	interfacial heat transfer coefficient for the k th phase
IN	total number of computational cells within the variable area section
P	pressure
p	period of the area generating function
T	temperature
t	time
TL	total length of test section including the half cells at the inlet and exit
u	velocity
V	volume
v'_i	velocity upstream of left cell boundary
v''_i	velocity downstream of left cell boundary
x	mixture quality used for steady state solution
z	spatial coordinate in the axial direction

Greek Symbols

Γ_l^g	interfacial mass transfer per unit volume out of the liquid
Θ	area ratio
μ	dynamic viscosity
ρ	density
ρ'	smeared density

τ	wall or interfacial drag force
ϕ	two phase friction multiplier

Subscripts

av	average
e	exit of test section
i	present computational cell
$i+1$	downstream cell
$i-1$	upstream cell
in	interface between liquid and vapor
IN	last computational cell in variable are section
g	vapor phase
k	k th phase, denotes liquid or vapor
l	liquid phase
m	mixture
o	entrance to test section
r	relative
ref	reference
s	saturation
T	total
w	wall

Superscripts

j	present time
$j+1$	future time
$'''$	denotes a quantity per unit volume
g	vapor phase

Acknowledgements

The authors wish to thank Mr. Amin Espat who developed the computer program for the homogeneous model. The authors also wish to thank Mrs. Muriel Murphy for typing the manuscript.

REFERENCES

1.   Tetner, A.M. and Wilder, H.U., "Pressure Drop Modeling in Variable Area Multiphase, Transient Flow," Multiphase Transport, Vol.2, Hemisphere Publishing Co., pp. 1137-1153, 1980.

2.   Espat, A.D., "Numerical Simulation of Two-Phase Bubbly Flow Across Orifices and Convergent-Divergent Nozzles, "M.S.   Thesis, University of New Orleans, May 1987.

3.   Akyuzlu, K.M., "Mathematical Modeling of Two-Phase Flow Oscillations," Ph.D. Thesis, University of Miami, November 1979.

4.   Wallis, G.B., "Annular Two-Phase Flow, Part 1: Simple Theory," Journal of Basic Engineering, March 1970.

5. Guillot, M.J., "Numerical Simulation of Two-Phase Flow Across Orifices and Converging - Diverging Nozzles using a Separated Flow (UVUT) Model," M.S.   Thesis, University of New Orleans, December 1988.

An Analysis of Void Wave Propagation in Bubbly Flows

ARTHUR E. RUGGLES, RICHARD T. LAHEY, JR.,
and DONALD A. DREW
Rensselaer Polytechnic Institute
Troy, New York 12180-3590, USA

ABSTRACT

A two-fluid model is presented which can be used for the prediction of the celerity and attenuation of various state variable perturbations in two-phase flows. This model has been used previously [1], [2] to predict the measured celerity and attenuation of both standing and propagating pressure perturbations in bubbly two-phase flow. In this paper the same model is used to analyze the existing data for void wave propagation and is related to other more simplified models for void wave propagation. The importance of interfacial drag, virtual mass and Reynolds stress is thereby demonstrated. Further, it is shown that the analysis of void waves represents an excellent means of assessing laws for interfacial momentum transfer, however improved void wave data is needed to facilitate such efforts.

INTRODUCTION

A two-fluid model consists of the averaged (in time and space) conservation equations for each phase. These equations are closed using postulated constitutive relationships that are deduced from considerations of heat, mass and momentum transfer. This modeling approach has been discussed in detail by many investigators [3]-[6].

A transient two-fluid model can be evaluated in a number of ways. One way is to solve the set of non-linear partial differential equations numerically, subject to appropriate initial and boundary conditions. Using this approach nonlinear phenomena can be investigated. Unfortunately, the direct numerical approach is cumbersome and not well suited to the assessment of the postulated interfacial transfer laws.

Another technique is to linearize the equation set, recognizing that the resultant linear model is only applicable for small perturbations around a steady-state, and to analyze the resultant dispersion relation [7], [1], [2]. A third approach is to evaluate the eigenvalues and eigenvectors of the equation set, noting that the eigenvalues represent the characteristic velocities of the two-fluid model [7], [2], [8]. These two techniques are commonly used to analyze wave propagation phenomena in two-phase flows, and are employed in this study to determine the sensitivity of void wave propagation to the assumed closure conditions.

In particular, the two-fluid model presented herein was used to predict the void wave celerities, and these predictions were compared with the data

175

taken in bubbly air/water flows by Pauchon & Banerjee [9], Bernier [10] and Mercadier [11]. While the two-fluid model is in basic agreement with these data, it was found that significant data scatter and insufficient documentation of the pertinent flow variables prevents complete assessment of the interfacial transfer laws in two-fluid models. Nevertheless, it appears that the analysis of carefully taken void wave data is an excellent way to assess models for interfacial momentum transfer.

The two-fluid model presented herein was also compared with simplified analytical models for void wave propagation due to Pauchon and Banerjee [9], [12] and by Wallis [13]. These comparisons indicate that, as expected, the interfacial drag law has a significant effect on the predicted void wave celerities in bubbly air/water systems. Moreover, the relationship between the dispersion relation and kinematic drift-flux models for void wave propagation [13] shows how typical drift-flux parameters correspond to specific models for interfacial momentum transfer (ie: interfacial drag, virtual mass, etc).

The Two-Fluid Model

A one-dimensional two-fluid model for bubbly air/water flow in a constant area duct can be written as the conservation of mass, momentum and energy equations for both the liquid (k=L) and gas (k=G):

CONSERVATION OF MASS

$$\frac{\partial}{\partial t} (\alpha_k \rho_k) + \frac{\partial}{\partial z} (\alpha_k \rho_k u_k) = 0 \tag{1a}$$

CONSERVATION OF MOMENTUM

$$\frac{\partial}{\partial t} (\alpha_k \rho_k u_k) + \frac{\partial}{\partial z} (\alpha_k \rho_k u_k^2) = - \alpha_k \frac{\partial p_k}{\partial z} + (p_{k_i} - p_k) \frac{\partial \alpha_k}{\partial z} - \tau_{ki} \frac{\partial \alpha_k}{\partial z}$$

$$+ \frac{\partial}{\partial z} \left[\alpha_k \tau^T_{zz_k} \right] - \alpha_k \rho_k g \cos \theta + M_k - \tau_{k_w}/D_H \tag{1b}$$

CONSERVATION OF ENERGY

$$\frac{\partial}{\partial t} (\alpha_k \rho_k h_k) + \frac{\partial}{\partial z} (\alpha_k \rho_k u_k h_k) = \alpha_k \left(\frac{\partial p_k}{\partial t} + u_k \frac{\partial p_k}{\partial z} \right)$$

$$- u_k \tau_{k_w} + \frac{q''_{ki}}{L_s} \tag{1c}$$

Appropriate constitutive equations must be used to model the interaction between the fluids. The momentum transfer between the gas and liquid phases

176

can be written as the sum of three forces:

$$M_L = - M_G = \alpha [F_D + F_{VM} + F_R] \qquad (2)$$

where $\alpha \equiv \alpha_g$.

The drag force, F_D, for bubbly flows is normally modeled as,

$$F_D = \frac{3}{8} \rho_L \frac{C_D}{R_b} (u_G - u_L) |u_G - u_L| \qquad (3)$$

where a typical drag coefficient, C_D, for bubbly air/water flows is given as as [14] ,

$$C_D = \frac{4}{3} R_b \left[\frac{g(\rho_L - \rho_G)}{g_c \ \sigma \ (1-\alpha)} \right]^{1/2} \qquad \text{(distorted bubbles)} \qquad (4)$$

The virtual mass force, F_{VM}, can be written as,

$$F_{VM} = \rho_L C_{VM} a_{VM} \qquad (5)$$

where the virtual volume coefficient, C_{VM}, is equal to 0.5 for a single spherical bubble in an infinite liquid media. The virtual mass acceleration, a_{VM}, is given by [15],

$$a_{VM} = \left[\frac{\partial u_G}{\partial t} + u_G \frac{\partial u_G}{\partial z} \right] - \left[\frac{\partial u_L}{\partial t} + u_L \frac{\partial u_L}{\partial z} \right] \triangleq \left[\frac{D_G u_G}{Dt} - \frac{D_L u_L}{Dt} \right] \qquad (6)$$

The force due to radial bubble pulsations, F_R, results from the interaction of an oscillating spherical bubble with the flow field around the bubble. This force is due to both bubble translation relative to the liquid phase and to radial bubble pulsations [7], and is given by,

$$F_R = \frac{3}{R_b} C_{VM} \rho_L (u_G - u_L) \frac{D_G R_b}{Dt} \qquad (7)$$

Since the gas phase is assumed to be dispersed within the liquid phase, the wall shear stress on the gas phase can be taken as:

$$\tau_{G_w} = 0 \qquad (8)$$

In contrast, the liquid phase wall shear stress is given by:

$$\tau_{L_w} = \frac{f}{2} \, \rho_L \, u_L \, |u_L| \tag{9}$$

For disturbances having time constants much greater than R_b/c_g, the pressure in the gas phase is essentially uniform, thus,

$$p_{Gi} - p_G \cong 0 \tag{10}$$

The interfacial pressure in the liquid phase can be calculated in essentially the same manner as for the pulsating bubble force, F_R, and yields:

$$p_{Li} - p_L = \Delta p_{Li} = -\frac{\rho_L}{4} \, (u_G - u_L)^2 \tag{11}$$

This expression has also been deduced by other authors [16].

In addition, one can consider the influence of Reynolds stress on void wave propagation. The Reynold's stress is negligible in the gas phase but not in the liquid phase. It is beyond the state-of-the-art to model this term in general, however, the Reynolds stress in the liquid phase due to "bubble-induced" turbulence has been given by Nigmatulin [17] as,

$$\underset{=\ell}{\tau^T} = -\alpha \rho_\ell \left[C_1 \, |\underline{u}_G - \underline{u}_L|^2 \, \underline{\underline{I}} + C_2 (\underline{u}_G - \underline{u}_L) \, (\underline{u}_G - \underline{u}_L) \right] \tag{12a}$$

It can be shown [18] C_1 and C_2 are given by,

$$C_1 = \frac{3}{20} \quad , \quad C_2 = \frac{1}{20}. \tag{12b}$$

It should be noted that for one-dimensional pipe flow, Eqs. (12) reduce to:

$$\tau^T_{zz_\ell} = -\frac{1}{5} \, \alpha \, \rho_\ell \, (u_G - u_L)^2 \tag{13}$$

Hence, the Reynolds stress gradient term in Eq. (1b) can be written as,

$$\frac{\partial}{\partial z} \left[(1-\alpha) \, \tau^T_{zz_\ell} \right] = -\frac{\partial}{\partial z} \left[\frac{1}{5} \, \alpha \, (1-\alpha) \, \rho_L \, (u_G - u_L)^2 \right] \tag{14}$$

This is basically the same expression as developed by Biesheuvel and Van Wijngaarden [18] and used by Pauchon and Banerjee [12].

The interfacial shear, τ_{k_i}, is often set to zero. However, one may also assume that it is equal to $\tau^T_{zz_\ell}$. When this is done the third and fourth terms on the right hand side of Eq. (1b) combine to yield,

178

$$-\tau^T_{zz_k} \frac{\partial \alpha_k}{\partial z} + \frac{\partial}{\partial z} \left[\alpha_k \; \tau^T_{zz_k} \right] = \alpha_k \frac{\partial}{\partial z} \left[\tau^T_{zz_k} \right] \tag{15}$$

A matrix form of the two-fluid model can be written as [2]:

$$\underline{\underline{A}} \; \frac{\partial \underline{\psi}}{\partial t} + \underline{\underline{B}} \; \frac{\partial \underline{\psi}}{\partial z} = \underline{c}(\underline{\psi}) \tag{16a}$$

where,

$$\underline{\psi} = \left[\alpha, p_L, u_G, u_L, h_L, R_b, \frac{D_G R_b}{Dt} \right]^T \tag{16b}$$

Equations (16) can be perturbed as follows,

$$\left[\underline{\underline{A}}_o + \delta \underline{\underline{A}} \right] \left[\frac{\partial \delta \underline{\psi}}{\partial t} \right] + \left[\underline{\underline{B}}_o + \delta \underline{\underline{B}} \right] \left[\frac{\partial \underline{\psi}_o}{\partial z} + \frac{\partial \delta \underline{\psi}}{\partial z} \right] = \underline{c}_o + \frac{\partial \underline{c}}{\partial \underline{\psi}} \Big|_o \; \delta \underline{\psi} \tag{17}$$

The steady-state equation describing the unperturbed fully developed two-phase flow is given by,

$$\underline{\underline{B}}_o \; \frac{\partial \underline{\psi}_o}{\partial z} = \underline{c}(\underline{\psi}_o) \equiv \underline{c}_o \tag{18}$$

Assuming that the spatial derivatives of the steady-state solution are of order δ, the linearized equation set describing the response of the system to small perturbations can be expressed as,

$$\underline{\underline{A}}_o \; \frac{\partial \delta \underline{\psi}}{\partial t} + \underline{\underline{B}}_o \; \frac{\partial \delta \underline{\psi}}{\partial z} = \underline{\underline{C}}_o \; \delta \underline{\psi} \tag{19a}$$

where,

$$\underline{\underline{C}}_o \overset{\Delta}{=} \frac{\partial \underline{c}}{\partial \underline{\psi}} \Big|_o \tag{19b}$$

Let us assume that the perturbation of the state variables is assumed of the form:

$$\delta \underline{\psi} = \underline{\psi}' \; e^{i(kz - \omega t)} \tag{20}$$

This transforms Eq. (19a) into the algebraic equation,

$$\{ \underline{\underline{A}}_o(\underline{\psi}) \; [-i\omega] + \underline{\underline{B}}_o(\underline{\psi}) \; [ik] - \underline{\underline{C}}_o(\underline{\psi}) \} \; \underline{\psi}' = 0 \; . \tag{21}$$

Equation (21), in conjunction with the requirement that $\underline{\psi}'$ be nontrivial, implies a linear dispersion relation of the form,

$$\det \{(\underset{=o}{A} - (i/\omega)\underset{=o}{C}) - k/\omega\ \underset{=o}{B}\} = 0 \quad . \tag{22}$$

Equation (22) has seven roots (k/ω), each giving a (dispersion) relationship between angular frequency, ω, and wavenumber, k. As shown by Ruggles et al. [1], [2], two of these roots predict the propagation of pressure perturbations (ie, the sound speed) in two-fluid flows. Similarly, another root predicts the propagation of void perturbations in two-phase flows.

Comparison of the Dispersion Model to Void Wave Data

The void wave propagation data of Pauchon and Banerjee [9] and Bernier [10] are compared with the dispersion relation, Eq. (21), in Figure-1. For consistency with previous investigations, [9], [12], the virtual volume coefficient, C_{VM}, was taken to be 0.5 in these evaluations.

Two interfacial drag coefficients have been used in conjunction with the two-fluid model, the first being the distorted bubble model of Harmathy [14] given in Eq. (4). The second was the so-called Newton's regime drag model [19], given by,

$$C_D = 0.45 \left[\frac{1 + 17.67\ [f(\alpha)]^{6/7}}{18.67\ f(\alpha)} \right]^2 \tag{23a}$$

where,

$$f(\alpha) = (1-\alpha)^{1/2}\ (1-\alpha/\alpha_m)^{2.5\alpha_m(\mu_G+0.4\mu_L)/(\mu_G+\mu_L)} \tag{23b}$$

and, α_m is the maximum packing void fraction, which for bubble flows can be taken as unity [19]. This drag coefficient is recommended for bubble radii, R_b, which satisfy:

$$R_b \leq 34.65 \left[\frac{\mu_L}{\rho_L\ g\ (\rho_L - \rho_G)} \right]^{1/3} \tag{24}$$

That is, $R_b \leq 16$ mm for a bubbly air/water flow at the STP conditions which characterize the conditions of the data under consideration here. Predictions are also shown on Figure-1 for the dispersion model with no interfacial drag. It can be seen that the drag models for distorted bubbles and zero interfacial drag ($C_D = 0$) tend to bracket the data, while the Newton's regime drag model falls in between. It is evident that the predictions are quite sensitive to the interfacial drag law used.

The linear dispersion model, Eq. (22), is compared to data of Mercadier [11] in Figure-2. Only the distorted bubble drag law is shown since it was

Figure 1. Void Wave Data, λ^* vs $\langle\alpha\rangle$

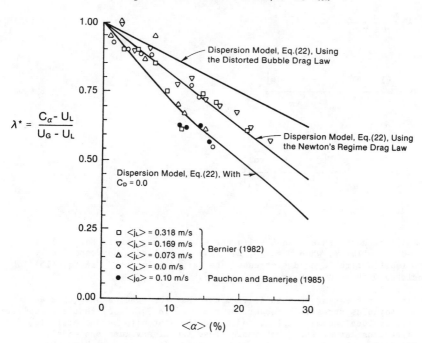

considered to be the most appropriate for the test conditions examined. This
choice of drag law is further supported by comparing the interfacial drag data
from Mercadier [11] given in Figure-3 with the two interfacial drag models.
Note that while neither of these drag laws is appropriate for the entire range
of void fractions investigated, the distorted bubble drag law does the best
job for $(\langle \alpha \rangle) \geq 10\%$.

Comparison of the Dispersion Model to other Void Wave Models

The models of Pauchon and Banerjee [9], [12] are based on the continuity
and momentum equations for liquid and gas, however, phasic compressibility,
surface tension, bubble pulsation induced forces (F_R) and gas phase inertia
are neglected. The resulting expression for the eigenvalues is [12]:

$$\lambda^* = \frac{c_\alpha - u_L}{u_G - u_L} = V \pm \sqrt{\nu/\tau} \tag{25}$$

where,

$$V = \frac{(1 - \alpha)\left[C_{vm} - \xi - \kappa\alpha\right]}{\alpha(1 - \alpha) + C_{vm}} \tag{26a}$$

$$\tau = \alpha(1 - \alpha) + C_{vm} \tag{26b}$$

$$\nu = (1 - \alpha)^2 \frac{\left[C_{vm} - \xi - \kappa\alpha\right]^2}{\left[\alpha(1 - \alpha) + C_{vm}\right]} + \alpha(1 - \alpha)\left[\xi + \kappa - C_{vm}\right]$$

$$+ 2(1 - \alpha)^2 \left[\xi - C_{vm}/2\right] \tag{26c}$$

It should be noted that when $\xi = 1/4$ the expression given in Eq. (11) is
implied. Similarly, when $\kappa = 1/5$, Eq. (14) is implied. Correspondingly,
$\kappa = 0$ implies zero Reynold's stress. The larger value of λ^* in Eq. (25) is
associated with void wave propagation [12].

It is interesting to note in Figure-4 that the results are quite sensitive to Reynolds stress. Moreover, we can note in Eq. (25c) that the two void
wave eigenvalues become complex, and cause the system to become elliptic, for
void fractions larger than that where these two eigenvalues are equal.
Significantly, the point at which they coincide (ie, the "nose" of the
eigenvalue curve in Figure-4) shifts dramatically with the addition of
Reynolds stress. This change in nature of the system of differential
equations indicates that the assumed constitutive relationships may not be
appropriate for void fractions beyond where the eigenvalues are real (ie,
beyond where the system model is well-posed). This is not surprising since
the models used for virtual mass effects and Reynolds stress are based on
single bubble analyses. Appropriate models for closely packed bubbly flows
may require improved constitutive models in order to be well-posed (ie,

Figure 2. Void Wave Data Due to Mercadier (1981)

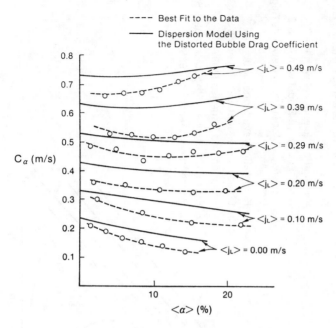

- - - - Best Fit to the Data
——— Dispersion Model Using
the Distorted Bubble Drag Coefficient

C_α (m/s)

$\langle j_L \rangle$ = 0.49 m/s
$\langle j_L \rangle$ = 0.39 m/s
$\langle j_L \rangle$ = 0.29 m/s
$\langle j_L \rangle$ = 0.20 m/s
$\langle j_L \rangle$ = 0.10 m/s
$\langle j_L \rangle$ = 0.00 m/s

$\langle \alpha \rangle$ (%)

Figure 3. Drag Coefficient vs. Data

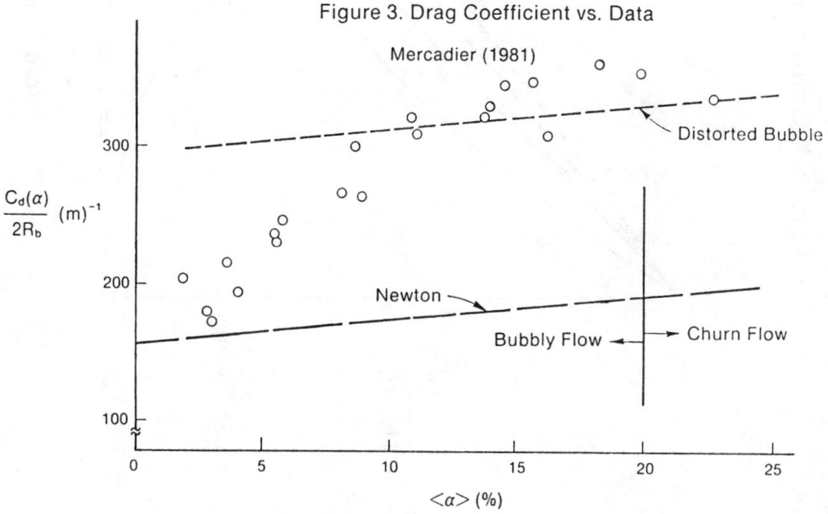

Mercadier (1981)

$\dfrac{C_d(\alpha)}{2R_b}$ (m)$^{-1}$

Distorted Bubble

Newton

Bubbly Flow ◄—— ——► Churn Flow

$\langle \alpha \rangle$ (%)

183

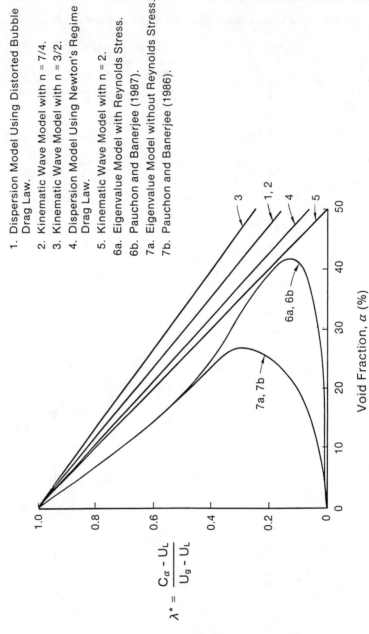

Figure 4. Comparison of Void Wave Models, $C_{VM} = 0.5$

1. Dispersion Model Using Distorted Bubble Drag Law.
2. Kinematic Wave Model with n = 7/4.
3. Kinematic Wave Model with n = 3/2.
4. Dispersion Model Using Newton's Regime Drag Law.
5. Kinematic Wave Model with n = 2.
6a. Eigenvalue Model with Reynolds Stress.
6b. Pauchon and Banerjee (1987).
7a. Eigenvalue Model without Reynolds Stress.
7b. Pauchon and Banerjee (1986).

$$\lambda^* = \frac{C_\alpha - U_L}{U_g - U_L}$$

Void Fraction, α (%)

hyperbolic). Thus, it should not be assumed that the mathematical transition from hyperbolic to non-hyperbolic systems is necessarily indicative of flow transition. Indeed, flow regime transition involves phenomena (eg, bubble coalescence) which are not in the model presented herein. Thus flow regime transition should not necessarily be inferred from the results presented herein.

The two-fluid model, Eqs. (16), can also be treated as an eigenvalue problem,

$$\det \left\{ \underline{\underline{A}} v - \underline{\underline{B}} \right\} = 0 \tag{27}$$

where, v, is a system eigenvalue.

These eigenvalues include the effect of compressibility, vapor inertia, and bubble pulsation induced forces (F_R). It is seen in Figure-4 that the two-fluid model gives virtually identical values for (c_α) as the simplified eigen-value models due to Pauchon and Banerjee [9], [12]. This is a useful check on consistency and verifies that, as expected, the effects of compressibility, bubble pulsation and gas inertia are negligible for void wave propagation.

A kinematic drift-flux model for void waves has been proposed by Wallis [13]. Significantly, this model is based entirely on steady-state consider-ations. The celerity of void perturbations is given by,

$$c_\alpha = \left(\frac{\partial j_G}{\partial \alpha} \right)_j = j + \frac{\partial j_{GL}}{\partial \alpha} \tag{28}$$

since,

$$j = \alpha u_G + (1 - \alpha) u_L$$

we have,

$$\lambda^* \overset{\Delta}{=} \frac{c_\alpha - u_L}{u_G - u_L} = \alpha + \frac{1}{(u_G - u_L)} \frac{\partial j_{GL}}{\partial \alpha} \tag{29}$$

Wallis [13] also proposed a drift-flux relation of the form,

$$j_{GL} = \alpha (1 - \alpha) (u_G - u_L) = u_\infty \alpha (1 - \alpha)^n \tag{30}$$

Thus Eqs. (29) and (30) yield,

$$\lambda^* = 1 - n\alpha \tag{31}$$

The value of the drift-flux parameter (n) depends on the flow conditions. For example, for steady, fully-developed vertical bubbly two-phase flow in the free stream, Eqs. (1b) & (3) yield:

$$-\frac{\partial p_G}{\partial z} - \rho_G g - \frac{3}{8} \rho_L \frac{C_D}{R_b} (u_G - u_L)^2 = 0 \tag{32a}$$

$$-(1 - \alpha) \frac{\partial p_L}{\partial z} - (1 - \alpha) \rho_L g + \alpha \frac{3}{8} \rho_L \frac{C_D}{R_b} (u_G - u_L)^2 = 0 \tag{32b}$$

Assuming the axial pressure gradients are equal, Eqs. (32) can be combined to yield:

$$(u_G - u_L) = [\frac{4}{3} \frac{g(\rho_L - \rho_G)(1 - \alpha)D_b}{\rho_L C_D}]^{1/2} \tag{33}$$

For the distorted bubble regime, Eq. (4) can be combined with Eq. (33) to yield:

$$(u_G - u_L) = 1.414 \ [\frac{g g_c (\rho_L - \rho_G) \sigma}{\rho_L^2}]^{1/4} (1 - \alpha)^{3/4}$$

That is,

$$(u_G - u_L) = u_\infty (1 - \alpha)^{3/4} \tag{34}$$

Hence Eqs. (30) and (34) imply n = 7/4 for distorted bubbly flow. Similarly, for other flow conditions the value of the drift-flux parameter recommended by Wallis [13], n = 2, may be appropriate.

Finally, it should be noted that, using an entirely different analytical approach, Pauchon and Banerjee [12] deduced a result for the case of a constant interfacial drag coefficient which also yields Eq. (31). In particular, they found n = 3/2, which is consistent with the analysis presented above for a constant C_D, and the results proposed previously by Zuber and Hench [20]. Moreover, they also derived the void dependence of neutrally stable void waves and found agreement with the Newton regime drag model [19] up to a void fraction of about 30%. It will be shown later that this model corresponds to an 'n' of about 2.

The linear dispersion model, Eq. (22), is compared with the simplified eigenvalue models due to Pauchon and Banerjee [9], [12] and the kinematic drift-flux model of Wallis [13] in Figure-4. It can be seen that the dispersion model gives results which are larger than the eigenvalue models, with the deviation increasing as the interfacial drag coefficient (C_D) increases. Also, it was found that for zero interfacial drag (ie, C_D = 0) that the dispersion model gave results which were close to the eigenvalue model. This implies that the other terms (eg, wall shear) in the matrix $\underset{=o}{C}$ of Eq. (22) are relatively small compared to interfacial drag.

Finally, it is interesting to note in Figure-4 that the kinematic drift-flux model for distorted bubbly flow (n = 7/4) agrees well with the corresponding dispersion results up to fairly large void fractions. Interestingly, the drift-flux parameter n = 2 agrees reasonably well with the dispersion results for Newton's regime drag law. Thus we find that the void wave celerities given by a drift-flux model (which are based only on steady-state considerations) are closely linked (via the interfacial drag law) to the corresponding linear dispersion results of the full two-fluid model. This clearly shows the importance of interfacial drag on void wave propagation phenomena. Moreover, it implies that a lot of physics associated with interfacial momentum transfer (eg: virtual mass effects, turbulence, etc.) is implicit in the drift-flux parameter, n.

CONCLUSION

It has been shown that the linear dispersion relation of a typical two-fluid model appears to be capable of predicting void wave propagation. In contrast, the void eigenvalues underpredict such data since these results imply zero interfacial drag, among other things (ie, setting $C_{=o}$ = 0). It has also been shown that kinematic drift-flux models of void wave propagation are closely linked to the dispersion relation in which the corresponding interfacial drag law is used.

It can be noted in Figure-5 that the eigenvalue model yields increasing values for λ^* as the void fraction approaches unity. This occurs because the primary contribution to the liquid pressure gradient is the hydrostatic head due to the mixture density. Thus, the relative velocity, $u_G - u_L$, approaches zero as the mixture density approaches that of the gas phase (ie, bubble bouyancy goes to zero). This result is somewhat artificial since the Newton's regime drag law used in the eigenvalue model predictions shown in Figure-6 is not appropriate for very high void fractions, nor are some of the other closure laws which have been used. While the same comments apply, this behavior is not seen in the model of Pauchon and Banerjee [12] since no explicit expression for relative velocity was used.

Finally, it should be noted that all the two-fluid models shown on Figure-4 were run with the virtual volume coefficient, C_{VM}, equal to 0.5 in order to be consistent with the eigenvalue models of Pauchon and Banerjee [9], [12]. However, it can be noted in Figure-5 that the eigenvalue results are significantly affected by the value of C_{VM}. In particular, the domain of hyperbolicity of the two-fluid model is reduced when C_{VM} is greater than 0.5 and increased when it is less. Including the interfacial shear stress, τ_{L_i}, also reduces the domain of hyperbolicity. Indeed, a significant change can be noted when $C_{VM} \leq 0.5$. It is also interesting to note that for $C_{VM} < 0.5$, the lower branch of λ^* can be negative.

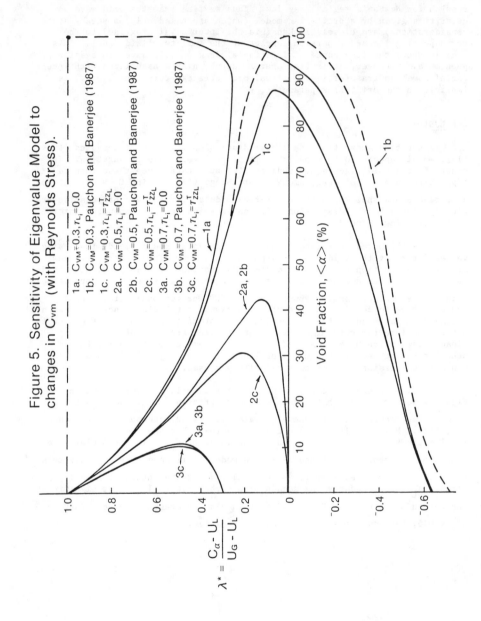

Figure 5. Sensitivity of Eigenvalue Model to changes in C_{VM} (with Reynolds Stress).

1a. $C_{VM}=0.3, \tau_{L_i}=0.0$
1b. $C_{VM}=0.3$, Pauchon and Banerjee (1987)
1c. $C_{VM}=0.3, \tau_{L_i}=\tau_{zz_L}^I$
2a. $C_{VM}=0.5, \tau_{L_i}=0.0$
2b. $C_{VM}=0.5$, Pauchon and Banerjee (1987)
2c. $C_{VM}=0.5, \tau_{L_i}=\tau_{zz_L}^I$
3a. $C_{VM}=0.7, \tau_{L_i}=0.0$
3b. $C_{VM}=0.7$, Pauchon and Banerjee (1987)
3c. $C_{VM}=0.7, \tau_{L_i}=\tau_{zz_L}^I$

Void Fraction, $\langle\alpha\rangle$ (%)

$$\lambda^* = \frac{C_\alpha - U_L}{U_G - U_L}$$

Fortunately the linear dispersion model with representative interfacial drag included is not sensitive to C_{VM}, Reynolds stress, or the interfacial shear stress. This indicates that the models for interfacial drag can be assessed through void wave measurements. Unfortunately, the existing void wave propagation data exhibits significant scatter and does not include the measurement of the attenuation coefficient, η_α. Furthermore, the existing data for c_α is not accompanied by sufficient supporting information on the flow state, or the nature of the void perturbation (ie: frequency, amplitude, etc.), thus, complete two-fluid model assessment is not possible. Nevertheless, it appears that carefully taken, harmonically-forced, void wave propagation data can be a powerful tool for assessing two-fluid interfacial drag laws due to the relatively large sensitivity in the void wave speed, c_α, to interfacial drag. Indeed, it appears that the void fraction dependence of the drag coefficient could be verified in this way.

NOMENCLATURE

a	Acceleration
a_{VM}	Virtual mass acceleration
c	Speed of propagation (celerity)
C_{VM}	Virtual volume coefficient
D	Diameter
$\dfrac{D_k(\ )}{Dt}$	Material derivative, $\dfrac{\partial(\ )}{\partial t} + u_k\dfrac{\partial(\ )}{\partial z}$
f	Frequency; friction factor
F	Force
g	Gravitational acceleration
g_c	Gravitational constant
h	Enthalpy
i	Imaginary number, $\sqrt{-1}$
j_k	Superficial velocity of phase-k
j_{GL}	Drift-flux
k	Wavenumber
M_k	Interfacial momentum transfer
n	Drift-flux parameter
$1/L_S$	Interfacial area density
p	Pressure
q_{ki}''	Interfacial heat transfer rate
R	Radius
STP	One atmosphere, 23°C
T	Temperature
t	Time
u	Velocity
v	Eigenvalue

| V | Parameter defined in Eq. (26a) |
| z | Axial location |

Greek

α	Void fraction
$\delta(\ )$	Perturbed quantity
η_α	Attenuation coefficient
λ	Wave length
μ	Dynamic viscosity
ω	Angular frequency
ρ	Density
κ	Reynolds stress parameter
ψ	State vector
σ	Surface tension
θ	Angle of inclination of flow from vertical
ν	Kinematic viscosity or parameter defined in Eq. (26c)
τ	Stress or parameter defined in Eq. (26b)
ξ	Interfacial pressure distribution parameter

Subscripts

b	Bubble
2ϕ	Two-phase
1ϕ	One-phase
G	Gas
H	Hydraulic
i	Interfacial
k	Phase indicator (G = gas, L = liquid)
L	Liquid
o	Equilibrium (steady-state) value
VM	Virtual mass
w	Wall
∞	Terminal rise (velocity)

Symbols

$\delta(\ )$	Perturbation
$(\ )^T$	Transpose of a vector
$\overset{\Delta}{=}$	Defined as
∇	Gradient
$\dfrac{\partial(\ )}{\partial(\ )}$	Partial derivative
$\dfrac{d(\ )}{d(\ )}$	Total derivative

REFERENCES

[1]   Ruggles, A.E., Scarton, H.A., and Lahey, R.T., Jr., 1988, "An Investiga-
      tion of Pressure Perturbations in Bubbly Air-Water Flows", Journal of
      Heat Transfer, Vol-110.

[2]   Ruggles, A.E., Drew, D.A., and Lahey, R.T., Jr., 1987, "The Relationship
      Between Standing Waves, Pressure Pulse Propagation and Critical Flow
      Rate in Two-Phase Mixtures", ANS Symposium Volume—*Waves in Multiphase
      Systems*, 24th ASME/AIChE National Heat Transfer Conference, Pittsburgh,
      PA.

[3]   Ishii, M., 1975, "Thermo-Fluid Dynamic Theory of Two-Phase Flow",
      Eyrolles.

[4]   Delhaye, J.M. and Achard, J.L., 1978, "On the Use of Averaging Operators
      in Two-Phase Flow Modelling", ASME Symposium Vol., *Thermal and Hydraulic
      Aspects of Nuclear Reactor Safety*, Vol. 1 - Light Water Reactors, (Eds.
      O.C. Jones, Jr. and S.G. Bankoff).

[5]   Banerjee, S., 1986, "Current Approaches to Modeling Multicomponent
      Multiphase Flows: Problems and Potential", Symposium Vol., *Multiphase
      Fluid Transients*, ASME Winter Annual Meeting, Anaheim, CA.

[6]   Drew, D., 1983, "Mathematical Modeling of Two-Phase Flow", Annual Rev.
      of Fluid Mech., 15: 261-91.

[7]   Cheng, L-Y., Drew, D.A. and Lahey, R.T., Jr., 1983, "An Analysis of Wave
      Dispersion, Sonic Velocity, and Critical Flow in Two-Phase Mixtures",
      NUREG-CR/3372.

[8]   Boure, J.A., Fritte, A.A., Giot, M.M. and Reocreux, M.L., 1975, "A
      Contribution to the Theory of Critical Two-Phase Flow: On the Links
      Between Maximum Flow Rates, Choking, Sonic Velocities, Propagation and
      Transfer Phenomena in Single and Two-Phase Flow", Energie Primaire,
      Vol-11, pp. 1-27.

[9]   Pauchon, C., and Banerjee, S., 1986, "Interphase Momentum Interaction
      Effects in the Averaged Multifield Model, Part I: Void Propagation in
      Bubbly Flows", Int. J. of Multiphase Flow, 16, 4, pp. 559-573.

[10] Bernier, R.N.J., 1982, "Unsteady Two-Phase Flow Instrumentation and
      Measurement", Report E200.4, Division of Engineering and Applied
      Science, California Institute of Technology.

[11] Mercadier, Y., 1981, "Contribution a L'Etudes des Propagations de
      Perturbations de Taux de Vide dans les Ecoulements Diphasiques Eau-Air a
      Bulles", Docteur Engenieur, Thesis, L'Universite Scientifique et
      Medicale & L'Institut National Polytechnique de Grenoble.

[12] Pauchon, C. and Banerjee, S., 1987, "Interphase Momentum Interaction Effects in the Averaged Multifield Model, Part II: Kinematic Waves and Interfacial Drag in Bubbly Flows", Int. J. of Multiphase Flow, 14, 3, pp. 253-263.

[13] Wallis, G.B. 1969, "One-Dimensional Two-Phase Flow", McGraw-Hill.

[14] Harmathy, T.Z., 1960, "Velocity of Large Drops and Bubbles in Media of Infinite or Restricted Extent", AIChE Journal, Vol. 6, No. 2.

[15] Drew, D.A. and Lahey, D.R., Jr., 1987, "The Virtual Mass and Lift Force on a Sphere in Rotating and Straining Inviscid Flow", Int. Journal of Multiphase Flow, Vol. 13, No. 1.

[16] Stuhmiller, J.H., 1977, "The Influence of Interfacial Pressure Forces on the Character of Two-Phase Flow Model Equations", Int. J. Multiphase Flow, Vol.-3, 551-560.

[17] Nigmatulin, R.I., 1979, "Spacial Averaging in the Mechanics of Heterogeneous and Dispersed Systems", Int. J. of Multiphase Flow, Vol-5, 353-385.

[18] Biesheuvel, A. and Van Wijngaarden, L., 1984, "Two-Phase Flow Equations for a Dilute Dispersion of Gas Bubbles in Liquid", J. Fluid Mech., 168, pp. 301-318.

[19] Ishii, M. and Zuber, N., 1979, "Relative Motion and Interfacial Drag Coefficient in Dispersed Two-Phase Flow of Bubbles, Drops and Particles", AIChE Journal, Vol. 25, No. 5.

[20] Zuber, N. & Hench, J., 1962, "Steady-State and Transient Void Fraction for Bubbling Systems and Their Operating Limit: Part I. Steady-State Operation, General Electric Report No. 62 GL 100.

A Novel Two-Fluid Model for Predicting Transient Subcooled Two-Phase Flows

H. A. KHATER
Department of Mechanical Power Engineering
Cairo University
Giza, Egypt

Abstract

A one-dimensional Lagrangian two-velocity model for predicting transient subcooled two-phase flow problems which preserves the advantages of the FAST approach is presented. The equation set consists of five equations, two representing mass conservation, two for energy conservation, and one momentum equation for the velocity difference between the phases. Analytical solutions which are valid over finite time and space intervals are derived and used to advance the solution through time. The model predictions indicate that the results are both accurate and physically plausible.

1. INTRODUCTION

The FAST solution method was shown [1] to be an economical tool for solving problems involving transient, non-equilibrium, two-phase flows where either the velocities of the two phases are equal, or a drift velocity is specified. The FAST method uses analytical solutions to the one-dimensional Lagrangian drift-flux formulation which are valid over finite time and space intervals. The method is particularly useful for situations in which the inlet velocity and the pressure level at one point in the channel are specified.

In some problems the difference in phase velocities is an important feature, and the above approach no longer suffices (e.g., counter-current flows, annular dispersed flows, etc.). In this paper a model which preserves the advantages of the FAST method and at the same time predicts the individual phase velocities is presented.

Several systems of equations for handling two-velocity, two-phase flows have appeared in the literature, differing in the way of averaging the instantaneous equations [2-6]. A good survey of the available systems is presented in Ref. [7].

In the present model an additional conservation equation was introduced to the FAST system of equations. This equation was obtained by combining the momentum equations for the separate phases in such a way that the axial pressure gradient was eliminated. For the special case in which the interphase drag is

193

large, this extra equation asserts that the phase velocities will be nearly equal, and the model reduces to the original FAST model. The conservation equations are, as in the original FAST model, transformed into a Lagrangian formulation but with a propagation velocity equal to the liquid phase velocity. Analytical solutions to these conservation equations which are valid over restricted time and space intervals are then found, and these solutions form the basis of the numerical solution.

2. FORMULATION OF THE CONSERVATION EQUATIONS

The proposed analytical scheme assumes that: (1) the flow channel cross-sectional area is constant, (2) the thermodynamic and transport properties are uniform over the portion occupied by each phase, (3) kinetic and potential energies and interfacial shear work are all ignored in the energy equations, (4) the effect of the difference between the pressures of each phase is ignored in the energy equation, instead one pressure is specified, p*, the interface pressure, and (5) the interface velocity is equal to the liquid velocity. The one-dimensional transient formulation of the conservation equations subject to the above mentioned assumptions is described below.

2.1 Governing Equations

Vapour Mass Conservation

$$\frac{\partial}{\partial t}(\alpha\rho_v) + \frac{\partial}{\partial z}(\alpha\rho_v V_v) = \Gamma \tag{1}$$

Liquid Mass Conservation

$$\frac{\partial}{\partial t}((1-\alpha)\rho_L) + \frac{\partial}{\partial z}((1-\alpha)\rho_L V_L) = -\Gamma \tag{2}$$

Vapour Momentum Conservation

$$\frac{\partial}{\partial t}(\alpha\rho_v V_v) + \frac{\partial}{\partial z}(\alpha\rho_v V_v^2) = -\alpha\frac{\partial p^*}{\partial z} - \alpha\rho_v g\sin\theta - a^* \alpha\rho_v g\cos\theta \frac{\partial h'}{\partial z}$$

$$+ \Gamma V_L - \frac{\tau_i P_i}{A_c} - \frac{\tau_{wv} P_{wv}}{A_c} \tag{3}$$

Liquid Momentum Conservation

$$\frac{\partial}{\partial t}((1-\alpha)\rho_L V_L) + \frac{\partial}{\partial z}((1-\alpha)\rho_L V_L^2) = -(1-\alpha)\frac{\partial p^*}{\partial z} - (1-\alpha)\rho_L g\sin\theta$$

$$-a^*(1-\alpha)\rho_L g\cos\theta \frac{\partial h'}{\partial z} - \Gamma V_L + \frac{\tau_i P_i}{A_c} - \frac{\tau_{wL} P_{wL}}{A_c} \tag{4}$$

194

Vapour Energy Conservation

$$\frac{\partial}{\partial t}(\alpha \rho_v (i_v - \frac{p^\star}{\rho_v})) + \frac{\partial}{\partial z}(\alpha \rho_v V_v i_v) = \frac{q_w'' P_{hv}}{A_c} - \frac{q_{vi}'' P_i}{A_c}$$

$$+ \Gamma i_G - p^\star \frac{\partial \alpha}{\partial t} \qquad (5)$$

Liquid Energy Conservation

$$\frac{\partial}{\partial t}((1-\alpha)\rho_L (i_L - \frac{p^\star}{\rho_L})) + \frac{\partial}{\partial z}((1-\alpha)\rho_L V_L i_L) = \frac{q_w'' P_{hL}}{A_c} + \frac{q_{iL}'' P_i}{A_c}$$

$$- \Gamma i_f - p^\star \frac{\partial(1-\alpha)}{\partial t} \qquad (6)$$

In the conservation of momentum equations a* is unity for stratified flows, and is zero for other flows. a* is the coefficient of a term which accounts for gravity forces due to the inclination of the interface in stratified flows.

Global energy conservation

From a global energy balance the following equivalent equations are obtained:

R.H.S. of equation (5) + R.H.S of equation (6) = $\frac{q_w'' P_h}{A_c}$ (7)

L.H.S. of equation (5) + L.H.S. of equation (6) = $\frac{q_w'' P_h}{A_c}$ (8)

2.2 Approximations

For the sake of convenience but not necessity two approximations are made in the study

(i) the wall heat flux is specified.
(ii) the effect of the axial pressure change on the thermo-dynamic and transport properties is neglected. The advantage of this is to decouple one momentum equation from the set of equations. The disadvantage is the limitation of the practical use of the method to problems where vapour generation due to axial pressure drop is small compared to that due to average pressure change with time and/or heat flux. Also, it precludes the analysis of problems involving shock waves.

2.3 Lagrangian Formulation

As mentioned above the liquid velocity is chosen to be the propagation velocity, thus the differential operator D/Dt is defined as

$$\frac{D}{Dt} \triangleq \frac{\partial}{\partial t} + V_L \frac{\partial}{\partial z} \tag{9}$$

The Resultant Equations

Upon substituting the operator (9) into the conservation equations and with some algebraic manipulation the following Lagrangian formulation results.

Conservation of Mass

The Void Propagation Equation

$$\frac{D\alpha}{Dt} = \frac{\Gamma}{\rho_L} + (1-\alpha) \frac{\partial V_L}{\partial z} + \frac{(1-\alpha)}{\rho_L} \frac{D\rho_L}{Dt} \tag{10}$$

The Mixture Continuity Equation

$$\frac{\partial j}{\partial z} = \frac{\Delta\rho \, \Gamma}{\rho_L \rho_V} - \frac{(1-\alpha)}{\rho_L} \frac{D\rho_L}{Dt} - \frac{\alpha}{\rho_V} \frac{D\rho_V}{Dt} - \frac{\alpha}{\rho_V} (V_V - V_L) \frac{\partial \rho_V}{\partial z} \tag{11}$$

where j is the mixture velocity defined by

$$j \triangleq \alpha V_V + (1-\alpha) V_L \tag{12}$$

Conservation of Energy

Vapour Enthlpy Propagation Equation

$$\alpha\rho_V \frac{D i_V}{Dt} = \frac{q_w'' P_{hv}}{A_c} - \frac{q_{vi}'' P_i}{A_c} - \Gamma(i_V - i_G) + \alpha \frac{Dp^\star}{Dt} - \alpha\rho_V (V_V - V_L) \frac{\partial i_V}{\partial z} \tag{13}$$

Liquid Enthalpy Propagation Equation

$$(1-\alpha)\rho_L \frac{D i_L}{Dt} = \frac{q_w'' P_{hL}}{A_c} + \frac{q_{iL}'' P_i}{A_c} - \Gamma (i_f - i_L) + (1-\alpha) \frac{Dp^\star}{Dt} \tag{14}$$

Conservation of Momentum

As mentioned above, the second approximation permits one momentum equation to be replaced by a specification of the

196

pressure-time history $p^*(t)$. The second momentum equation is obtained by eliminating $\partial p^*/\partial z$ through manipulating equations (5) and (6), which yields a conservation equation governing the relationship between the vapour and liquid velocites. Before writing down the resultant equation, constitutive relations are required to describe the shear stresses in terms of the physical properties. The following relationships are used

$$\tau_i = 1/2 \; \rho_v \, C_{D_{vL}} \, (V_v - V_L) \, \left| V_v - V_L \right| \tag{15}$$

$$\tau_{wv} = 1/2 \; \rho_v \, C_{D_{wv}} \, V_v \, \left| V_v \right| \tag{16}$$

$$\tau_{wL} = 1/2 \; \rho_L \, C_{D_{wL}} \, V_L \, \left| V_L \right| \tag{17}$$

Thus, the resultant conservation equation for the relationship between the two velocities is

$$\rho^* \frac{DV_r}{Dt} = - V_r \left\{ \frac{\Gamma}{\alpha} + \rho_v \frac{\partial V_v}{\partial z} + \frac{1}{\alpha(1-\alpha)} \frac{1}{2} C_{D_{vL}} \, \rho_v \, \left| V_v - V_L \right| \, \frac{P_i}{A_c} \right\}$$

$$+ \Delta\rho \frac{DV^*}{Dt} + g\Delta\rho \, (\sin\theta + a^* \frac{\partial h'}{\partial z} \cos\theta)$$

$$- \frac{1}{\alpha} \frac{1}{2} C_{D_{wv}} \, \rho_v \, V_v \left| V_v \right| \frac{P_{wv}}{A_c} + \frac{1}{(1-\alpha)} \frac{1}{2} C_{D_{wL}} \rho_L V_L \left| V_L \right| \frac{P_{wL}}{A_c} \tag{18}$$

where ρ^* and V^* are given by

$$\rho^* = \rho_L \quad , \qquad V^* = V_v \qquad\qquad\qquad \text{or} \tag{19}$$

$$\rho^* = \rho_v \quad , \qquad V^* = V_L \tag{20}$$

and where

$$V_r = V_v - V_L \tag{21}$$

The relationships (19) and (20) provide two equivalent forms. Depending on the range of α, one of the forms will lead to better convergence properties of the numerical solution. This will be discussed later.

Global Energy Conservation

Equation (7) yields the rate of vapour generation in terms of interface parameters which are suitable for use in enthalpy propagation equations.

$$\Gamma = \frac{P_i}{A_c} (q''_{vi} - q''_{iL}) \tag{22}$$

where

$$q_{iL}'' = \begin{cases} h_{iL}(i_f - i_L)/\overline{c_{p_L}} & , \quad i_L < i_f \quad (23) \\[3mm] \dfrac{A_c}{P_i} \{ \dfrac{q_w''P \, hL}{A_c} + (1-\alpha) \dfrac{Dp^\star}{Dt} (1 - \rho_f \dfrac{di_f}{dp}) \} & , \quad i_L = i_f \quad (24) \end{cases}$$

and

$$q_{vi}'' = \begin{cases} h_{vi}(i_v - i_G)/\overline{c_{p_v}} & , \quad i_v > i_G \quad (25) \\[3mm] \dfrac{A_c}{P_i} \{ \dfrac{q_w''P \, hv}{A_c} + \alpha \dfrac{Dp^\star}{Dt} (1 - \rho_G \dfrac{di_G}{dp}) \} & , \quad i_v = i_G \quad (26) \end{cases}$$

Equation (8) gives the rate of vapour generation in terms of global parameters which is suitable for use in the void propagation equation

$$\Gamma = \{ \frac{q_w''P \, h}{A_c} + \frac{Dp^\star}{Dt} - \alpha\rho_v \frac{Di_v}{Dt} - (1-\alpha)\rho_L \frac{Di_L}{Dt} - \alpha\rho_v (V_v - V_L) \frac{\partial i_v}{\partial z} \}/(i_v - i_L) \tag{27}$$

2.4 A Mathematical Approximation

In order to render the conservation equations amenable to analytical solutions, a mathematical approximation is introduced. This is, that over the limited interval in time and space for which the solution to equation for is sought, all the other unknowns in this equation are assumed to vary linearly with time. For example, when solving an equation for ϕ_1 in which the variable ϕ_2 appears, ϕ_2 is replaced by

$$\phi_2 = a_i^{\phi_2} + a_{i+1}^{\phi_2} t \tag{28}$$

Here, $a_i^{\phi_2}$ and $a_{i+1}^{\phi_2}$ are the two constants defining the linear variation of ϕ_2 with time when ϕ_2 appears in the ϕ_1 equation. With t=0 at the begining of the time interval,

$$a_i^{\phi_2} = \phi_2^0 \tag{29}$$

$$a_{i+1}^{\phi_2} = \frac{\phi_2^n - \phi_2^0}{\Delta t} \tag{30}$$

This approximation requires that iteration be used to solve the set of equations for all dependent variables. Table 1 lists the variables used in this procedure and their corresponding coefficients a_i and a_{i+1}

198

VARIABLE	CORRESPONDING COEFFICIENTS	VARIABLE	CORRESPONDING COEFFICIENTS	VARIABLE	CORRESPONDING COEFFICIENTS
V_L	a_1, a_2	$C_{D_{wv}}$	a_{21}, a_{22}	$\dfrac{\Gamma}{\alpha}$	a_{41}, a_{42}
V_v	a_3, a_4	$C_{D_{wL}}$	a_{23}, a_{24}	i_G	a_{43}, a_{44}
α	a_5, a_6	$\dfrac{\partial V_v}{\partial z}$	a_{25}, a_{26}	i_f	a_{45}, a_{46}
i_L	a_7, a_8	$\dfrac{\partial i_v}{\partial z}$	a_{27}, a_{28}	$\dfrac{P_i h_{vi}}{A_c \lambda \bar{c}_{p_v}}$	a_{47}, a_{48}
i_v	a_9, a_{10}	$\dfrac{\partial h'}{\partial z}$	a_{29}, a_{30}	$\dfrac{P_i h_{iL}}{A_c \lambda \bar{c}_{p_L}}$	a_{49}, a_{50}
P_{hL}	a_{11}, a_{12}	$\dfrac{\partial V_L}{\partial z}$	a_{31}, a_{32}	$\dfrac{P_i h_{iL}}{A_c \bar{c}_{pL}}$	a_{51}, a_{52}
P_{hV}	a_{13}, a_{14}	$\dfrac{\partial \alpha}{\partial z}$	a_{33}, a_{34}	$\dfrac{P_i h_{vi}}{A_c \bar{c}_{p_v}}$	a_{53}, a_{54}
P_i	a_{15}, a_{16}	$\dfrac{P_i}{\alpha(1-\alpha)}$	a_{35}, a_{36}	P_{wv}	a_{55}, a_{56}
Γ	a_{17}, a_{18}	$\dfrac{P_{wv}}{\alpha}$	a_{37}, a_{38}	P_{wL}	a_{57}, a_{58}
$C_{D_{vL}}$	a_{19}, a_{20}	$\dfrac{P_{wL}}{(1-\alpha)}$	a_{39}, a_{40}	ρ_L	a_{59}, a_{60}

2.5 The Final Form of the Conservation Equations

The final form of the conservation equations is obtained by substituting the above mentioned approximations in the relevant equations.

The Rate of Vapour Generation

Here the equation governing the rate of vapour generation as a function of α, and as a function of interface parameters under different energy transfer models (discussed later), is presented.

The Rate of Vapour Generation as a Function of α

Equation (27) is used to yield

$$\Gamma = C_1 + C_2\,\alpha + C_3\,\alpha t + C_4\,\alpha t^2 \tag{31}$$

where the coefficients C_1 through C_4 are listed in Table 2.

The Rate of Vapour Generation as a Function of the Energy Transport Models

The term 'energy transport models' is used here to refer to the assumptions made regarding the temperature of the vapour and the liquid. A detailed discussion of the energy transport models is given by Khater et al [8]. For the purpose of this paper we present a summary of these models.

1. General thermal non-equilibrium model, where both the liquid and the vapour are not at saturation. This model may be applied under diabatic subcooled liquid stratified flow conditions where the vapour may be superheated.

2. Subcooled liquid-saturated vapour model. In this case the vapour is assumed to remain at saturation. This may occur under flow boiling situations.

3. Saturated liquid-superheated vapour model. In this case the liquid is assumed to remain at saturation. This may be suitable for post-dryout situations.

4. Thermal equilibrium model. Here both phases remain at saturation. This will occur under well mixed flow conditions.

Now one can proceed to derive the equations for Γ under the different energy transport models just mentioned above. However, for the purpose of illustration of the method only one energy transport model will be considered; namely the saturated vapour-subcooled liquid model. The derivation of the final equations for the other models will be published elsewhere.

200

The Saturated Vapour-Subcooled Liquid Model

In this case we need to derive Γ as a function of i_L. This is obtained by substituting the appropriate linea variations (Table 1), equation (23) and equation (26) into equation (22). Thus we obtain

$$\Gamma = C_5 + C_6 i_L + C_7 i_L t + C_8 t + C_9 t^2 \tag{32}$$

where the coefficients C5 through C9 are listed in Table 2 (note that $c = \frac{di_G}{dp}$).

The Void Propagation Equation

Equation (10) yields, after substituting the appropriate linear approximations

$$\frac{D\alpha}{Dt} = e_1 + e_2\, \alpha + e_3\, \alpha t + e_4\, \alpha t^2 + e_5 t \tag{33}$$

where the coefficients e_1 through e_5 are listed in Table 3.

The Mixture Velocity

The equation governing the mixture velocity can be obtained by integrating equation (11)

$$j = j_i + \int_0^z \left(\frac{\Delta\rho\Gamma}{\rho_L \rho_v} - \frac{(1-\alpha)}{\rho_L} \frac{D\rho L}{Dt} - \frac{\alpha}{\rho_v} \frac{D\rho_v}{Dt} - \frac{\alpha}{\rho_v}(V_v - V_L) \frac{\partial\rho_v}{\partial z} \right) dz \tag{34}$$

The Liquid Enthalpy Propagation Equation

Again, here only the case for the saturated vapour-subcooled liquid energy transport model will be shown. Substituting equations (23) and (32), and the appropriate linear variations (Table 1), into equation (14), and assuming that the quantity $(1-\alpha)\rho_L$ can be replaced by an average value $\overline{(1-\alpha)\rho_L}$ during the time interval under consideration, one obtains.

$$\frac{Di_L}{Dt} = e_6 + e_7 i_L + e_8 i_L t + e_9 i_L t^2 + e_{10} i_L^2 + e_{11} i_L^2 t$$

$$+ e_{12} t + e_{13} t^2 + e_{14} t^3 \tag{35}$$

where the coefficients e6 through e14 are listed in Table 3.

The Vapour Enthalpy Propagation Equation

Under saturation conditions equation (13) yields

$$\frac{Di_v}{Dt} = \frac{Di_G}{Dt} = \rho_G C_G \frac{Dp^*}{Dt} = e_{15} \tag{36}$$

TABLE 2: THE COEFFICIENTS C_1 THROUGH C_9

$$C_1 = (\frac{q_w''P_h}{A_c} + \frac{Dp^*}{Dt} - \rho_L a_8)/(\overline{i_v - i_L})$$

$$C_2 = [- \rho_v a_{10} + \rho_L a_8 - \rho_v a_{27}(a_3 - a_1)]/(\overline{i_v - i_L})$$

$$C_3 = [- \rho_v(a_{27}(a_4 - a_2) + a_{28}(a_3 - a_1))]/(\overline{i_v - i_L})$$

$$C_4 = [- \rho_v a_{28}(a_4 - a_2)]/(\overline{i_v - i_L})$$

$$C_5 = \frac{q_w'' \, a_{13}}{A_c \, \bar{\lambda}} + (1 - \rho_G C_G) \frac{a_5}{\bar{\lambda}} \frac{Dp^*}{Dt} - a_{49} \, a_{45}$$

$$C_6 = a_{49}$$

$$C_7 = a_{50}$$

$$C_8 = \frac{q_w'' a_{14}}{A_c \, \bar{\lambda}} + (1 - \rho_G C_G) \frac{a_6}{\bar{\lambda}} - (a_{49} \, a_{46} + a_{50} \, a_{45})$$

$$C_9 = - a_{50} \, a_{46}$$

TABLE 3: THE COEFFICIENTS e_1 THROUGH e_{22}

$$e_1 = \frac{c_1}{\rho_L} + a_{31} + \frac{a_{60}}{\rho_L}$$

$$e_2 = \frac{c_2}{\rho_L} - a_{31} - \frac{a_{60}}{\rho_L}$$

$$e_3 = \frac{c_3}{\rho_L} - a_{32}$$

$$e_4 = \frac{c_4}{\rho_L}$$

$$e_5 = a_{32}$$

$$e_6 = (\frac{q_w'' a_{11}}{A_c} + a_{45} (a_{51} - c_5) + (1-a_5) \frac{Dp^*}{Dt}) / (\overline{(1-\alpha)\rho_L})$$

$$e_7 = (- a_{45} c_6 - a_{51} + c_5) / (\overline{(1-\alpha)\rho_L})$$

$$e_8 = (- a_{45} c_7 - a_{46} c_6 - a_{52} + c_8) / (\overline{(1-\alpha)\rho_L})$$

$$e_9 = (- a_{46} c_7 + c_9) / (\overline{(1-\alpha)\rho_L})$$

$$e_{10} = c_6 / (\overline{(1-\alpha)\rho_L})$$

$$e_{11} = c_7 / (\overline{(1-\alpha)\rho_L})$$

$$e_{12} = (\frac{q_w'' a_{12}}{A_c} + a_{45} (a_{52} - c_8) + a_{46} (a_{51} - c_5) - a_6 \frac{Dp^*}{Dt}) / (\overline{(1-\alpha)\rho_L})$$

TABLE 3 Continued

$$e_{13} = (- a_{45} \, c_9 + a_{46} \, (a_{52} - c_8))/((1-\alpha)\rho_L)$$

$$e_{14} = (- a_{46} \, c_9)/((1-\alpha)\rho_L)$$

$$e_{15} = \rho_G c_G \frac{Dp^*}{Dt}$$

$$e_{16} = (- a_{41} - \rho_v a_{25} + p_3 \, a_{35} \, a_{19})/p_1$$

$$e_{17} = (- a_{42} - \rho_v \, a_{26} + p_3 \, (a_{35} \, a_{20} + a_{36} \, a_{19}))/p_1$$

$$e_{18} = (p_3 \, a_{36} \, a_{20})/p_1$$

$$e_{19} = (p_2 + \Delta\rho g \, (\sin\theta + a^* \, a_{29} \, \cos\theta) + p_4 a_3 a_{37} a_{21} + p_5 a_1 a_{39} a_{23})/p_1$$

$$e_{20} = (a^* \, \Delta\rho g \, a_{20} \, \cos\theta + p_4 \, (a_{21} \, p_6 + a_3 \, a_{37} \, a_{22})$$
$$\quad + p_5 (a_{23} p_7 + a_1 a_{39} a_{24}))/p_1$$

$$e_{21} = (p_4 \, (a_{21} \, a_4 \, a_{38} + a_{22} \, p_6) + p_5 \, (a_{23} \, a_2 \, a_{40} + a_{24} \, p_7))/p_1$$

$$e_{22} = (p_4 \, a_{22} \, a_4 \, a_{38} + p_5 \, a_{24} \, a_2 \, a_{40})/p_1$$

The Equation of Conservation of Momentum

Two forms of the equation of propagation of the relative velocity, V_r, are considered here. These forms are obtained by substituting the relationship (19) or (20) into equation (18). Using the appropriate linear approximations one obtains

$$\frac{DV_r}{Dt} = V_r \ (e_{16} + e_{17}t + e_{18}t^2) + e_{19} + e_{20}t + e_{21}t^2 + e_{22}t^3 \tag{37}$$

where the coefficients $e16$ through $e22$ are listed in Table 3. These coefficients are functions of other coefficients $P1$ through P_7, which are listed in Table 4.

3. ANALYTICAL SOLUTIONS

In this section, the analytical solutions to the governing equations, and the numerical method of finding the solution over the whole time and space domain are presented.

3.1 Solution of the Equation of Propagation of the Void and the Relative Velocity

Both the void propagation equation and the equation of propagation of the relative velocity are first order, linear, non-homogeneous ordinary differential equations. The general form of both equations is

$$\frac{dx}{dt} - x \ (\tilde{a}_1 + \tilde{a}_2t + \tilde{a}_3t^2) = \tilde{a}_0 + \tilde{a}_4t + \tilde{a}_5t^2 + \tilde{a}_6t^3 \tag{38}$$

The boundary condition is

$$x = x^0 \quad \text{at} \quad t = 0 \tag{39}$$

The solution to equation (38) subject to the boundary condition (39) is

$$x = x^0 \ \{\exp \ [\tilde{a}_7(t^n) + \frac{\tilde{a}_2}{2} \ (t^n)^2 + \frac{\tilde{a}_3}{3} \ (t^n)^3]\}$$

$$+ \int_0^{t^n} (\tilde{a}_0 + \tilde{a}_4t + \tilde{a}_5t^2 + \tilde{a}_6t^3)\{\exp[\tilde{a}_1(t^n-t) + \frac{\tilde{a}_2}{2} \ ((t^n)^2-t^2)$$

$$+ \frac{\tilde{a}_3}{3} \ ((t^n)^3 - t^3)]\} \ dt \tag{40}$$

where t^n is the new t level.

3.2 Solution of the Enthalpy Propagation Equations

Subject to the boundary condition $iv = iv^0$ at $t=0$, equation (36) yields

$$i_v = i_v^0 + e_{15} \ t^n \tag{41}$$

TABLE 4: THE COEFFICIENTS p_1 THROUGH p_7

$$p_1 = \begin{cases} \rho_v & ; \text{ Eqn. [20]} \\ \\ \rho_L & ; \text{ Eqn. [19]} \end{cases}$$

$$p_2 = \begin{cases} \Delta\rho \; a_2 & ; \text{ Eqn. [20]} \\ \\ \Delta\rho \; a_4 & ; \text{ Eqn. [19]} \end{cases}$$

$$p_3 = -1/2 \; \rho_v \; \frac{\overline{|V_v - V_L|}}{A_c}$$

$$p_4 = 1/2 \; \rho_v \; \frac{\overline{|V_v|}}{A_c}$$

$$p_5 = 1/2 \; \rho_L \; \frac{\overline{|V_L|}}{A_c}$$

$$p_6 = (a_3 \; a_{38} + a_4 \; a_{37})$$

$$p_7 = (a_1 \; a_{40} + a_2 \; a_{39})$$

TABLE 5: SPECIFICATIONS OF THE SEPARATED TWO-PHASE TRANSIENT FLOW PROBLEM

Geometry

Horizontal rectangular channel

Width (W)   = 0.00261 m

Height (H) = 0.0255 m

Length   = 0.686   m

Boundary Conditions

System pressure   = 13.79 MPa

Inlet subcooling = 12.7°C

Wall heat flux   = 946 KW/m^2

Inlet velocity   = 1.83 m/sec

Inlet void   = 0

Interface Perimeter

P_i = $2(1-\alpha)(W+H)$

Heat Transfer Coefficient from Interface to Liquid

h_{iL} = $0.4 \mu_L c_{pL} (Re)^{0.662}/D_h$

The equation of propagation of the liquid enthalpy (35) is a Ricatti equation which can be solved analytically using Frobenius series solution method, after transforming the equation to a second order differential equation. Because of space limitations the derivation of the solution will not be presented here, only the final solution is given. For those readers who are interested in more details they can refer to Khater et al [8].

The solution to equation (35) subject to the boundary condition

$$i_L = i_L^0 \quad \text{at } t = 0 \tag{42}$$

is given by

$$i_L = - \frac{\sum\limits_{n=1}^{\infty} n \, B_n t^{n-1}}{(e_{10} + e_{11} t) \sum\limits_{n=0}^{\infty} B_n t^n} \tag{43}$$

where B_n, $n=0,1,2,\ldots$etc. are the series coefficients given by

$$B_0 = 1 \tag{44}$$

$$B_1 = - e_{10} \, i_L^0 \quad \text{(from the boundary condition)} \tag{45}$$

$$B_n \atop n \geq 2 = \frac{[(n-1)(n-2)d_2 + (n-1)d_3]B_{n-1} + [(n-2)d_4 + d_7]B_{n-2} + [(n-2)d_5}{n(n-1) \, d_1}$$

$$\frac{+ d_8]B_{n-3} + [(n-4)d_6 + d_9]B_{n-4} + d_{10}B_{n-5} + d_{11}B_{n-6} + d_{12}B_{n-7}}{} \tag{46}$$

where in equation (46), in the case of a negative value of the subscript m of Bm is equal to zero. The coefficients d1 through d_{12} are given by

$$d_1 = e_{10} \quad , \quad d_2 = e_{11} \quad , \quad d_3 = - e_7 \, e_{10} \, e_{11} \quad ,$$

$$d_4 = - (e_8 \, e_{10} + e_7 \, e_{11}) \quad , \quad d_5 = - (e_9 \, e_{10} + e_8 \, e_{11}) \quad , \quad d_6 = e_9 \, e_{11} \quad ,$$

$$d_7 = e_6 \, e_{10}^2 \quad , \quad d_8 = e_{10}^2 \, e_{12} + 2 \, e_{10} \, e_{11} \, e_6 \quad ,$$

$$d_9 = e_{10}^2 e_{13} + 2 e_{10} e_{11} e_{12} + e_{11}^2 e_6 \quad ,$$

$$d_{10} = e_{10}^2 e_{14} + 2 e_{10} e_{11} e_{13} + e_{11}^2 e_{12} \quad ,$$

$$d_{11} = 2 e_{10} e_{11} e_{14} + e_{11}^2 e_{13} \quad , \quad d_{12} = e_{11}^2 e_{14}$$

3.3   The Axial Position of a Fluid Element

The axial position of a fluid element can be found by integrating the equation

$$\frac{dz}{dt} = V_L = a_1 + a_2 t \tag{47}$$

subject to the boundary condition

$$z = z^0 \quad \text{at} \quad t = 0 \tag{48}$$

The solution is thus given by

$$z = z^0 + a_1 t + \frac{a_2}{2} t^2 \tag{49}$$

3.4   Solution of the Mixture Velocity Equation

The mixture velocity at any position is found by numerically integrating equation (34).

3.5   The Method of Advancing the Solution in the Time and Space Domain

The following steps are carried out to obtain the solution in the time and space domain of interest.

1. The flow in the flow channel is divided into a number of fluid elements.
2. A time step t is chosen. The number of fluid elements has to be sufficiently large, and at the same time sufficiently small to ensure that the solutions given are accurate.
3. For each fluid element the solutions derived are used to find , Vr, iL, iv, j and z. Iteration is neccessary for each element to find the solution for that element. During the time step under consideration, one or more additional fluid elements may be introduced at the channel inlet.
4. Since in step 3 the dependent variables for each element were determined independently of the changes that occur to the rest of the elements, step 3 is repeated to couple the fluid elements until the solution converges at the end of the time step.

5. The solutions obtained in step 4 are now used as the initial conditions for the next time step, and steps 3 and 4 are repeated.

6. Step 5 is repeated as many times as the problem requires to cover the time domain of interest.

4. RESULTS AND DISCUSSION

To demonstrate the ability of the above mentioned model to predict two-phase flow problems in which the velocities of the two phases are different, three problems have been chosen. The first problem involves vapour generation due to wall heat flux, and was chosen to study global mass conservation characteristics of the model, and to determine whether reasonable dependence on interfacial shear was exhibited. The other problems, involve wave phenomena in two component flows where analytical solutions are available and they were used to study the accuracy of the predictions of the model.

4.1 A Separated One-Component Two-Phase Flow Transient

This problem is a wall heat flux transient type. The initial condition is steady adiabatic flow of subcooled liquid at high pressure in a rectangular channel. At the start of the transient the wall heat flux undergoes a step increase over the whole channel. The inlet velocity, void and liquid enthalpy and the system pressure are kept constant throughout the transient. Table 5 shows the problem specifications. The interfacial drag coefficient was varied from 0 to 100 and its effect on phase velocities, void and global mass conservation was examined.

Figure 1 shows the final new steady flow axial distribution of both the liquid and vapour velocities under various conditions of interfacial drag coefficient. It is seen that the results are qualitatively plausible in that the two velocities approach each other as the interfacial drag coefficient is increased. At the highest considered value, the phase velocities become practically identical.

Figure 2 shows the axial void distribution for this problem. It is seen that for the lower values of interfacial drag the void is less at any position which is , again, at least qualitatively correct.

In addition to the above plausible results, it has been shown by Mathers et al [9] that this problem is a critical test of the ability of a code to conserve mass. In this study a mass gain ranging from 0.8% and 2% occurred.

This problem was also helpful in the exploration of some important features of the equation of propagation of the relative velocity. It was mentioned earlier in this paper that two equivalent forms for this equation were used and that depending on the range of α , one of the forms will lead to better

210

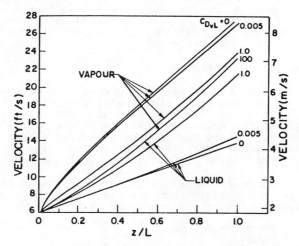

Figure 1: Steady liquid and vapour velocity axial
profiles for various values of $C_{D_{vL}}$

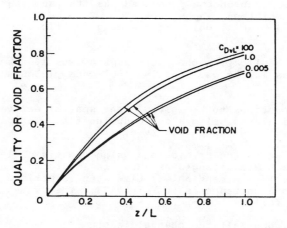

Figure 2: Steady axial void profile for various
values of $C_{D_{vL}}$

211

convergence properties of the numerical solution. It was found, in the above problem, that using equation (18) with the substitution of the relationships (19) for the whole range of α, a large number of iteations was required in the lower range of ($\alpha < 0.05$) to obtain a converged solution. Beyond this value of α, the number of iterations required was reduced to the acceptable value. This procedure was then adopted for all cases and was considered a feature of the model.

4.2 Propagation of Dynamic Waves

This problem considers the propagation of waves in an incompressible two-component (water-air) flow, in a plane channel where both ends of the channel are closed. An initial discontinuity in the water level at the middle of the channel is assumed. Figure 3 shows the channel geometry, dimensions and the initial fluid discontinuity. In this problem, the wave shape and speed are examined using the suggested two-fluid model.

Wallis [10] reported an analytical solution to this problem, where all the drag coefficients are assumed to be zeros. The wave speed is given by (small amplitude wave):

$$c = \pm \sqrt{\frac{gH(\rho_L - \rho_v)}{\dfrac{\rho_L}{1-\alpha} + \dfrac{\rho_v}{\alpha}}} \qquad (50)$$

The purpose of examining this problem is to compare the results of the predicted wave speed with the analytical value, and to examine the effect of the number of the elements chosen and the time step on the predicted results so that a "grid independent" solution can be obtained. To make comparisons with the analytical solution, all the drag coefficients were set equal to zero.

In this application the water density was 1000 kg/m^3 and the air density was 1.2 kg/m3. Equation (50) thus yields a wave speed of 2.21 m/s.

In the equations to be solved, the spatial derivatives of V and h are calculated using a cubic spline fit. The distribution of the elements is shown in Figure 4.

Figure 5 shows the predictions using ten fluid elements (the centres of each are arranged as shown in Figure 4), and a time step of 0.1 s. If a dimensionless time step is defined as $\Delta\tau = \Delta t c / \Delta z$, then the above corresponds to $\Delta\tau = 0.177$. The wave speed was calculated as follows:

$$c_{calc} = \frac{\text{one-half the channel length}}{\text{time for the interface to be flat over the entire channel}} \qquad (51)$$

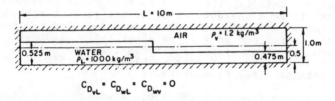

$$C_{D_{vL}} = C_{D_{wL}} = C_{D_{wv}} = 0$$

Figure 3: Closed-end channel with water discontinuity at the middle

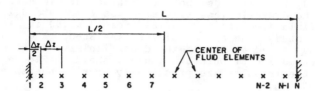

Figure 4: Distribution of fluid elements

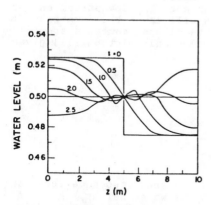

Figure 5: Wave propagation in closed-end channel (N = 10, ∠t = 0.1 s)

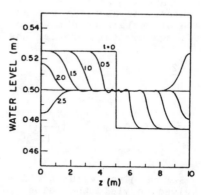

Figure 6: Wave propagation in closed-end channel (N = 50, Δt = 0.025 s)

The computed wave speed in this case was 2.37 m/s, an errror of 7.13% compared to the analytical value. As for the shape of the wave, it is seen from Figure 5 that some under-and over-shoots occur, the maximum of which is at t= 1.0 s; these decay for larger t. This is attributed to the coarse grid used, and it also seems to be a characteristic of the present method in such wave problems. However, the over-and under-shoots almost vanish as the grid is refined as will be shown below. Another feature to be noticed is that symmetry between the wave heights at the channel ends is good, at least untill the interface completely flattens for the first time at t=2.2 s.

Using a very fine grid (50 fluid elements and a time step of 0.02 s corresponding to $\Delta \tau$ =0.21) gives the results shown in Figure 6. The wave speed computed using equation (51) yields a value of 2.22 m/s, which is 0.45% larger than the analytical value. It is seen that very small over and under-shoots still occur untill 0.5 s, after which they vanish and the interface becomes, for all practical purposes, flat at 0.5 s.

To study the effect of both the number of elements and the time step, the dimensionless time step was fixed at 0.2, and the number of fluid elements was varied from 10 to 50 in increments of 10. To study the effect of the time step, the number of elements was fixed at 30 and the dimensionless time step was varied. The results of both studies are shown in Figure 7. It is seen that the results are practically "grid-independent" for N=30 and $\Delta \tau$ =0.6.

4.3 Stratified Flow in a Plane Channel which is Abruptly Closed at One End

The configuration for this problem is shown in Figure 8 where a stratified flow of water of density 1000 kg/m3 and air of density 1 kg/m3 moves at a constant speed of 1 m/s through a plane channel. At the start the transient, the downstream end of the channel is abruptly closed . This will cause the formation of a wave of certain height which will travel upstream at a specific speed. Again, this problem has an analytical solution [11] given by

$$\frac{h_2}{h_1} = 1 + \frac{V_{Li}}{c} \tag{52}$$

$$V_{Li} + c = (gh_1)^{\frac{1}{2}} \ \{\frac{h_2}{2h_1} \ (1 + \frac{h_2}{h_1})\}^{\frac{1}{2}} \tag{53}$$

For a given value of h1 and V_{Lo} the above two equations can be solved to yield the values of both the wave height, h2, and the wave speed, c. Under the conditions given in Figure 8 (V_{L1}=1 m/s and h =0.5 m) the analytical solution gives
1

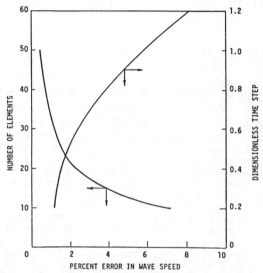

Figure 7: Error in predicted wave speed as a function of
the number of elements, and the dimensionless
time step

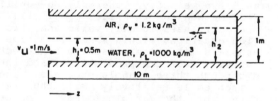

Figure 8: Stratified flow in a plane channel
which is abruptly closed at one end

$$c = 2.023 \text{ m/s} \tag{54}$$

$$h_2 = 0.75 \text{ m}$$

To solve this problem using the FAST code, the following boundary conditions were used

$$V_L(o,t) = 1 \text{ m/s}$$

$$V_L(L,t) = V_v(L,t) = 0 \text{ m/s} \tag{55}$$

$$\frac{\partial}{\partial z}(V_L, V_v, \alpha)(L,t) = 0$$

The derivatives along the channel were calculated as in the previous problem.

Figure 9 shoes the predictions using 10 fluid elements and a time step of 0.1 s. Again, over-shoots occur at the start of the transient ;the maximum value occurs at t=0.5 s but they soon die out. The wave height for t>1 s is approximately 0.76 m. To compute the wave speed, the wave front was assumed to be located at middle plane,i.e. where the liquid level has risen by a value equal to one-half of the maximum level risen. The computed speed was 2.04 m/s, i.e an error less than 1% exists compared to the analytical value. A total of 50 s of CPU time was required for executing this problem on an IBM 360-75 computer. Refining the grid to 20 fluid elements and a time step of 0.05 s reduces the wave front smearing considerably. Again, over-and under-shoots occur, but they die out more quickly than in the first case. The predicted wave height was 0.765 m, and the computed wave speed was 2.02 m/s. The results are shown in Figure 10.

5. CONCLUDING REMARKS

A Lagrangian, two-velocity, thermal non-equilibrium model based on the FAST method due to Khater et al [1], has been presented. The procedure uses locally valid analytical solutions to the conservation equations within specified time and space intervals, and these solutions are used to construct the global solution. The plausibility of the predictions, and their accuracy were demonstrated through three example problems. The execution time for a given problem is relatively small thus retaining the economical feature of the original FAST procedure.

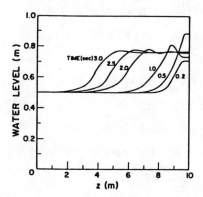

Figure 9: Wave propagation
N = 10, and Δt = 0.1 s

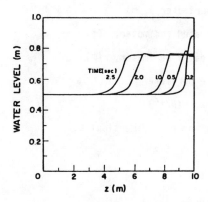

Figure 10: Wave propagation
N = 20, and Δt = 0.05 s

217

NOMENCLATURE

A_c = cross-sectional area, (m^2)

c = rate of change of saturation enthalpy with pressure (J/kg MPa)

C_D = drag coefficient, (dimensionless)

c_p = specific heat, (J/kg K)

c_{pL} = $(i_f-i_L)/(T_S-T_L)$, (J/kg K)

c_{pv} = $(i_G-i_v)/(T_S-T_v)$, (J/kg K)

D_h = hydraulic diameter, (m)

g = gravitational acceleration, (m/s^2)

h = coefficient of heat transfer, (W/m^2K)

h' = liquid height, (m)

H = channel height, (m)

i = enthalpy, (J/kg)

j = $\alpha V_v+(1-\alpha)V_L$, superficial velocity, (m/s)

p = pressure, (MPa)

P_h = heated perimeter, (m)

P_{hL} = liquid heated perimeter, (m)

P_{hv} = vapour heated perimeter, (m)

P_i = interface perimeter, (m)

q'' = heat flux, (W/m^2)

Re = Reynolds number, (dimensionless)

t = time, (s)

Δt = t^n-t^o, (s)

T = temperature, (K)

V = velocity, (m/s)

V_r = V_v-V_L, (m/s)

z = axial position, (m)

Greek Letters

α =   void fraction, (dimensionless)

Γ =   vapour generation rate, (kg/m^3 s)

μ =   viscosity, (kg/m s)

ρ =   density, (kg/m^3)

$\Delta\rho$ = $\rho_L - \rho_v$, (kg/m^3)

θ =   angle of inclination of the channel with the horizontal, (rad)

τ =   shear stress, (kg/m^2)

Subscripts

f   =   saturated liquid

G   =   saturated vapour

i   =   channel inlet

iL   =   interface -liquid

L   =   Liquid

s   =   saturation

v   =   vapour

vi   =   vapour-interface

vL   =   vapour-liquid

w   =   wall

wL   =   wall-liquid or wetted liquid

wv   =   wall-vapour or wetted vapour

Superscripts

o   =   old time level

n   =   new time level

___   =   average during the time and space interval

REFERENCES

1. Khater, H.A., Nicoll, W.B. and Raithby, G.D., "The FAST Procedure for Predicting ransient Subcooled Two-Phase Flows", Vol 8, No. 3, pp. 261-278, Pergamon Press, June 1982.

2. Delhaye, J.M., "General Equations of Two-Phase Systems and their Application to Air-Water Bubble Flow and to Steam-Water Flashing Flow", ASME Preprint No. 69-HT-63, 1969.

3. Ishii, M.,"Thermo-Fluid Dynamic Theory of Two-Phase Flow", Eyrolles, Paris, 1975.

4. Vernier, P. and Delhaye, J.M.,"General Two-Phase Flow Equations Applied to Thermodynamics of Boiling Nuclear Reactors", Energie Primaire, 4, pp. 5-46, 1968.

5. Yadigaroglu, G. and Lahey, R.T.,"On the Various Forms of Conservation Equations in Two-Phase Flow", International J. of Multiphase Flow, 2, 1976.

6. Zuber, N. and Staub, F.W.,"The Propagation and the Wave Form of the Vapour Volumetric Concentration in Boiling, Forced Convection System Under Oscillating Conditions", International J. of Heat and Mass Transfer, 9, pp. 871-895, 1966.

7. Hughes, E.D., Lyckowski, R.W. and McFadden, J.H.,"An Evaluation of State-of-the-Art Two-Velocity Two-Phase Flow Models and their Applicability to Nuclear Reactor Transient Analysis", Electric Power Research Institute, Report No. EPRI NP-143, 2, 1976.

8. Khater, .A. and Raithby, G.D.,"Development of a Two-Velocity Model for Transient Non-Equilibrium Two-Phase Flow Based on the FAST Approach", Electric Power Research Institute, Report No. EPRI NP-1732, 1981.

9. Mathers, W.G., Zuzak, W.W., McDonald, H.H. and Hancox, W.T.,"On Finite Difference Solutions to the Transient Flow Boiling Equation", Invited paper presented at the OECD Committee on the Safety of Nuclear Installations Specialists Meeting on Transient Two-Phase Flow, Toronto, Ontario, 1976.

10. Wallis, G.B.,"One-Dimensional Two-Phase Flow", McGraw-Hill Book Company, 1969.

11. Streeter, V.L. and Wylie, E.B.,"Hydraulic Transients", McGraw-Hill Book Company, 1969.

Transient Phenomena under Force to Natural Circulation of Horizontal Heating and Vertical Cooling Figure-of-Eight and -Zero Flow Loops

JEN-SHIH CHANG and M. KIELA
Department of Engineering Physics
McMaster University
Hamilton, Ontario, Canada L8S 4M1

V. S. KIRSHNAN
AECL-RC
Manitoba, Canada

Abstract

In this paper, transient phenomena under the force to natural circulation of horizontal heating and vertical cooling figure-of-zero and -eight flow loops are experimentally investigated. The MRD-2 loop was equipped with two orifice flow meters, 38 thermocouples and 12 pressure transducers and 6 capacitance void transducers. AGEMA IR thermo-image processing system is used for the transient tube wall surface temperature measurements. The results show that the flow rate and frequency of oscillation, under natural circulation in pump trip experiments show similar characteristics with natural circulation from stagnated water with gradual heating. Rewetting of the heating section under natural circulation is cyclic phenomena and significantly different wall temperature profile are observed with other force flow rewetting experiments.

I.   INTRODUCTION

The Three Mile Island small break loss of coolant accident (LOCA) demonstrate the necessity for long term unpowered cooling schemes to remove decay heat from the nuclear reactor core [3]. In a CANDU nuclear power plant, the TMI type of accident is less likely to occur because there are additional heat losses through the system piping and other components in the CANDU primary heat transport system (PHT). However, under LOCA conditions, the steam generators are expected to be primary heat sinks. Depending on the particulars of the accident scenario, different heat removal mechanisms may be occurring in each steam generator tube of the PHT system. These include reflux condensation, single- and two-phase natural circulations [11]. Here, thermosyphoning in a thermohydraulic loop may be defined as fluid flow caused primarily by density differences created by a heat source(s) and a heat sink(s) at a higher elevation with respect to the source. The phenomenon is governed by essentially the same physics as natural convection of a fluid in a container. Thermosyphoning is possible both without a change of phase of the fluid (single-phase thermosyphoning) and with a change of phase in the fluid (two-phase thermosyphoning). The combinations of several thermosyphoning of a thermalhydraulic loop to maintain the unidirectional flow without force convection, we called it the "natural circulation" [11].

A need for long term unpowered cooling schemes to remove decay heat from the nuclear reactor core has led to many studies of natural circulation phenomena. Some work has been performed, but is applicable only to U.S. nuclear reactors since different reactor designs can lead to different natural circulation mechanisms [1-8]. On CANDU system, some thermosyphoning tests have been conducted in actual power plants and simulation loops [9-12]. As an exception to previous study in a simulated or real CANDU systems, a recent study in RD-12 and -14 [12] offers some theoretical explanation of the mechanism of the phenomena using Zuber's void propagation theory [13]. In this work, transient phenomena observed under a force to natural circulation transition in a figure-of-eight and -zero flow loops are experimentally investigated.

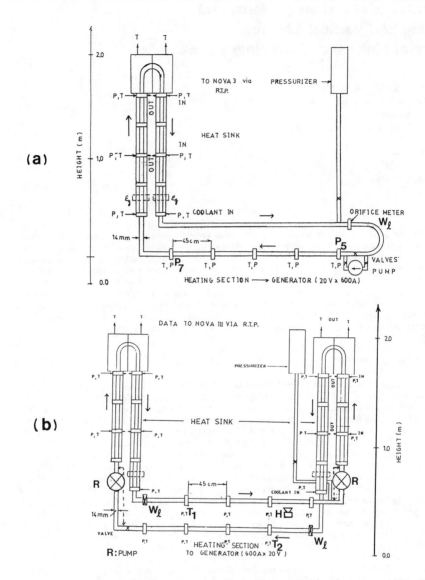

Fig. 1   Schematics of the test loops: P pressure gauges, T thermocouples, O orifice flowmeters, ε_g capacitance void meter, H IR thermoimage camera.
a) Figure-of-Zero flow loop   b) Figure-of-Eight flow loop

222

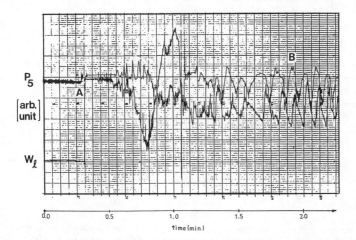

Fig. 2   Transient of mass flow rate and heating channel inlet pressures under the transition from force convection by pumps to the natural circulation with the presence of condensable gas and orifice in the figure-of-zero flow loops. A: pump trip, B: steady state condition.

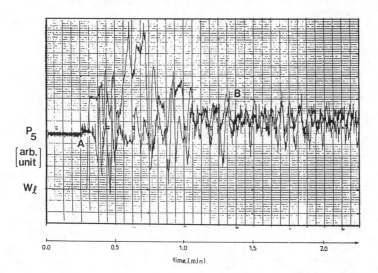

Fig. 3   Transient of mass flow rate and heating channel inlet pressure under the transition from force to natural circulation with orifice flowmeter but without noncondensable gases in the figure-of-zero flow loop. A: pump trip, B: steady state condition.

223

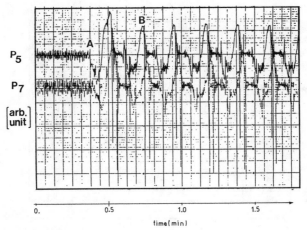

Fig. 4 Transient of heating channel exit and inlet pressure and the transitions from force to natural circulation without orifice and noncondensable gas in the figure-of-zero flow loop. A: pump trip, B: steady state condition.

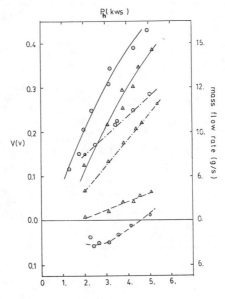

Fig. 5 Time averaged maximum and minimum flow rates, measured by orifice meters at different heater power during quasi-steady state natural circulation for different investigated experimental conditions: —— maximum forward flow, --- maximum flow reversal, —·— time averaged, 000 without non-condensable gas, ΔΔΔ with non-condensable gas.

II. EXPERIMENTAL SETUP AND PROCEDURE

A figure-of-zero loops is used as a experimental setup in this study (Figure 1a). As the system is unsymmetric, it may be easier to establish two-phase natural circulation with a relatively low power input, so that study can be done on a small scale, experimental setup (i.e. system limitation is low). Also, possible interferences which may occur between competing phenomena in the two vertical condensing sections, a figure-of-eight loop (Figure 1b) is also investigated in the present study.

The heating section is electrically heated using a maximum $20V \times 600A$ regulated dc generator. It consists of four stainless steel heating elements about 45 cm long and 13 mm in inner diameter. These heating elements are electrically insulated from each other by Teflon spacers, so that individual elements can be individually heated (if so desired). At these Teflon spacers, absolute pressure near the wall and centre line temperature are monitored. The condensing section consists of eight separated 30 cm long coaxial Pyrex glass (four on each of the two legs). The two legs are connected at the top by an inverted glass U-tube. The inner tube (primary side) is also 12 mm in inner diameter so that cross-sectional flow area is the same for both the heating and cooling system.

Pressure is monitored using Validyne pressure transducer and the temperature is monitored using type T Omega thermocouple. The volumetric average void fraction at each of the eight glass sections can also be monitored by using a well calibrated capacitance void meter. These void fraction measurement along the U-tube section enables the measurement of an axial void fraction distribution in the condensing section. This axial distribution of void fraction is a valuable information in an attempt to construct a model for the phenomena. Flow rate in the primary loop is measured by an orific flow meter which is situated on the horizontal returning side of the loop. System pressure is controlled by a pressurizer which is located about the same elevation with the U-tube section of the system. The loop has a by-pass channel for a water pump therefore, both the force convection and transient to the natural circulation can be studied. A NOVA III mini-computer with a 60 channel RTP system was used as a data acquisition system.

The experimental procedure is as follows:
(1)   Turn on the pump and set the flow rate in the both primary and secondary side of the loop to 0.4 [L/min] and 200 [mL/min] respectively.
(2)   Turn on the electric heating section power to around 1.5 [kW].
(3)   Observe temperatures and flow rates until the system reaches steady state.
(4)   Observe the phenomena when pump power goes to zero. Open by-pass channel for pump.

III. EXPERIMENTAL RESULTS
III.1 Figure-of-Zero Loops

Typical experimentally observed the force to natural circulation transient for the figure-of-zero flow loops are shown in figures 2, 3 and 4 for the flow loop with the orifice flow meters and the existence of noncondensable gas (air), with the orifice flow meter but without noncondensable gas, and without both the orifice flowmeter and noncondensable gas, respectively, where in order to compare these flow transient without orifice flow meter, the absolute pressure transient at the inlet of the heating channel, P_5, for all three cases are compared in these figures. Since this work was only part of a series of studies on two-phase natural circulation, the amount of non-condensable gases in the system as not accurately measured, and only qualitative trends of experimental data would be given here concerning the presence of non-condensable gases. From this, the approximated percentage of amount of non-condensable gas in the system is 3%. By compared with figures 2, 3, and 4, it shows clearly that the on-set of steady natural circulation is slow down by the existence of orifice and non-condensables, since the presence of both the orifice and noncondensable gas bubbles in the

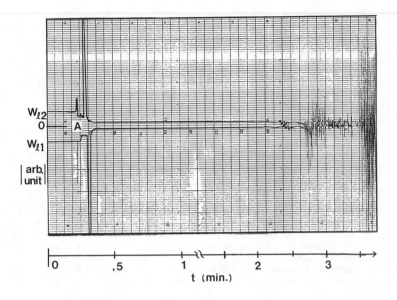

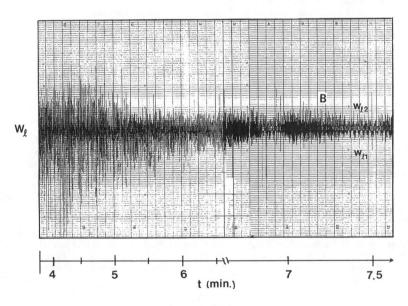

Fig. 6   Transient of mass flow rate under the transition from force to natural circulation in the figure-of-eight flow loop. A: pump trip, B: steady state condition.

226

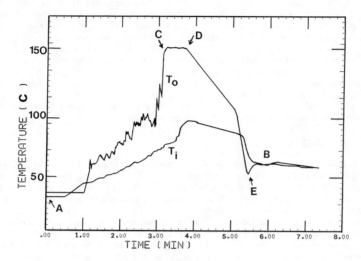

Fig. 7   Transient of heating channel inlet and outlet temperature under the transition from force to natural circulation in the figure-of-eight flow loop. A: pump trip, B: steady state condition, C: heater off, D: initiation of natural circulation, E: slug flow.

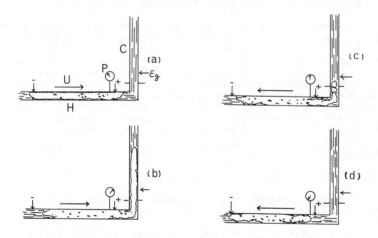

Fig. 8   The sequence of events observed the quasi-steady under two-phase natural circulations in the figure-of-eight flow loop.

227

flow loop will be acts as an extra flow resistance in the system as has been suggested by Tran and Chang [14].

The effect of noncondensable gas also tends to reduce steady natural circulation flow rate as shown in Figure 5, since the major driven forces of two-phase natural circulation, such as condensation in a U-tube section, and boiling in a heating section etc. are significantly influenced by noncondensable gases. However, thermosyphoning was observed to be established earlier than without initial pump force convection (establishment of natural circulation from gradual heating of system) as compared with experimental results of Tran and Chang [14]. Figure 2 and 3 also show that the flow without noncondensable gas has a more unstable and faster oscillation, since when noncondensable gases are present, it may act as a spring to damped down flow oscillation due to the compressibility of gas. These phenomena clearly can be observed from figure 5, since no flow reversal was observed with a presence of noncondensable gas. Here, we must note that the flow rate and frequency of oscillation in pump trip experiment show similar characteristics with natural circulation obtained from the stagnated water with gradual heating.

III.2   Figure-of-Eight Loop

Typical experimentally observed the force to natural circulation transient for figure-of-eight flow loop are shown in Figures 6 and 7 for flow rate, and the heating channel inlet and exit temperatures, respectively. Figures 6 and 7 both show that the transient phenomena from the force to natural circulation in a figure-of-eight loop is similar to figure-of-zero loop. It starts within a few minutes and has a highly oscillatory nature. For the quasi-steady state two-phase natural circulations, the sequence of events observed is shown in Figure 8, where the behaviour in the U-tube section was directly observed, and the heated section could only be inferred from the axial temperature measurements at the center of the tube and the real-time IR thermal image processing system observation of wall surface temperature as shown in Figures 9 to 11. The loop was still full with water except horizontal boiling section and the riser section of the U-tube were two-phase flows. The riser section of the U-tube was countercurrent annular wavy flow as shown in Figure 8b and only small amount of entrainments were observed in the present range of experiments. The behavior of the two-phase region was periodic , at time violent, and oscillatory in both time and space.

The thermal image of these sequence of events observed from the thermal image processing system are shown in Figures 9 to 11 at the exit location of the heated section. In Figure 9, the void inside the heating channel started to expand and the subcooled heated section at the exit become heating up. The shape of the two-phase interface can also clearly be observed with some existance of entrainment in the front. In figure 10, the heating section is full with void and some entrainment droplets and natural convection generate oscillatory temperature profiles inside. After two seconds, Figure 11 shows that flow reversal phase beginning, and the rewetting of the heating wall was intiated by entrainment droplets. Highly oscillatory nature of rewetting process can be more clearly observed in axial and radial temperature profiles in Figure 11b, and this oscillatory rewetting process may be due to the mixed convections and the existance of the significant amount of entrainment in the system.

IV.   CONCLUDING REMARKS

Transient phenomena observed in a horizontal heating vertical cooling figure-of-eight and figure-of-zero loops are experimentally investigated, and the results show that:

1.   In a pump trip experiment, thermosyphoning was observed to be established earlier than without initial pump force convection.

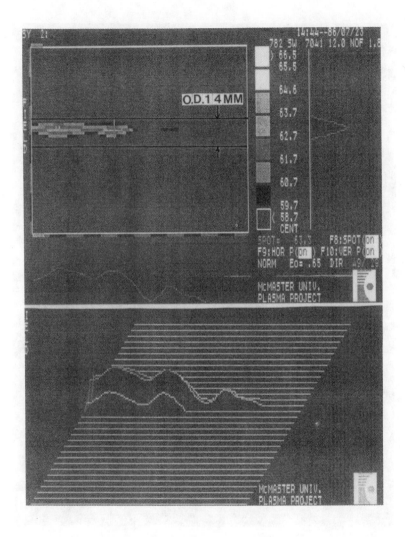

Fig. 9   Thermal image of heating section exit wall temperature profile - the void expansion phase.

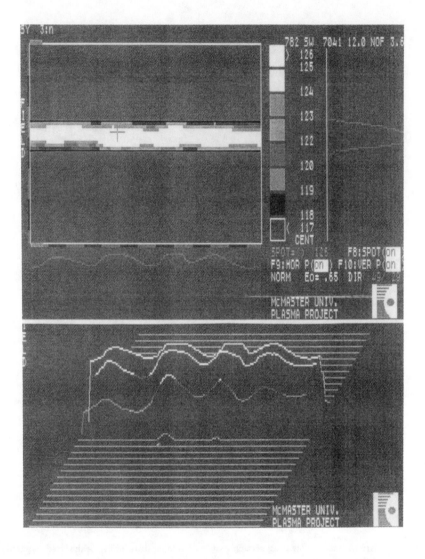

Fig. 10   Thermal image of heating section exit wall temperature profiles - the void fully expanded conditions.

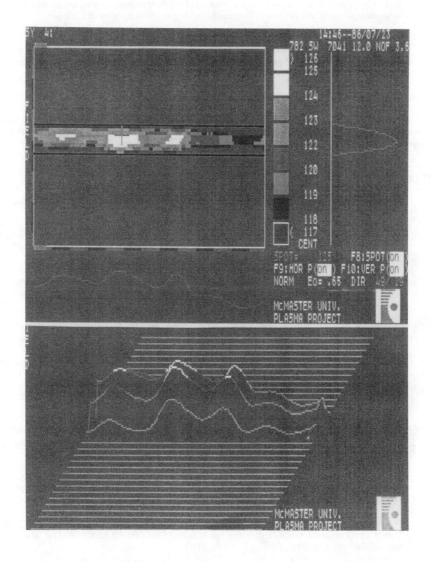

Fig. 11　　Thermal image of heating section exit wall temperature profiles - the flow reversal phase.

2.   The flow rate and frequency of oscillation in the pump trip experiment show similar charactersitics with natural circulation obtained from stagnated after with gradual heating.

3.   The flow observed in two-phase thermosyphoning is oscillating the flow and the frequency of oscillation which was observed to be a function of the heater input power, the secondary side water temperature, the presence of noncondensable gas and the orifice in the loops.

4.   Rewetting of the heating section under the natural circulation is cyclic phenomena and a significantly different wall temperature profile is observed with other rewetting experiments.

Acknowledgement

The authors wish to express their sincere thanks to J.C. Campeau, F.B.P. Tran and W.J. Garland for valuable discussions and comments. This work was supported by the Ontario Government, Ministry of Colleges and Universities under BILD grant, and AECL-RC.

REFERENCES

1.   Y. Zvirin, "A Review of Natural Circulation Loops in PWR and Other Systems", EPRI Report NP-1676-SR, January 1981.

2.   F.C. Engel, R.C. Markley, B. Minushkim, "Temperature Profiles in Natural and Forced Circulation of Sodium Through a Vertical LMFBR Blanket Test Assembly", AIChE Symposium Series 208, Vol. 77, pp. 265-272, 1981.

3.   D.J. Denver, J.F. Harrison, N.G. Trikouros, "TETRAN Natural Circulation Analyses During the TMI Unit 2 Accident", ANS/ENS Thermal Reactor Safety Meeting, Knoxville, Tennessee, April 1980.

4.   Y. Avirin, C.W. Sullivan, R.B. Duffey, P.R. Jenck III, "Experimental and Analytical Investigation of a PWR Natural Circulation Loop", EPRI Report NP-1364-SP, March 1981.

5.   P. Jenck III, L. Lennert, R.L. Kiang, "Single-Phase Natural Circulation Experiments on Small Break Accident Heat Removal", EPRI Report NP-2007, August 1981.

6.   R.L. Kiang, J.S. Marks, "Two-Phase Natural Circulation Experiments on Small Break Accident Heat Removal", EPRI Report NP-2007, August 1981.

7.   J.P. Adams, G.E. McCreery and V.T. Berta, "Natural Circulation Cooling Characteristics During PWR Accident Simulations", Thermal Hydraulics of Nuclear Reactors, M. Merilo, Ed., ANS Press, vol. II, pp. 825-832, 1983.

8.   D.J. Shimeck and G.W. Johnson, "Natural Circulation Cooling in PWR Geometry Under Accident-Induced Conditions", Thermal Hydraulics of Nuclear Reactors, M. Merilo, Ed., ANS Press, vol. II, pp. 849-857, 1983.

9.   P.J. Ingham, J.P. Mallory and V.S. Krishnan, "Two-phase Natural Circulation in a CANDU Type Heat Transport Loop", Int. Symp. on Natural Circulation, ASME Winter Meeting, Boston, Dec. 13-18, 1987.

10. F.W. Barclay, E.H. Hawley, K.O. Spitz, K.W. Demonline, E.P. Rawlings, T.A. Borgford, "Two-Phase Flow Instability Tests in RD-12", 8th Simulation Symposium on Reactor Dynamics and Plant Control, paper III-b-2, 1981.

11. V.S. Krishnan, J.S. Chang, S. Banerjee, "Two-Phase Natural Circulation in a Horizontal Boiling-Vertical U-tube Condensing Thermal-Hyraulics Loop", Proceeding of the Canadian Nuclear Society 3rd Annual Conference, pp. A1-A6, 1982.

12. K.H. Ardron, V.S. Krishnan, "Stability of a Two-Phase Natural Circulation Loop with Figure of Eight Symmetry", Proceedings of 3rd Miami International Conference on Multi-phase Flow, 1983.

13. N. Zuber, F.W. Staub, "The Propagation and Wave From of the Vapour Volumetric Concentration in Boiling, Forced Convection System Under Oscillatory Conditions", Int. J. Heat Mass Transfer, Vol. 9, pp. 871-895, 1966.

14. F. Tran, J.S. Chang, "The Role of Orifice and Noncondensable Gas on the Two-Phase Natural Circulation in a Horizontal Heating-Vertical Condensing Loop", Proc. 4th Canadian Nuclear Soc. Meeting, J.G. Charuk Ed., CNS Press, pp. 20-33 (1983).

Gas Holdup and Flow Heterogeneity in Air Lift Pumps of Small Height and Diameter

M. DÍAZ and A. SAENZ
Department of Chemical Engineering
University of Oviedo
33071 Oviedo, Spain

Abstract

An experimental study of the upward movement of water and bubble swarms in small height and diameter air lift pumps is firstly presented. Approximate relationships between flow heterogeneity, gas flow rate and riser tube length and diameter are developed. On the basis of this two-phase flow heterogeneity, an empirical equation is included to calculate the gas holdup in the column.

1. INTRODUCTION

The air lift pump is a system of great importance in two main fields of the chemical industry. Simple design, low initial costs, absence of moving parts, ability to handle small particles of solids in the liquid and simplicity of operation and maintenance are some of the reasons for the common use of airlifts for liquid (1) and solids pumping. Also, chemical reactions with interphase mass-transfer are often carried out in bubble columns and air lift loop reactors. Reactions typically take place in the liquid phase, but mass transfer of a reactant (hydrogenation, oxidation) from gas to liquid can limit overall reaction rate (2).

A major use of the air lift loop reactor is for fermentation, where it can be satisfactorily adapted to the requirements of the fermentation processes, since the fermentor design for aerobic processes is directed towards finding an inexpensive solution of the fundamental problems of transferring oxygen to the liquid phase and mixing of all the components with a minimum energy input. Besides, this pumping device has a great appeal because an analogy exists between two phase steam-water flow in evaporators, boilers and condensers and air-water flow in the riser tube.

On a conceptual basis, the magnitude of superficial liquid velocity in two-phase flow constitutes the main difference between bubble column and air lift loop reactor . Liquid throughput in bubble column is in general very small, and is applied only in continuous phase, whereas in air lift loop reactor the liquid circulates with a high velocity independent of the operation method. In contrast to bubble columns, superficial liquid velocity in air lift loop reactors is generally so high that two-phase flow is turbulent.

The characteristic flow pattern in bubble columns results from the superposition of multiple circulation cells, whilst in air lift loop reactors, the bubbles move upwards in a nearly uniform field of liquid flow and the gas phase is uniformly distributed across the column cross-section. Cocurrent flow of gas and liquid leads to an essentially different behaviour of important scaleup parameters, such as gas holdup, interfacial area between gas and liquid phase, mass transfer performance, mixing behaviour and longitudinal dispersion (3).

Gas flows in this study were chosen to cover conditions in both bubble columns and airlifts. The experimental points of this paper are shown plotted in Weyland's chart (Figure 1).

According to Weyland's (4) classification the conditions specified for D=0.020 m and D=0.080 m correspond to the air lift and bubble column operating conditions respectively, whereas for D=0.042 m an intermediate case can be seen.

Gas holdup, interfacial mass transfer coefficient, gas-liquid interfacial area, dispersion coefficients and heat transfer characteristics depend mainly upon the prevailing flow regime, bubble size distribution and coalescence characteristics. A thorough knowledge of all these interdependent parameters is also necessary for a proper scaleup of these re actors (5).

The two-phase flow depends on the amount of air introduced into the tube. The air can be distributed in small bubbles (bubble flow), in long rond nosed bubbles (s'ug flow) or, finally, in a central core surrounded by a liquid annulus (an nular flow) (6).

Slug flow is very persistent, and is the important regime for air lift pumps. Nicklin (7,8) gives a theory of slug flow based on the rate of rise of the bubbles. This theory showed that the rising velocity of such a bubble is:

$$v_s = 1.2 \ v_1 + 0.35 \ (g.D)^{1/2} \qquad |1|$$

where D is the diameter, g the acceleration due to gravity and v_1 the average liquid velocity due to the expansion of the slug as it rises. The knowledge of the behaviour of slugs in moving liquid streams will be applied to the problem of steady two-phase flow in air lifts.

A vast amount of research has been carried out in the field of two-phase flow in vertical tubes (9, 6, 4, 8). It is usually agreed that an individual flow regime, considered as an homogeneous column, settles down. This regime depends on the phase velocities, viscosities and densities, the vapour fraction, the flow direction, the heat flux in the fluid, the interphase heat transfer, the geometry of the confining walls and the transient convective momentum. In fact, the published flow regime maps are usually for cocurrent upward (or horizon tal) flow, one dimensional, steady state and isothermal flow, without taking into account all other variables (10).

In this study it has turned out that the two-phase flow

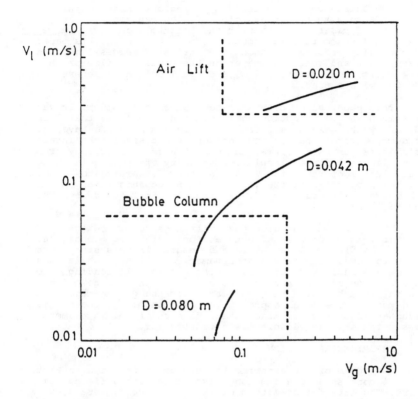

Figure 1.- Experimental conditions range in a Weyland's chart.

237

pattern can be expressed in two terms corresponding to single two-phase flow regimes for different zones of the riser tube . The validity of this assumption was experimentally verified under the conditions studied. Flow regime determinations carried out in small airlifts show that the main flow results in many cases from the superposition of two individual regimes (heterogeneous column). For instance, when $v_g = 0.10$ m/s there is bubble flow in the lower part of the D= 0.020 m. These separate bubbles become slug bubbles in the higher part of the column and slug flow appears.

The existence of two individual regimes in the column is in accordance with the observation of flow behaviour, and the influence of riser tube diameter and height, and liquid and gas superficial velocities is accounted for. The aim of this paper is to provide information about the influence of gas-liquid residence time and column diameter on the resulting flow pattern. Here, short riser tubes (H<2 m) and small column diameters (D<0.080 m) are used.

This paper also includes a section on gas holdup. It can be defined as the fraction of the channel volume that is ocupied by the gas phase. Gas holdup determines the average residence time of the gas and influences decisively the interfacial area per unit volume, the mass transfer efficiency from gas to liquid and the circulation velocity of the liquid phase in air lift loop reactors. It is very sensitive to the physical properties of the liquid (3) and depends mainly on the bubble swarm. Accordingly, it is strongly influenced by the flow regime.

Most published studies deal with high column height and diameters, in which wall effects are only of minor importance (11, 12, 3, 5). Small diameters and lengths in this study were chosen to cover slug flow hydrodynamic conditions, where gas bubbles are rapidly stabilized by the column wall leading to the formation of slug bubbles. As a result, the change of the flow in the tube (slug bubbles) can influence gas holdup. A new equation reveals an approximate relationship between gas holdup and flow regime. Also, the influence of column diameter and superficial gas velocity is accounted for.

2. EXPERIMENTAL SECTION

A diagram of the experimental equipment is shown in Figure 2. It consisted of a 2 m long riser tube with different diameters. The internal diameters of the three tubes, were 0.0020, 0.042 and 0.080 m. The columns were made of transparent plexiglass pipes. The water level in the recirculation tank was used to adjust the net lift of the pumping liquid and the submergence ratio of the pump ($\alpha = l_s/H$). The liquid temperature measured with an in-line thermometer, was maintained at 25° C ($\pm$ 1°C) by means of water circulating from a thermostat.

Water was used as the liquid phase in all the columns. Liquid flow rate was measured with a calibrated flowmeter. The water was lifted by an air flow, which was presaturated with

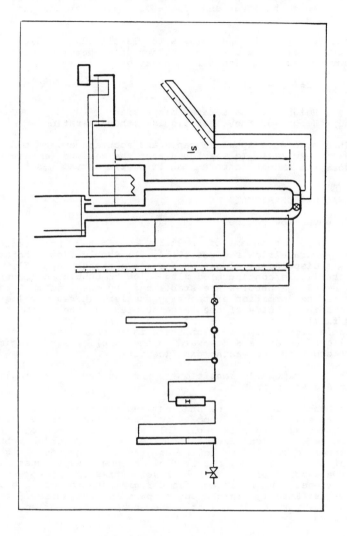

Fig. 2.– Schematic representation of an air lift pump.

water. A pressure reducer, a needle valve, an in-line thermometer and two manometers were installed in the pressurized air
feeding line in order to regulate the entering air conditions.
The air flow was measured with a rotameter. The air supply
temperature was maintained at 23°C (± 1°C). The air injection
piece was made of plexiglass with an orifice of 1.5 mm internal
diameter. The injection point was located at three different
positions above the lower end of the lift pipe for changing the
gas residence time.

The observed flow patterns were classified according to
visual evidence. The average gas holdup was determined directly
by measuring bed expansion during aeration:

$$\varepsilon = (h_1 - h_2)/h_1 \qquad\qquad |2|$$

h_1 - Height of the aerated dispersion
h_2 - Height of the clear liquid without aeration

This procedure appeared rapid and accurate owing to the
absence of foam (11). It was found that the maximum experimental
error does not amount to more than 5% of the actual gas holdup.

3.   ANALYSIS OF RESULTS

Air lift pump flow regimes

The upward movement of the bubble swarms has been reduced
to three separate flow regimes (6, 4, 3): Bubble flow (characterized by a dispersion, almost uniform in size bubbles, flowing
within the liquid continuum); slug flow (bubble colescence
occurs, and large bubbles are stabilized by the column wall
leading to the formation of the slug bubbles, whereas the liquid
flows down the outside of the large bubbles in the form of a
falling film); churn flow (a breakdown of the bubbles leads to
an unstable flow regime in wich there is an oscillatory motion
of the liquid upward and downward in the tube). These regimes
occur in order of increasing gas flow rate.

The experimental transitions and intermediate regimes
observed are shown in Figure 3.

In bubble flow (A) the bubbles keep their separate identity; from time to time two bubbles collide and may form a larger
bubble, but coalescence is negligible at the lower end of the
lift tube, where it can be located. The process of bubble
coalescence and collision increases rapidly in the higher
region of the tube. Thus, bubble flow becomes very unstable,
with large bubbles moving with high velocities in the presence
of smaller ones, and an intermediate regime due to coalescence
(D) prevails. Finally, in the upper part of the riser tube
large characteristically bullet-shaped bubbles are formed, leading to slug flow (B).

At higher gas velocities slug bubbles break down leading
to an unsteady flow pattern with channeling: churn flow (C) .

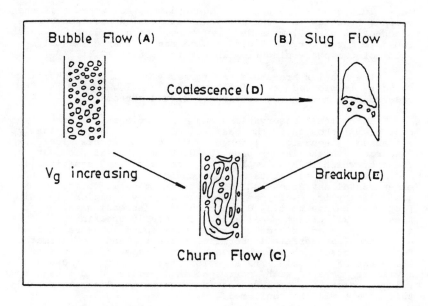

Fig. 3.- Hydrodynamic flow regimes of an air
lift pump.

There is an intermediate regime, in which the shape of the slugs cannot be maintained and air flowing upward as small bubbles in the separating liquid regions increases. When this occurs dispersed slug flow appears (E). At higher gas flow rates, the bubble flow-churn flow transititon is observed at the lower end of the riser tube. This boundary is due to both coalescence and breakup processes rapidly increasing with gas velocities.

On the basis of a theoretical and experimental study of slug flow in vertical flow, Nicklin's theory (7) can be applied to the problem of steady two-phase flow in vertical tubes. The absolute velocity of slug was taken as:

$$v_s = 1.2 \ (G+L)/A + 0.35 \ (g.D)^{1/2} \qquad\qquad |3|$$

This expression was then equated to $G/A\varepsilon$, where ε is the voidage fraction, and the resultant expression provides a mean of checking slug flow in the riser tube. These predictions were found to agree very well with a wide range of experimental results of two-phase slug flow in vertical tubes (7,8).

The results are shown in Figures 4 and 5, in which rising velocity is plotted as a function of gas flow rates in the case of both moving and stationary liquid streams.

Both visual observations and inspections of these two charts will show that the gas velocity is satisfactorily expressed by equation |1| for D= 0.042 m. and stable slug flow is formed. In the larger pipe (D= 0.080 m) bubbles rise at a velocity which is higher than that predicted by equation |1| and bubble flow is maintained. In the case of D= 0.020 m the experimental deviations showed that slug flow is formed whereas gas velocity is lower than that predicted by equation |1|.This fact must be due to flow distortion with increasing gas veloci ties and wall effect. Wall effect, generally speaking, in a system of fixed fluid properties is found to cause bubble elongation, a decrease in terminal velocity, and the diameter becomes controlling length governing the velocity and the fron tal shape of a bubble. The aerations measurements for D=0.020 m will be compared with corresponding values calculated by using predictive procedures recommended in the literature and this effect will be confirmed.

From Figure 5 for stationary liquid, the existence of a gas velocity minimum in the vicinity of the bubble flow-slug flow transition can be seen.

Heterogeneous column

The prevailing flow regime depends upon the characteristics of coalescence and breakup of the bubbles. The transition occurs because of complex physical interactions; the nature of these interactions varying from transition to transition. In order to study the influence of the gas velocity, different tube lengths, diameters and degrees of submergences were used and the height of each regime in the tube was measured. Figure

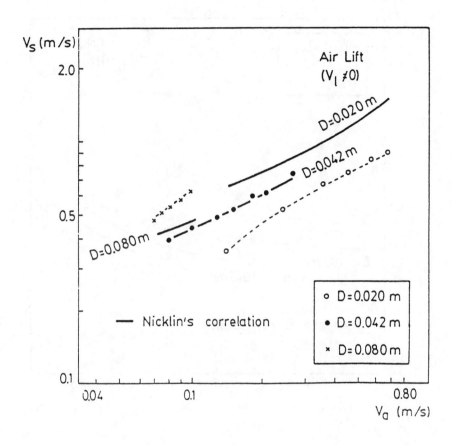

Fig. 4.- Gas rising velocity v.s. gas superficial velo
city in air lift operating conditions.

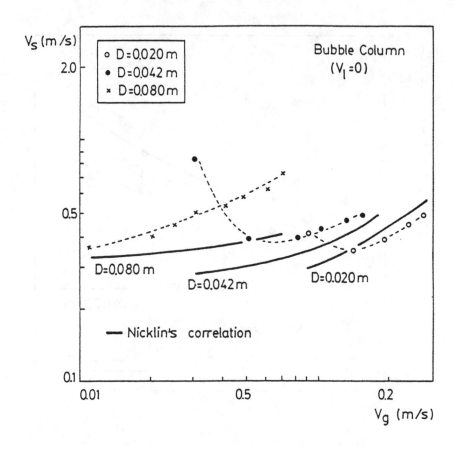

Fig. 5.- Gas rising velocity v.s. gas superficial velocity in bubble column operating conditions.

6 gives H_j as a function of G, for H, D and α as parameters for some of the conditions studied.

If the superficial gas velocity is small, bubble flow is maintained in the lower part of the tube while in the higher part slugs are formed by means of an intermediate coalescence region. At higher gas flow rates, the transition from bubble to churn flow occurs in the lower part. In addition, more distoted slugs are formed at the end of the tube. If the superficial gas velocity is higher than 0.40 m/s slugs are wholly dispersed , and churn flow covers all the column. Figure 7 reveals the approximate relationship between flow regime and pipe diameter using gas flow rate as parameter.

It is significant that previous workers have disagreed regarding the dependence of flow regime on tube diameter. Some of these workers (3, 5, 7) found an approximate relationship between flow regime and diameter, though many other authors reported flow regime maps without including column diameter as a parameter. It is generally admitted that vertical two-phase flow behaviour is influenced by pipe size for diameters up to 0.15 m.

It can be seen from Figure 7 that heterogeneous flow pattern is strongly affected by the pipe diameter owing to wall effects and longitudinal mixing within the limits of tube areas studied. Ohki and Inoue (13) reported that the column diameter had a significant effect on the dispersion coef - ficient: The larger the diameter, the greater the dispersion coefficient. On the other hand, wall effects, disminishing as diameter increases, were reviewed by Clift (14). In practice, the larger the diameter (from 0.020 to 0.042 m) the longer the height of bubble flow, whereas there is no transition to slug flow for D=0.080 m.

At gas flow rates greater than 0.16 m/s the transition of bubble flow to churn flow is observed. At gas flow rates of about 0.25 m/s the slugs are increasingly distorted up to 0.40 m/s. Once the air flow rate exceeds 0.40 m/s a churn flow pattern is formed.

In order to state the dependence of flow regime on tube length the injection level was varied. A distance of 0.77 and 1.27 m between the bottom orifice and the injector was used . The results are shown plotted in Figure 7. Using different lengths there is a marked difference between the boundary lines due to variation of volumetric gas flow rates along the tube . Because the average volumetric flow rate of gas increases towards the top of the column, the local average voidage fraction, ε, and the frequency of coalescence will also increa se, leading to slug flow at the end of short riser tubes.

Gas holdup

Gas holdup depends mainly on the gas flow rate and it is strongly influenced by flow regime. Figure 8 shows gas holdup data measured in the upflow column of the air-lift as a function of gas flow rates and diameters as parameters, within

245

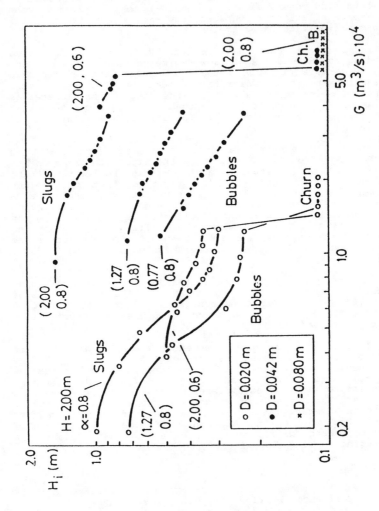

Fig. 6.– Experimental structure of two phase flow at different gas flow and operating conditions, H and D.

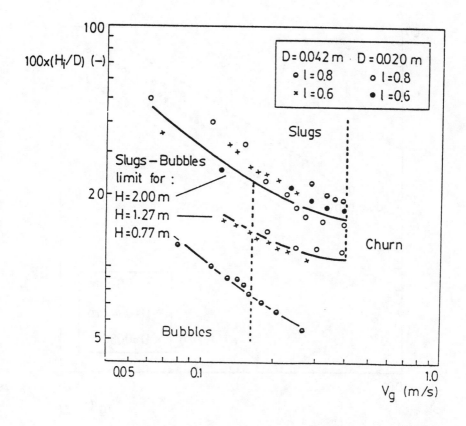

Fig. 7.- Heterogeneous flow regime as a function
of superficial gas velocity and H/D.

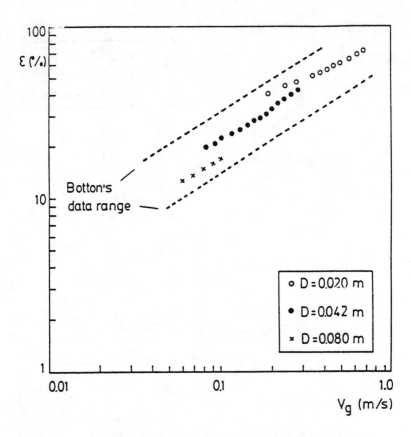

Fig. 8.- Gas holdup experimental data in air lift
pumps.

the limits reported by Botton (15).

The results seem likely to depend upon both gas and heterogeneous flow pattern simultaneously. Weyland (4) reported that there exists a typical relationship between two-phase homogeneous structure and gas holdup. If there is a two-phase flow transition, the holdup profile versus superficial gas velocity is modified. This could be indicated by means of a summation that would include the contributions due to specific regimes.

In the case of homogeneous flow pattern, the dependence of gas holdup on gas velocity can generally be assumed to be of the form (3) $\varepsilon \sim v_g^n$, where n depends on the flow regime, varying from 0.4 to 1.0.

In the case of heterogeneous column we propose an equation for the relationship between gas holdup and two-phase flow pattern. We need to introduce a correlation term, K, given by $K = \Sigma h_i . a_i$, extended to the three main regimes: Bubble flow (A), slug flow (B) and churn flow (C). In this equation h_i is the height (%) of the column covered by the regime i, and it can be obtained from Figure 7.

The most accurate linear fit of experimental points yields an adjusted multiple correlation coefficient of 0.995.

$$\varepsilon = v_g^{0.5}/ D^{1/3} . (0.225\ h_A + 0.251\ h_B + 0.241\ h_C) \quad |4|$$

Equation $|4|$ constitutes an empirical expression of all the results given in Figure 8. The equation was tested for adequacy by applying Fisher's test (on the basis of Fisher's variance ratio). The coefficients of the regression equation were tested for significance for the 95% selected significance level and gave the values of 0.225 ± 0.004, 0.251 ± 0.004 and 0.241 ± 0.004 respectively. The range for which this equation is valid is $0.06 < v_g < 0.60$ m/s and $0.020 < D < 0.080$ m.

The dependence of gas holdup on superficial gas velocity and diameter is given by the values of the equation coefficient, K. If homogeneous flow pattern prevails (e.g. $v_g > 0.40$ m/s) the equation coefficient is $a_i/D^{1/3}$. In this case, the individual coefficient is greater for churn flow than the one for bubble flow. In practice it is assumed that the heterogeneous flow regime prevails if slug flow appears. In this case, the coefficient depends on both diameter and superficial gas velocity and increases rapidly with increasing air flow rate. After reaching a maximum, it finally decreases owing to the formation of churn flow. The smaller the diamter, the smaller the range of heterogeneity and the greater the equation coefficient.

The influence of riser tube diameter is accounted for by means of equation $|4|$ within the range D <0.080 m. Such an influence is due to the dependence of flow regime on pipe diamete. Various authors have demonstrated that gas holdup is not significantly affected as diameter is further increased above 0.15 m (5,3).

Comparison with previous results

As a final step in correlating results, holdup measurements
were compared with corresponding values calculated using predic
tive procedures recommended in the literature. There are a
number of these, all differing substantially. Some of the
important correlations referring to bubble columns have been
reported in Table I from the admirably unified reviews
published by Shah (3) and Miller (5).

The average deviation of the correlated data are shown
plotted in Figure 9 as a function of diameter.

Gas flow rates in this study were chosen to afford
maximum versatility in studying both airlift (D=0.020 m) and
bubble column (D=0.080 m) hydrodynamics. For D = 0.020 and
D= 0.042 m the correlations predicted a much lower gas holdup
than experimental values. Because the correlations agree quite
closely with results in a 0.080 m tube at the flow rates concerned, it would appear that this disagreement must be caused
by wall effects and high liquid and gas velocities (airlift
hydrodynamic conditions). Thus, it is clear that small column
diameters and air lift hydrodynamic conditions play an important part in the determination of gas holdup.

CONCLUSIONS

On the basis of flow regime observations, with H up to 2
metres, the coexistence of two different regimes in the riser
tube (heterogeneous column) was stated for D=0.020 m. The para
meter (H_i/D) for each regime accounts for the influence of pipe
diameter within this range. At v_g < 0.16 m/s bubble flow in the
lower part is replaced by slug flow in the higher part of the
column. At v_g > 0.16 m/s churn flow in the lower part becomes
slug flow in the higher fraction of the column. At vg> 0.25 m/s
slug bubbles are progressively distorted up to 0.40 m/s. Once
v exceeds 0.40 m/s a single column pattern is formed (homogeneous column). For D= 0.080 m there is no transition to slug
flow, bubble flow becomes increasingly distorted but slug
bubbles are not stabilized by wall effects. Also, Fig.7 shows
the effect of smaller riser tube lengths leading to shorter
bubble zones, and can be used to estimate two-phase flow
pattern in air-lifts, with column dimensiones (H and D) and v_g
parameters, under the conditions specified.

- Based on gas holdup observations an empirical correla-
tion for ε is proposed |4|. This equation takes account for
the strong influence of small diameters, which stabilizes
slugs. The coefficient of the regression equation reveals the
effect of flow pattern and can be computed from Fig.7.

- Comparing the hold-up correlated data with those
computed using procedures recommended in the literature (for
greater values of diameter and small v_1), the average devia -
tion shows, Fig. 9, that small column diameters and air lift
conditions (such as those used in this study) play an important role in increasing gas holdup.

TABLE I: HOLDUP CORRELATIONS FOR BUBBLE COLUMN.

Fig.9	Authors	Gas-liquid	v (m/s)	v (m/s)	D(m)	H(m)
1	Hershey (1935)	Air-Water	-	-	-	4.57
2	Hughmark (1967)	Air-Aqueous/ Organic Liq.	.004-.45	-	0.1	-
3	Kim (1972)	Air-Water	.26	-014-.102	-	-
4	Akita, Yoshida (1973)	Air-Aqueous Solutions	.003-40	.044	.152-.600	1.26-3.5
5	Hikita, Kikuka-wa (1974)	Air-Aqueous Solutions	.042-.38	0	.10-.19	.60-1.35
6	Botton (1974)	Air-Aqueous Solutions	14	.05	.02-.48	.32-2.0
7	Mersmann (1978)	Gas-liquid	Semitheoretical		Equation	
8	Hikita (1980)	Gas-Aqueous/ Organic Liq.	.042-.38	-	.1	.65

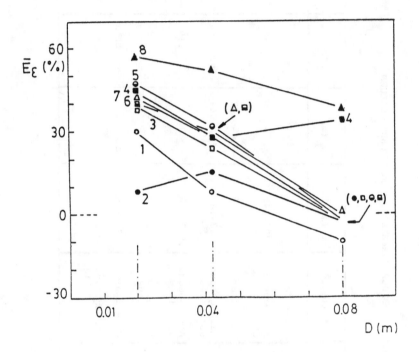

Fig. 9.– Average deviations of several correlations in the prediction of holdup data.

NOMENCLATURE

a_i Regression coefficient for regime i, defined by equation $|4|$.

A Channel cross-sectional area, m^2.

D Tube diameter, m.

E_ε Average gas holdup deviation between predicted and experimental data (-).

g Acceleration due to gravity, m/s^2.

G Gas volumetric flow, m^3/s at NTP.

h_i Height of column covered by regime i, %

H Tube height, m.

H_i Tube height covered by regime i, m.

h_1 Height of aerated water in the column, m.

h_2 Height of clear liquid without aeration in the column, m.

K Correlation term to consider the column heterogeneity ,

l_a Net lift, m.

L Liquid volumetric flow, m^3/s.

l_s Submergence, m.

n Factor defined by equation $|4|$.

v_g Gas superficial velocity, m/s.

v_l Liquid superficial velocity, m/s.

v_s Absolute gas rising velocity, m/s.

Greek letters

α Degree of submergence (-)

ε Gas holdup (%).

REFERENCES

(1) Witte, R.; "Ind. Chim. Belge", $\underline{30}$(7), 701-713, (1965)

(2) Halde, R.; Svensson, H., "Chem. Eng. J.", $\underline{21}$, 223-227, (1981)

(3) Shah, Y.T.; Kelkar, B.G.; Godbole, S.P.; Deckwer, W.D.;
"A. I. Ch. E. Journal", $\underline{28}$(3), 353-379, (1982).

(4) Weyland, P.; Onken, U.; "Ger. Chem. Eng.",$\underline{4}$, 174-181, (1981).

(5) Miller, D.N.; "Ind. Eng. Chem. Process Des. Develop.", $\underline{19}$,
371-377, (1980).

(6) Hetsroni, G.; "Handbook of Multiphase Systems", Mc Graw-
Hill, New York, (1982).

(7) Nicklin, D.J.; Wilkes, J.D.; Davidson, J.F.; "Trans. Instn.
Chem. Engrs.", $\underline{40}$, 61-68, (1962).

(8) Nicklin, D.J.; "Trans. Instn. Chem. Engrs.", $\underline{41}$, 29-39,(1963)

(9) Perry, R.H.; Chilton, H.C.; "Chemical Engineer's Handbook"
5th End., Mc Graw-Hill, New York, (1973).

(10) Mujumdar, A.S.; Mashelkar, R.A.; "Advances in Transport
Processes", Vol II, Wiley Eastern Limited, New Delhi,(1982)

(11) Nöttemkanper, R.; Steiff, A.; Weinspach, P.N.; "Ger. Chem.
Eng.", $\underline{6}$, 147-155, (1983).

(12) Lewis, D.A.; Davidson, J.F.; "Chem. Eng. Sci", $\underline{38}$(1),
161-167, (1983).

(13) Ohki, Y.; Inoue, H.; "Chem. Eng. Eng. Sci.", $\underline{25}$, 1-16, (1970)

(14) Clift, R.; Grace, J.R.; Weber, M.E.; "Bubbles, Drops and
Particles", Academic Press, New York, (1978)

(15) Botton, R.; Cosserat, D.; Charpentier, J.C.; "Chem. Eng.
J.", $\underline{16}$, 107-115, (1978).

Unsteady Motion of a Conducting Viscous Fluid through a Porous Medium in Circular Pipes with Heat Transfer

ADEL A. MEGAHED and M. H. KAMEL
Department of Engineering Physics and Mathematics
Faculty of Engineering
Cairo University, Egypt

ABSTRACT

The present work is devoted to the study of unsteady motion of an electrically conducting, viscous and incompressible fluid through a porous medium, in a circular pipe of infinite length. The walls of the pipe is considered non-conducting. The motion is subjected to an external transverse magnetic field which is assumed uniform and undisturbed. The equations governing this type of motion are the momentum equation and Maxwell's equations. They are written in the nondimensional form following the usual magnetohydrodynamic model. A term expressing the effect of the porous medium is included according to Darcy's model which is based on experimental studies.
An exact analytical solution of the problem is obtained and the unsteady velocity distribution is represented graphically with the variation in space and time. The effects of magnetic field and porocity on the flow are interpreted graphically with the aid of computations. Heat transfer characteristics are also studied taking account of viscous and Joule dissipations. The energy equation is solved using an implicit numerical method to study the time history of the temperature distribution inside the pipe. The resuls are depicted graphically.

1. FORMULATION OF THE PROBLEM

Consider the unsteady motion of an electrically conducting, incompressible and viscous fluid through a porous medium in an infinte circular pipe under the external action of a transverse magnetic field. The system of magnetohydrodynamic equations are constructed in cylindrical coordinates (r,θ,x). The direction of motion is axial and the direction of the magnetic field is normal to that of flow. So we have;

The velocity vector $\underline{V}=(0,0,u)$ and

The magnetic field vector $\underline{B}=(B0 \cos\theta, -B0 \sin\theta, b)$ where b is the

induced magnetic field whitch can be neglected for small values of velocities and the magnetic field is assumed constant.

255

i.e $R_{em} \ll 1$ where ;

R_{em} is the Magnetic Reynold's number.

The Governing Equations

$$\partial f/\partial t + f \, \nabla \cdot \underline{V} = 0 \qquad \text{------------------------------(1)}$$

$$f \partial \underline{V}/\partial t = -\nabla p + \mu \, \nabla^2 \underline{V} - \mu \, \underline{V}/k + \underline{J} \times \underline{B} \qquad \text{----------------(2)}$$

and Ohm's law;

$$\underline{J} = \sigma \, (\underline{E} + \underline{V} \times \underline{B}) \qquad \text{----------------------------(3)}$$

From equation (3) we can deduce Lorentz force and the above

equations reduce to;

$$f \, \partial u/\partial t = -\partial p/\partial x + \mu \, \nabla^2 u - \mu u/k - \sigma B_0^2 \, u \qquad \text{-----------(4)}$$

$$-1/f \, \partial p/\partial r = 0 \qquad \text{------------------------------(5)}$$

$$\partial u/\partial x = 0 \qquad \text{------------------------------------(6)}$$

With the initial and boundary conditions;

u=uo for t = 0 (The fluid initially is pushed with this

velocity.)

u=0   for r=a ,   t $\geq$ 0

u= finite value   for r=0 ,   t $\geq$ 0   -----(7)

Where;

σ is the conductivity of the fluid,

μ is the viscosity of the fluid.   and

k   is the porocity of the porous medium.

Let us introduce the following nondimensional quantities;

$\bar{U} = u/u_o$

$\bar{X} = x/a$

$\bar{B} = B/(B_o * R_{em})$

$\bar{t} = t\mu/f a^2$

256

$\bar{P} = pa/\mu u_o$

$\bar{K} = k/a^2$

By using these quantities in equations (4) and (7) and drop the primes for simplicity we get;

$$\partial U/\partial t = -P_x + \nabla^2 U - U/K - H^2_a U \quad ----------------(8)$$

With the initial and boundary conditions;

$U = 1$ for $t = 0$

$U = 0$ for $r = 1$, $t \geq 0$

$U =$ finite value for $r = 0$,   for $t \geq 0$ $-------(9)$

2. METHOD OF ANALYSIS

It is clear that these equations which describing the un-steady velocity distribution belong to the parapolic type of partial differential equations .   In the proposed method of analysis, the method of separation of variables is followed to solve equation (8) with the initial and boundary coditions (9) and the dependence on the time is expressed by an exponential function which tends to zero as time tends to infinity.

At first we make the substitution ;

$$U = -P_x/(H^2_a + 1/K) + F \quad ------------------(10)$$

where F is a new function of (r,θ,t). equation (8) will be;

$$\partial F/\partial t = -\nabla^2 F - (H^2_a + 1/K)F \quad ------------(11)$$

with B.C ;

$$F(r,\theta,t) = P_x/(H^2_a + 1/K) = F_1(\theta) \quad \text{for } r = 1 \; --(12)$$
$$F(r,\theta,t) = \text{finite value} \quad \text{for } r = 0 \quad -------(13)$$

and I.C ;

$$F(R,\theta,t) = 1 + P_x/(H^2_a + 1/K) = F_o(r,\theta) \quad -----(14)$$

The general solution of these equations take the form;

$$F(r,\theta,t) = F^{(1)}(r,\theta) + F^{(2)}(r,\theta,t) \quad ----------(15)$$

Where $F^{(1)}$ is the solution of the equation ;

$$\nabla^2 F^{(1)} = (H^2_a + 1/K) \; F^{(1)} \quad \text{------------------(16)}$$

which satisfies the B.C;

$$F^{(1)} = F^{(1)}(\theta) \qquad \text{for } r=1 \quad \text{------------------(17)}$$

And the function $F^{(2)}$ is the solution of the equation;

$$\delta F^{(2)}/\delta t = \nabla^2 F^{(2)} - (H^2_a + 1/K) \; F^{(2)} \quad \text{---------(18)}$$

with the   B.C $F^{(2)} = 0$ for $r=1$ and; -------(19)

I.C $F^{(2)} = F(r,\theta) - F^{(1)}_o(r,\theta)$ --------(20)

Following the steps of the separation of variables technique ; The final solution of the problem will be;

$$U(r,t) = P_x/(H^2_a +1/K) \left[\frac{I_o\{\sqrt{H^2_a + 1/K}\,r\}}{I_o\{\sqrt{H^2_a + 1/K}\}} - 1 \right]$$

$$+ 2 \, (1+ P_x/(H^2_a +1/k)) \sum_{m=1}^{\infty} \frac{J_o(Z_m^o \; r)}{Z_m^o \cdot J_1(Z_m^o)} \; \text{EXP}-\{H^2_a +1/K+Z_m^{o2}\}t$$

$$\text{------------(21)}$$

Where;

J_o and J_1 are BESSEL functions and I_o is the modified BESSEL function.   and;

Z_m^o are the roots of J_o

The effects of magnetic field and the porocity of the medium are shown in the figures (1),(2),(3) and (4)

The coefficient of skin effect on the pipe surface is given by;

$$\tau = \mu \, (\delta u/\delta r) \mid r=a \quad \text{---------------(22)}$$

Refering to the nondimensional quantities defined before we get the coefficient of skin friction in the nondimensional form ;

$$\tau = 1/Re \, (\delta U/\delta r) \mid r=1 \quad \text{--------------(23)}$$

Where Re is the Reynold's number . Differentiating equation (21) the skin friction will be ;

258

$$\tau = \frac{P}{X} / Re(H_a^2 + 1/K) \left[\frac{\overline{I}_o (\sqrt{(H_a^2 + 1/K)} \mathbf{r} \mid r=1}{I_o (\sqrt{H_a^2 + 1/K})} \right]$$

$$+ 2(1 + P_X / (H_a^2 + 1/K)) \sum_{m=1}^{m=\infty} \frac{J_o (Z_m^o \ r) \mid r=1}{Re(Z_m^o J (Z_m^o))} EXP-(H_a^2 + 1/K + Z_m^{o 2}) t$$

$$-------------------------(24)$$

3. HEAT TRANSFER

We now consider the heat transfer characteristics in the MHD pipe flow when the temperature of the outer surface is considered constant and T and the temperature of the flow is considered maximum at the axis of the cylinder. The energy equation for the unsteady MHD pipe flow in cylinderical form is given by;

$$\rho C \, \delta T / \delta t = (C \mu / P_r) \nabla^2 T + \mu (\delta U / \delta r)^2 + J^2 / \sigma \quad --(25)$$

Where C is the specific heat at constant pressure, T is the temperature of the fluid taken as a function of the time t and the pipe radius r and P_r is the Prandtl number. The last two terms are the viscous and Joule dissipations respectively.

The boundary conditions for T are

T = T_o at r = a (a is the radius of the pipe) and

T = T_m at r = 0 ($\delta T / \delta r = 0$ at r = 0) $-----------(26)$

The initial condition is

T = T_o for all r $------------(27)$

Introducing non-dimensional quantities

$$\theta = (T - T_o)/(T_{max} - T_o)$$

$$\tau = t \mu / \rho a^2$$

$$\overline{r} = r / a$$

$$E \text{ (Eckert number)} = U^2 / C(T_{max} - T_o)$$

Equations (25),(26) and (27) reduce to

$$P \, \delta\theta / \delta\tau = \nabla^2 \theta + E \, P_r \, (\delta U / \delta r)^2 + E \, P_r \, (H^2 U^2) \quad ---------(28)$$

With the boundary conditions

$$\theta = 0 \text{ at } r = 1 \quad \text{and} \quad \theta = \theta_{max} \text{ at } r = 0 \qquad \text{------(29)}$$

and the initial condition

$$\theta = 0 \quad \text{for all } r \qquad \text{--------(30)}$$

Equations (28), (29) and (30) are solved numerically using CRANK-NICOLSON method and the results are shown in figure (5).

4. DISCUSSIONS

The effects of the magnetic field (which is expressed by Hartmann number H) and the porocity of the medium (which is expressed by K) on the flow are depicated in figures (1) and (2) for different values of Hartmann H and $1/K$. It is evident from fig.1 that; the increasing of the magnetic field acts as an increasing resistans against the flow. Also from fig. (2) the increasing of $1/K$ decrease the velocity of the flow. Figure (3) indicates the effect of the initial velocity given to the flow but this effect (the increasing of the velocity) decreases with time and vanishes completely at steady state at which the flow is forced only by the pressure gradient.
The change of the velocity of the flow with time is indicated in fig.(4). This figure shows that the velocity reaches the steady state value after approximately $t=0.5$.
Heat transfer characteristics results are shown in fig. (5) at time level 100 where each time step represents $1/110$ from the charateristic time t. The figure shows the effect of Joule and viscous dissipations for different values of Hatmann number. It is clear that the effect of viscous dissipation is too small compared with Joule , also as H increased, Joule dissipation increased through the pipe.
The results of this work are of great importance in the study of of motion of crude oil through transmission pipes and many other useful engineering applications.

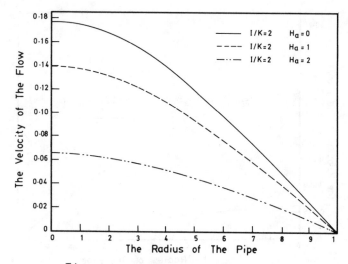

Fig. 1.– THE EFFECT OF THE MAGNETIC FIELD ON THE FLOW

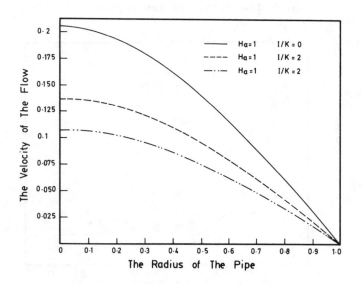

Fig. 2– THE EFFECT OF THE POROCITY ON THE VELOCITY FIELD

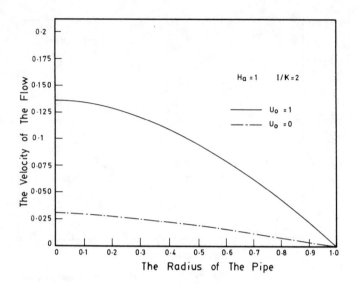

Fig. 3- The effect of the initial velocity

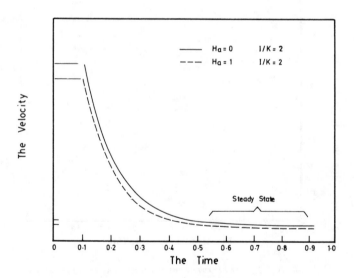

Fig. 4- THE CHANGE OF THE VELOCITY WITH TIME

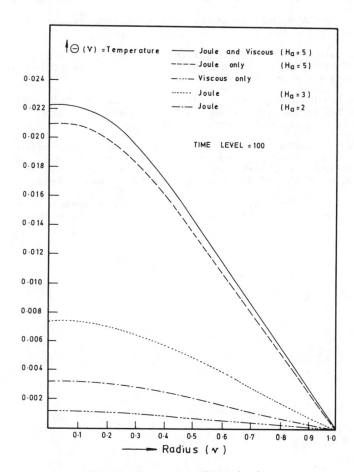

Fig. 5 – Temperature Distrubution

REFERENCES

[1] SCHLICHTING, H. (1968) " Boundary Layer Theory", 6th ed.
Mc Graw-Hill Book Co., Inc., New York.

[2] Cramer, K.R. and Pai, S.I (1973)"Magnetofluid Dynamics for
Engineers and applied Physicists" Scripta P.C,Washington,D.C

[3] Tikhonov.A.N,and Samarsky .A.A (1966)"Mathematical Physics
Equations", moscow.

[4] C.L.Varshney(1979)"Fluctuating Flow of Viscous Fluid through
a Porous medium bounded by a porius Plate" Dept. of math.
, Agra College, Agra.

[5] G.S SETH and M.R.MAJTI.(1981) "MHD COUETTE FLOW AND HEAT
TRANSFER in a ROTATING SYSTEM" Dept.of Math. Indian
Institute of Technology, Kharagpur.

[6] J.R.MAHAPATRA. (1973) "A note on the unsteady motion of a
viscous conducting liquid between two porous concentric
circular cylinders acted on by a radial magnetic field"
Dept.of Math. R.B.C College, Naihati,West Bengal , India.

[7] Curtis F. Gerald. (1970) "Applied Numerical Analysis"
California State Polytechnic College ,San Luis Obispo.

MULTIPHASE
FLOW INSTABILITIES

The Influence of Gravity on Density-Wave Instabilities in Boiling Channels

A. CLAUSSE
Centro Atomico Bariloche
8400 Bariloche, Argentina

R. T. LAHEY, JR.
Rensselaer Polytechnic Institute
Troy, New York 12180-3590, USA

ABSTRACT

The effect of gravity in the momentum equation on the stability of a boiling channel is analyzed. Froude number is utilized to characterize the relative importance of this term.

Two systems were studied: the classical parallel channel boundary condition, where the boiling channel is subjected to constant pressure drop; and, a natural circulation loop having a single boiling channel.

For the case of parallel channels, a decrease in the Froude number destabilizes the system for higher inlet subcoolings and stabilizes it for lower inlet subcoolings.

For natural circulation conditions complicated unstable regions at lower powers were found. This result agrees with experimental data for such systems.

1. INTRODUCTION

The dynamics of boiling flows is a complicated problem where generally more than one mechanism of instability is present. Friction pressure drops were found to play an important role, and therefore, in the past, a considerable amount of effort was spent on the study of systems where the frictional terms are predominant.

On the contrary, not too much attention has been paid to the influence of the gravity term in the stability of two-phase boiling flows. Achard et al. [1] found that for cases having low friction interesting "islands of instability" can appear. However, the parameters used in this analysis make it difficult to interpret their results from a practical point of view.

In this paper a linear stability analysis of a boiling channel is presented, in which the influence of the gravity term in the momentum equation was studied. The Froude number is utilized to characterize the relative importance of this term. Two systems were studied: the classical parallel channel boundary condition, where the channel is subjected to constant pressure drop, and a natural circulation loop having a single boiling channel.

2.   THE BOILING CHANNEL MODEL

Pressure Drop Transfer Functions

Let us consider the boiling channel shown in Fig. 1. This simple system has a rather complicated dynamics, which may lead to unstable situations for certain operating conditions. For the purpose of stability analysis, the following simplifying assumptions were used:

(1)   Homogeneous two-phase flow

(2)   Constant system pressure

(3)   No subcooling boiling

(4)   Constant and uniform heat flux

(5)   Constant inlet subcooling.

The one-dimensional conservation equations are:

Mass Conservation

$$\frac{\partial \rho}{\partial t} + \frac{\partial \rho u}{\partial z} = 0 \tag{1a}$$

Energy Conservation

$$\frac{\partial \rho h}{\partial t} + \frac{\partial \rho h u}{\partial z} = q \tag{1b}$$

Momentum Conservation

$$\frac{\partial \rho u}{\partial t} + \frac{\partial \rho u^2}{\partial z} + f \rho u^2 + \rho g = - \frac{\partial p}{\partial z} \tag{1c}$$

For constant system pressure, the enthalpy and the density are related by the following equation of state:

$$\rho = \left[v_f + \left(\frac{h - h_f}{h_{fg}} \right) v_{fg} \right]^{-1} \qquad \text{for, } h > h_f \tag{2a}$$

$$\rho = \rho_f \qquad \text{for, } h \leq h_f \tag{2b}$$

Equations (1) and (2) can be perturbed around a steady-state, and Laplace transformed in time. Integrating the resulting equations, the transfer functions of the pressure drop along the channel can be found. Partitioning the different terms of the momentum equation, we obtain:

$$\delta \Delta p_I = \delta \int_0^L \frac{\partial \rho u}{\partial t}\, dz = \rho_f \lambda_o\, s\, \delta u_i + s\rho_f\, I(-1)\, \delta u_e$$

$$+ \Omega \frac{v_{fg}}{h_{fg}}\, \rho_f^2\, u_{oi}\, h_o'\, I\left(-\frac{s}{\Omega}\right)\delta\lambda - \frac{v_{fg}}{h_{fg}}\, \rho_f^2\, u_{oi}\, h_o'\, \frac{\left[I\left(-\frac{s}{\Omega}\right) - I(-1)\right]}{s - \Omega}\, \delta u_e$$

$$\tag{3a}$$

$$\delta \Delta p_a = \delta \int_0^L \frac{\partial \rho u^2}{\partial z}\, dz = -2\rho_f u_{oi} \delta u_i + 2\rho_f u_{oi} \delta u_e + u_{oe}^2\, \delta\rho_e \tag{3b}$$

$$\delta \Delta p_i = \delta(K_i \rho_f u_i^2) = 2K_i \rho_f u_{oi} \delta u_i \tag{3c}$$

$$\delta \Delta p_e = \delta(K_e \rho_e u_e^2) = 2K_e \rho_f u_{oi} \delta u_e + K_e u_{oe}^2\, \delta\rho_e \tag{3d}$$

$$\delta \Delta p_f = f\, \delta \int_0^L \rho u^2 dz = 2f\, \rho_f\, u_{oi}\, \lambda_o\, \delta u_i + 2f\, \rho_f\, u_{oi}\, (L - \lambda_o)\, \delta u_e \tag{3e}$$

$$+ f\, \frac{v_{fg}}{h_{fg}}\, \rho_f^2\, u_{oi}^2\, h_o' \left\{ I\left(1 - \frac{s}{\Omega}\right)\delta\lambda - \frac{\left[I\left(1 - \frac{s}{\Omega}\right) + \lambda_o - L\right]}{s - \Omega}\, \delta u_e \right\} \tag{3f}$$

$$\delta \Delta p_G = \delta \int_0^L g\rho dz = g\, \frac{v_{fg}}{h_{fg}}\, \rho_f^2\, h_o' \left\{ I\left(-1 - \frac{s}{\Omega}\right)\delta\lambda - \frac{\left[I\left(-1 - \frac{s}{\Omega}\right) - I(-2)\right]}{s - \Omega}\, \delta u_e \right\}$$

$$\tag{3g}$$

where:

$$\delta\lambda = \frac{1 - \exp(-st_1)}{s}\, \delta u_i \tag{4}$$

$$\delta u_e = \delta u_i - \Omega\delta\lambda \tag{5}$$

$$\delta\rho_e = \rho_{oe}^2\, \frac{v_{fg}}{h_{fg}}\, h_o' \left\{ \left(\frac{u_{oe}}{u_{oi}}\right)^{1 - \frac{s}{\Omega}}\delta\lambda + \left[\frac{\left(\frac{u_{oe}}{u_{oi}}\right)^{1 - \frac{s}{\Omega}} - 1}{s - \Omega}\right]\delta u_e \right\} \tag{6}$$

And the auxiliary function $I(x)$ is defined as:

$$I(x) \triangleq \int_{\lambda_o}^{L} \left(\frac{u_o(z)}{u_{oi}}\right)^x dz = \begin{cases} \left[\dfrac{\left(\dfrac{u_{oe}}{u_{oi}}\right)^{x+1} - 1}{x + 1}\right] \dfrac{u_{oi}}{\Omega} & \text{if, } x \neq -1 \\[20pt] \dfrac{u_{oi}}{\Omega}\, \ell n \left[1 + \dfrac{\Omega}{u_{oi}}(L - \lambda_o)\right] & \text{if, } x = -1 \end{cases} \tag{7}$$

Dimensionless Equations

The above equations can be nondimensionalized by defining the following reference values:

$$z_r = L \tag{8a}$$

$$\rho_r = \rho_f \tag{8b}$$

$$u_r = u_{oi} \tag{8c}$$

$$t_r = \frac{1}{\Omega} \tag{8d}$$

$$h_r = \frac{Q}{w} = h_{eo} - h_{io} \tag{8e}$$

$$\Delta p_r = \rho_f u_{oi}^2 \tag{8f}$$

The resulting dimensionless pressure drop transfer functions are:

$$\Delta p_I(s) = N_{sub} + I(-1)\, U_e + N_{pch}\, I(-s)\, \Lambda - \frac{[I(-s) - I(-1)]}{s - 1}\, U_e \tag{9a}$$

$$\Delta p_a(s) = -2 + 2U_e + (1 + N_{pch} - N_{sub})^2\, R_e \tag{9b}$$

$$\Delta p_i(s) = 2K_i \tag{9c}$$

$$\Delta p_e(s) = 2K_e U_e + K_e(1 + N_{pch} - N_{sub})^2\, R_e \tag{9d}$$

$$\Delta p_f(s) = \left\{ 2\,\frac{N_{sub}}{N_{pch}} + 2\left(1 - \frac{N_{sub}}{N_{pch}}\right)U_e + \Lambda\,I(1-s) \right.$$

$$\left. - \left[\frac{I(1-s) - 1 + N_{sub}/N_{pch}}{s - 1}\right]\frac{U_e}{N_{pch}} \right\} F \qquad (9e)$$

$$\Delta p_G(s) = \left\{ \Lambda\,I(-s-1) - \left[\frac{I(-s-1) - I(-2)}{s - 1}\right]\frac{U_e}{N_{pch}} \right\} Fr^{-1} \qquad (9f)$$

where:

$$\Lambda = \frac{1 - \exp(-s\,N_{sub})}{s\,N_{pch}} \qquad (10)$$

$$U_e = 1 - N_{pch}\,\Lambda \qquad (11)$$

$$R_e = \frac{N_{pch}}{(1 + N_{pch} - N_{sub})^2}\left[N_{pch}\,I(-s)\,U_e + \Lambda(1 + N_{pch} - N_{sub})^{1-s}\right] \qquad (12)$$

And the dimensionless auxiliary function $I(x)$ is defined by:

$$I(x) = \begin{cases} N_{pch}^{-1}\left[\dfrac{(1 + N_{pch} - N_{sub})^{x+1} - 1}{x + 1}\right] & \text{if, } x \neq -1 \\[4ex] \dfrac{\ell n(1 + N_{pch} - N_{sub})}{N_{pch}} & \text{if, } x = -1 \end{cases} \qquad (13)$$

The following dimensionless parameters are defined:

$$N_{sub} = \frac{\Delta h_{sub}}{h_{fg}}\frac{v_{fg}}{v_f} \qquad \text{(Subcooling number)} \qquad (14)$$

$$N_{pch} = \frac{Q/w}{h_{fg}}\frac{v_{fg}}{v_f} \qquad \text{(Phase change number)} \qquad (15)$$

271

$$Fr = \frac{u_{oi}^2}{g \, L} \qquad \text{(Froude number)} \qquad (16)$$

The subcooling number represents the inlet subcooling of the liquid, the phase change number is the nondimensional power-to-flow ratio, and the Froude number represents the ratio of inertial to gravity forces. It is interesting to note that in the case of zero gravity, the gravity term vanishes, and the flow rate only influences the instability as Q/w in the phase change number (N_{pch}). Thus one of the effects of the gravity term is the decoupling of the influences of the power input (Q) and the flow rate (w).

3.   RESULTS

Constant Pressure Drop Boundary Condition

The characteristic equation of the system is given by the boundary condition of the pressure drop, which can be determined by a pump, an external loop, natural circulation, etc.

Let us consider the simplest case of a channel subjected to a constant pressure drop boundary condition. Since the perturbation of a constant is zero, the characteristic equation is:

$$\Delta p_{channel}(s) = 0 \qquad (17)$$

Channel stability is assured if all roots of Eq. (17) have negative real parts. The Nyquist criterion was used to analyze the effect of the gravity term on the stability. Figure 2 shows the stability margins for different values of the Froude number. The friction parameters used correspond to those of a typical BWR channel. The effect of the gravity term is not appreciable for values of the Froude number greater than 0.05. However, as the Froude number decreases, that is for lower flow rates, the channel becomes more stable for low subcoolings and more unstable for higher subcoolings.

Stability of a Natural Circulating Loop

The best situation to analyze the effect of gravity is under natural circulation conditions, where the gravity head is the cause of motion. The system shown in Fig. 3 was analyzed. A boiling channel is coupled with a downcomer connected to a surge tank with constant liquid level.

In this case the transformed boundary condition is given by the pressure drop in the downcomer, i.e.:

$$\Delta p_D(s) = \rho_f \, L_D \, u_D s + \rho_f Lg = -\Delta p_{channel}(s) \qquad (18)$$

Since the flow rate is controlled by the power input, the Froude number is not an independent parameter, and is determined by the steady-state momentum balance:

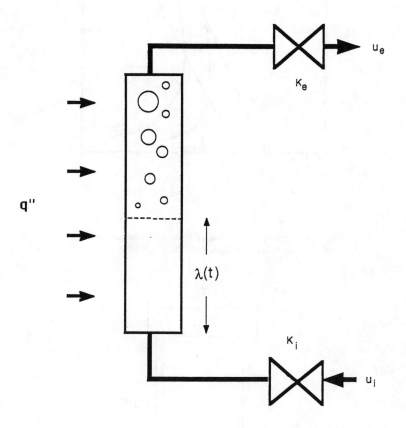

Fig. 1. Diagram of a single boiling channel system

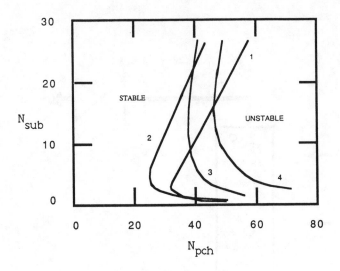

Fig. 2. Stability threshold of a boiling channel with constant pressure drop. 1. Fr=0.01; 2. Fr=0.001; 3. Fr=0.00013; 4. Fr=0.0001

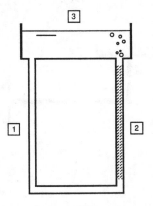

Fig. 3 : Natural circulation loop

1. Downcomer; 2. Heated section; 3. Surge tank

274

$$Fr = \left[\frac{N_{pch} - N_{sub} - \ell n(1 + N_{pch} - N_{sub})}{N_{pch}} \right] \left\{ K_i + K_e(1 + N_{pch} - N_{sub}) \right.$$

$$\left. + F \left[1 + \frac{(N_{pch} - N_{sub})^2}{2N_{pch}} \right] \right\}^{-1}$$

The stability threshold of the system is shown in Fig. 4. For high values of $N_{pch} - N_{sub}$, that is for high exit quality, the curve shows the same trend as in the "parallel channel" case with large Froude numbers. For lower powers, an interesting island of instability can be seen. This region extends for values of N_{sub} greater than 5. Figure-5 shows a sequence of Nyquist plots showing the island of instability at $N_{sub} = 10$. Figure-6 shows the same sequence for $N_{sub} = 5$. It can be noted that the region of instability becomes narrower as N_{sub} decreases. In Fig. 7 the case of $N_{sub} = 4$ is plotted. The island of instability no longer appears, but it is interesting to note its influence in the stable region.

At $N_{sub} = 50$ both regions of instability join as it is shown in Fig. 4. It is interesting to see what happens for higher N_{sub}. In Fig. 8 the sequence of Nyquist plots for $N_{sub} = 100$ is shown. The instability begins at $N_{pch} = 101.4$, but at $N_{pch} = 105$ the system becomes more stable, but becomes more unstable again for larger N_{pch}.

4. SUMMARY AND CONCLUSIONS

The effect of the gravity term in the momentum equation on the stability of boiling flows was analyzed. That term was found to be important under low flow rate conditions.

In the case of a boiling channel subjected to constant pressure drop, a decrease in the Froude number (lower flow rates) destabilizes the system for higher inlet subcoolings and stabilizes it for lower inlet subcoolings.

For natural circulating conditions, an island of instability was found at lower powers. This result agrees with experimental data for such systems [2,3].

5. NOMENCLATURE

f   Distributed friction per unit length
F   f·L

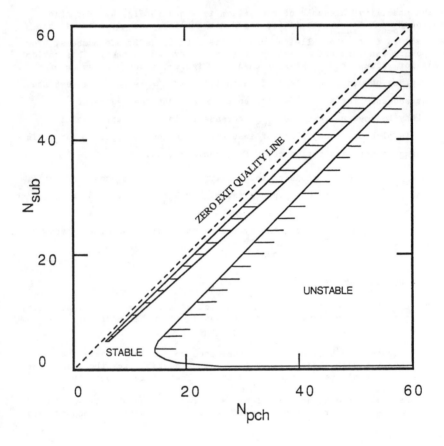

Fig. 4 : Stability threshold of a natural circulation system

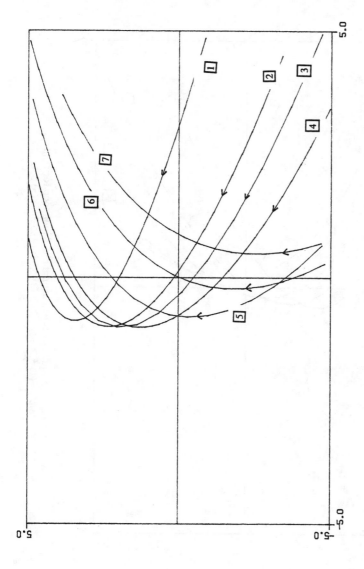

Fig. 5. Nyquist plots of a natural circulation loop for N_{sub}=10.
N_{pch}= (1) 10.26; (2) 10.36; (3) 10.4; (4) 10.46; (5) 10.625; (6) 10.9; (7) 11.15

277

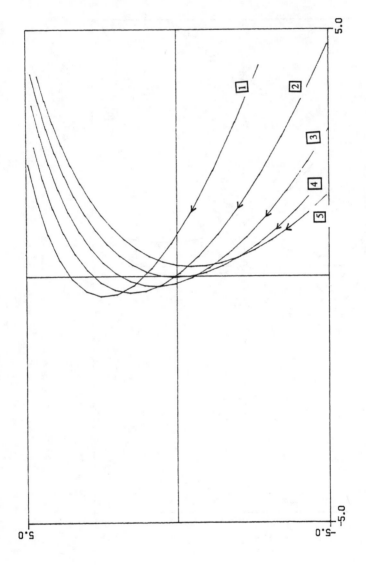

Fig. 6. Nyquist plots of a natural circulation loop for N_{sub}=5.
N_{pch} = (1) 5.2; (2) 5.225; (3) 5.31; (4) 5.365; (5) 5.42

278

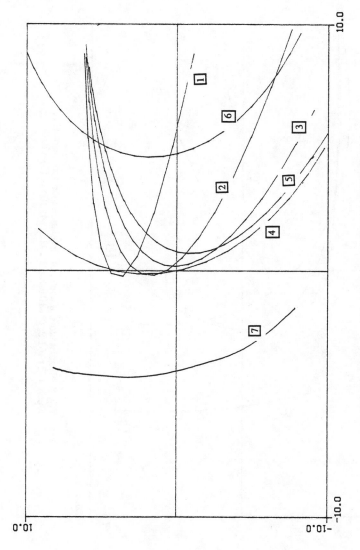

Fig. 7. Nyquist plots of a natural circulation loop for Nsub=4.
Npch = (1) 4.1; (2) 4.2; (3) 4.3; (4) 4.4; (5) 9.0; (6) 14.0; (7) 19.0

279

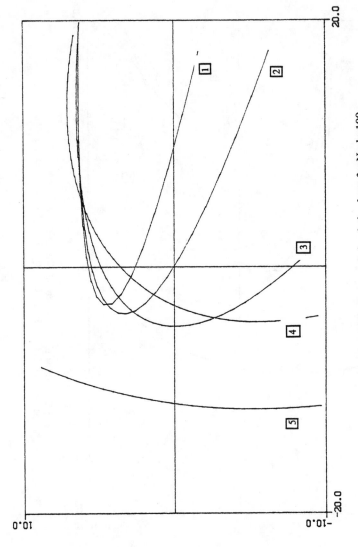

Fig. 8. Nyquist plots of a natural circulation loop for Nsub=100.
Npch = (1) 101.0; (2) 101.4; (3) 102.5; (4) 105.0; (5) 120.0

280

g	Gravity
h	Enthalpy
h_{fg}	Latent heat of evaporation
h'_o	Enthalpy increase per unit length
K	Local loss coefficient
L	Channel length
p	Pressure
q'''	Volumetric heat input
Q	Power input
R	Density transfer function
s	Time transform
t	Time
t_1	Single-phase residence time, $\dfrac{\lambda_o}{u_{oi}}$
u	Velocity
U	Velocity transfer function
v	Specific volume
v_{fg}	$v_g - v_f$
w	Flow rate
z	Space variable

Greek Letters

δ	Perturbation around the steady state
Δp	Pressure drop
λ	Subcooling length
Λ	Subcooling length transfer function
Ω	$q''' \, v_{fg}/h_{fg}$

Subscripts

a	Acceleration
D	Downcomer
E	Exit
f	Distributed function
g	Gas phase
G	Gravity head
i	Inlet
I	Inertia
o	Steady state

REFERENCES

1. Achard, J.L., Drew, D.A. and Lahey, R.T., Jr., "The Effect of Gravity and Friction on the Stability of Boiling Flow in a Channel", Chem. Engng. Commun., 11, 59-79, 1981.

281

2. Fukuda, K. and Kobori, T., "Classification of two-Phase Flow Instability by Density Wave Oscillation Model", _J. Nucl. Sci. Technol._, 16, 95-108, 1979.

3. Clausse, A. and Converti, J., "Flow Oscillations in a Boiling Natural Circulating Loop", Argentine Nuclear Technology Assoc. Meeting, Cordoba, Argentina, 1986.

Thermal Stability Investigations of Two Advanced Synthetic Based Fluids

VIJAY K. GUPTA
Chemistry Department
Central State University
Wilberforce, Ohio 45384, USA

Abstract

Thermal stability characteristics of two advanced synthetic based fluids a mixed silahydrocarbon and a polyalphaolefin (PAO) have been investigated. Three mixtures were prepared from above fluids with the following concentrations: 25% silahydrocarbon + 75% PAO, 50% silahydrocarbon + 50% PAO, and 75% silahydrocarbon + 25% PAO. Thermal decomposition studies of above mentioned base stock fluids and their mixtures were carried out as a function of time and temperature. Decompostion products were analyzed using viscosity measurements, capillary gas chromatography, gas chromatography coupled with mass spectrometry, and infra red spectroscopy. It was found that the silahydrocarbon fluid was more thermally stable than the polyalphaolefin. Both fluids did not show any appreciable degradation below 343°C, but above this temperature rate of degradation increased significantly. The fluid PAO decomposed almost completely when thermally stressed at 371°C for 48 hours, whereas the silahydrocarbon fluid exhibited same level of thermal degradation when heated at 371 °C for 72 hours. Mixing the two fluids did not provide any siginificant advantage as far as the thermal stability was concerned.

1. INTRODUCTION

Recent effort has led to the development of several functional fluids for use in aerospace applications such as jet engine oils, greases, and hydraulic fluids. Since hydraulic fluids are not expected to operate in oxidative environments, therefore thermal and hydrolytic stabilities are the two main areas of concern (1). In order to meet the need for a high temperature fluid, MIL-H-27601, a -40°C to 288°C hydraulic fluid was developed. A highly refined deep-waxed paraffinic mineral oil was selected as the base stock for the above application, and the fluid provided satisfactory service for over ten years in the temperature range of -40°C to 288°C.

The development of the supersonic missiles, like the advanced strategic air launched missile (ASALM) has created the need for a hydraulic fluid useable over the temperature range of -54°C to 315°C. Synthetic hydrocarbons based on hydrogenated polyalphaolefin oligomers have also been found to be deficient both in thermal stability and viscosity temperature properties. However, decene oligomers have emerged as a rahter unique class of synthetic hydrocarbon lubricants. These fluids are distinguished from their petroleum derived counterparts by their excellent low temperature properties, improved thermal stability, and significantly improved viscosity index. As a representative, an example of this type of fluid is a Gulf 4 cSt containing C_{30} oligomer. Perfluorinated fluids which have excellent thermal and oxidative stabilties, have several disadvantages such as high density, poor bulk modulus, elastomer incompatibility, and lack of suitable additives (3).

In response to the need for hydrocarbon type fluids with improved properties, an Air Force program has led to the development of a class of compounds called silahydrocarbons. These compounds have excellent viscosity temperature properties and thermal stability and are expected be hydrocarbon-like in their physical and chemical properties. Therefore, silahydrocarbons are good candiadates for hydraulic fluids useable over the temperature range of $-54^{\circ}C$ to $315^{\circ}C$. The studies on silahydrocarbon class of materials so far indicate that they are excellent candidate fluids that can be used over the temperature range of $-54^{\circ}C$ to $315^{\circ}C$, because of their excellent thermal stability coupled with desired flow properties down to $-54^{\circ}C$ (4-7). However, very little effort has been focussed on the kinetics and mechanisms of the thermal degradation processes occuring in the perfluorinated fluids, synthetic hydrocarbons, and silahydrocarbons. This report describes the effort to understand the thermal stability characteristics of two synthetic based advanced hydraulic fluids. One of the fluids is a mixture of silahydrocarbon compounds labelled as SHC and the other is a polyalphaolefin labelled as PAO.

2. RESULTS AND DISCUSSION

Thermal stabilities of the two experimental fluids SHC and PAO and their mixtures were conducted as a function of time and temperature. The representative structures of the two fluids are given in Figure 1. The two fluids were analyzed by capillary gas chromatography coupled with mass spectrometry (GC/MS) and their chromatograms are given in Figure 2(A) and 2(B). Specific GC conditions used for the analysis are given in Table 1. SHC was found to be a mixed silahydrocarbon consisting of the following four components:

Component	% Concentration
A. $CH_3 Si (n-C_8H_{17})_3$	29.3
B. $CH_3 Si (n-C_8H_{17})_2 n-C_{10}H_{21}$	46.4
C. $CH_3 Si n-C_8H_{17} (n-C_{10}H_{21})_2$	20.1
D. $CH_3 Si (n-C_{10}H_{21})_3$	3.1

The polyalphaolefin PAO was a Gulf 4 cSt $C_{30}H_{62}$ oligomer and its isomers (99%) as indicated by the chromatogram in Fig. 2(B). Thermal stabilty studies of SHC, PAO, and the mixtures (25% SHC + 75% PAO, 50% SHC + 50% PAO, and 75% SHC + 25% PAO), were investigated using the following procedure. The thermal stability test bombs were constructed of 304 stainless steel with 316 stainless steel high pressure end fittings. The bombs were 20 cm (8") long X 6 mm (0.25") diameter (unless otherwise specified). The bombs were cleaned with a suitable solvent (usually naphtha) and dried in an oven at $100^{\circ}C$ for one hour. Approximately 2 cc of the test fluid was added to the partially assembled bomb (one high pressure end fitting attached), and 99% pure nitrogen was bubbled through the fluid for 5 minutes to remove the air. The bomb was then quickly capped with the other high pressure end fitting and the whole assembly was weighed to the nearest mg. The bomb was then placed in an oven at the specified temperature (controlled to within $2^{\circ}C$) for the specified time. After the bomb was heated in the oven for the specified time, the bomb was removed, allowed to cool, and then reweighed. If the weight of the assembly changed more than 0.1 g, the test was considered invalid and rerun. If the weight change was less than 0.1 g, the bomb was placed in a bath at $-54^{\circ}C$ for at least 30 minutes to liquify the gaseous components. The bomb was then removed from the cold bath and one of the cap was replaced with a septum cap which was merely a 1/4" end cap with a septum in the space usually occupied by the tubing. The sample was then allowed to warm to room temperature and a headspace sample was taken through the septum top with a gas tight syringe. The gas sample was then injected into a gas chromatograph and analyzed for gaseous components. The bomb was then opened and the sample was

A. Silahydrocarbon

$$
\begin{array}{c}
R \\
| \\
CH_3 - Si - R_1 \\
| \\
R_2
\end{array}
$$

where   R, R_1, and R_2 groups can be either $n\text{-}C_8H_{17}$ or $n\text{-}C_{10}H_{21}$ or
          in combination

B. Polyalphaolefin

$$
\begin{array}{c}
HH \\
|| \\
CH_3 - (CH_2)_7 - C - CH_2 - C - (CH_2)_9 - CH_3 \\
|| \\
CH_3(CH_2)_7 \\
| \\
CH_3
\end{array}
$$

Figure 1. Chemical Structures of Representative Alpha-Olefin-Based Synthetic
           Hydrocarbon and Silahydrocarbons.

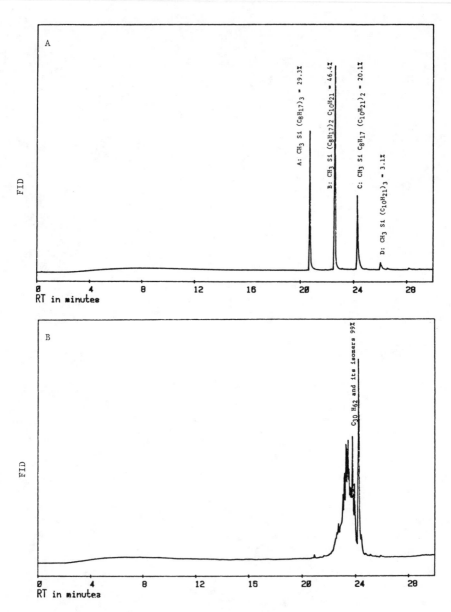

Figure 2. Capillary Gas Chromatograms of the Unstressed Fluids.
A. Silahydrocarbon (SHC) B. Polyalphaolefin (PAO).

286

Table 1.   Gas Chromatographic Conditions.

Gas Phase

Model: Perkin Elmer 900
Detector: FID
Column: Fused Silica Capillary Column
Length: 25 meters   Diameter: 0.22 mm
Liquid Phase: Methyl sicone carbowax deactivated
Carrier Gas: 1ml/min He
Auxilliary Gas: 40 ml/min He
Chart Speed: 1 cm/min   Attenuation: 10 X 16
Temperatures:
Injector: 220°C   Detector: 250°C
Column Temperature: -50°C to 200°C   Program Rate: 8°C/min
Initial Hold: 2 min   Final Hold: 20 min
Sample Size: 2 ml

Liquid Phase

Model: Hewlett-Packard 5710A
Detector: FID
Column: Fused Silica Capillary Column
Length: 12 meters   Diameter: 0.22 mm
Liquid Phase: Methyl Silicone Carbowax Deactivated
Carrier Gas: 1 ml/min He
Auxiliary Gas: 40 ml/min He
Chart Speed: 10 cm/hr.   Attenuation: 10 X 32
Temperatures:
Injector: 300°C   Detector: 350°C
Column: 70°C to 270°C   Program Rate: 8°C/min
Initial Hold: 2 min   Final Hold: 32 min
Sample Size: 1 Microliter

Table 2.   Viscosities of the Hydraulic Fluids as a Function of Temperature.

Hydraulic Fluid	Viscosities (cSt) at Different Temperatures in °C					Evis (KJ/mole)
	98.9	37.8	25.0	-54	37.8*	
PAO	3.79	17.26	27.56	11048.45	4.3	23.79
75% PAO + 25% SHC	3.54	15.08	23.60	7349.25	5.8	22.70
50% PAO + 50% SHC	3.23	12.91	19.99	5095.71	6.3	21.74
25% PAO + 75% SHC	2.98	11.16	17.18	3130.34	6.9	20.71
SHC	2.75	9.87	14.78	2216.83	8.6	20.05

---

* Viscosity data for the fluids when stressed at 371°C for 6 hours.

transferred to a glass vial for analysis of liquid components using GC analysis, viscosity measurements, and infrared spectroscopy.

The viscosities of the two fluids and their mixtures as a function of temperature are given in Table 2. Viscosity can be expressed as a function of temperature by the equation given below:

$$1/\phi = \eta = A \exp^{E_{vis}/RT}$$

where ϕ is the fluidity
η is the viscosity
A is the pre-exponential factor
and E_{vis} is the activation energy for viscous flow, the energy barrier that must be overcome before the elementary flow process can occur.

The above equation in the logarithmic form can be written as

$$\log_{10} \eta = \log_{10} A + E_{vis}/2.303 RT$$

Figure 3 gives the plot of $\log_{10} \eta$ as a function of 1/T. From the slopes, the activation energies have been computed and are given in Table 2. Both the fluids and their mixtures have the desired viscosities in the temperature range -54°C to 98.9°C and exhibit linear relationship within the limited temperature range.

The data in tables 3 and 4 and the chromatograms in figure 4 represent gas chromatography coupled with mass spectrometry (GC/MS) analysis of the gaseous phase of the two fluids stressed at 371°C. The % concentration values are the percent area, and the values have been averaged when more than one data point was obtained. Based on the above data, hydrocarbons containing 1-carbon atom to 10-carbon atoms were detected in the gaseous phase of both the fluids, but in PAO the hydrocarbons containing 11-carbon atoms to 15-carbon atoms were also detected, however their amounts were very small. The presence of these hydrocarbons in the gas phase of the stressed fluids indicates that the alkyl groups in the silahydrocarbon molecule and the hydrocarbon chain in PAO are breaking randomly as the molecules are heated to high temperature and lower molecular weight hydrocarbons are produced as the degradation products. The data in tables 5 and 6 and chromatograms in figure 5 represent the GC/MS analysis of the liquid phase of the two fluids stressed at 371°C. Various degradation products formed as reaction products under above stress conditions are listed in table 5 and 6. In the silahydrocarbon fluid, hydrocarbons with 5-carbon atoms to 8-carbon atoms, the silahydrocarbons originally present in the fluid, and the silahydrocarbons formed as a result of the degradation of one of the alkyl group of the molecule were found. The silahydrocarbon molecules formed as the degradation products contained one methyl group and two of the alkyl groups present in the original molecule and the fourth group formed as a result of the degradation of one of the alkyl group $-C_8H_{17}$ or $-C_{10}H_{21}$ present in the starting molecule. However, when the stress conditions are severe, other alkyl groups are also degraded. In the PAO fluid, the hydrocarbons containing 5-carbon atoms to 22-carbon atoms were found alongwith the isomers of the molecule $C_{30}H_{62}$. As siginificant concentrations of small molecules are produced, the hydrocarbon chains could preferentially be breaking away from the silicon atom in the silahydrocarbon molecule and away from the central carbon atom in the polyalphaolefin molecule (see the structures of the two compounds in figure 1).

The infra red spectra of the two experimental fluids stressed at 371°C are given in figure 6. A weak band was observed in the region 1630-1650 cm^{-1} in the silahdrocarbon fluid whereas a strong band was observed in the spectra of the

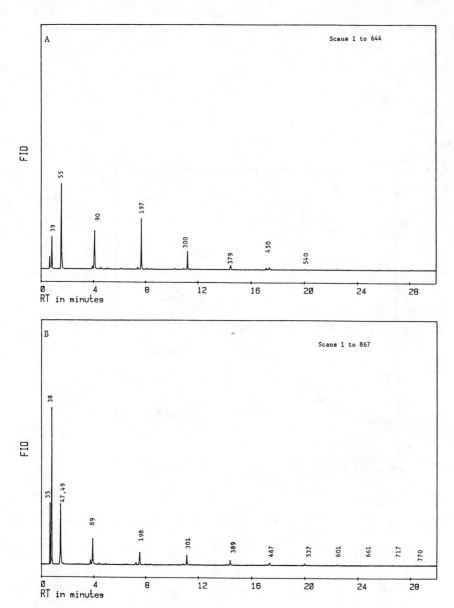

Figure 4.   Capillary Gas Chromatograms of the Two Fluids Stressed at 371°C.
from Headspace.
A.   Silahydrocarbon for 24 Hours
B.   Polyalphaolefin for 6 Hours

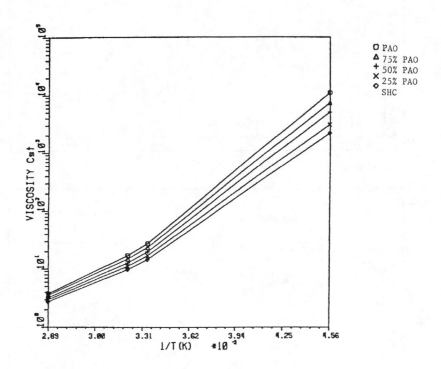

Figure 3.   Viscosities of the Two Fluids and Their Mixtures as a Function
of Temperature.

290

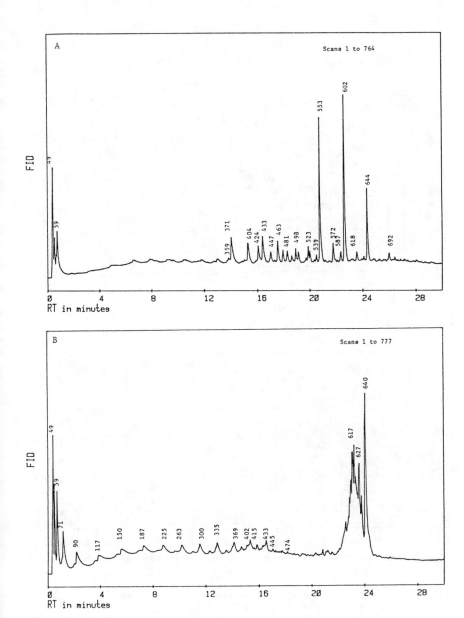

Figure 5.   Capillary Gas Chromatograms of the Two Fluids Stressed at 371°C in the liquid phase.
A.   Silahydrocarbon for 24 Hours
B.   Polyalphaolefin for 6 Hours

Fluid A. Silahydrocarbon Stressed at 371°C for 24 Hours

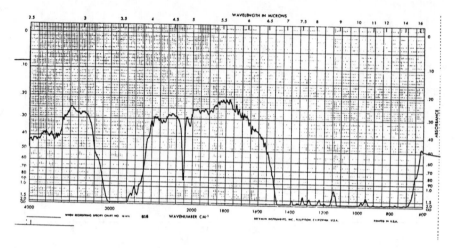

Fluid B. Polyalphaolefin PAO Stressed at 371°C for 6 Hours

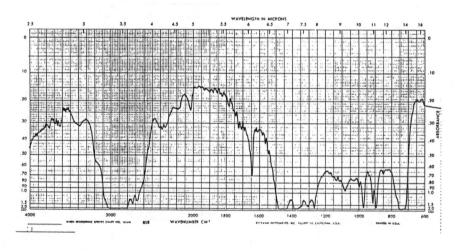

Fig. 6. Infra Red Spectra of the two Experimental Fluids Stressed at 371°C,
using 0.5 mm Path-length KBr Cell.

Table 3. GC/MS Analysis of SHC Stressed at $371^{\circ}C$ for 24 Hours in the Gaseous Phase.

Scan No.	Components	% Concentration
35,36	CH_4, C_2H_4, and C_2H_6*	9.85
39	C_3H_6 and C_3H_8*	26.17
49,55	C_4H_8 and C_4H_{10}*	23.48
83	C_5H_{10}	0.90
90	C_5H_{12}	13.91
148	$(CH_3)_4$ Si**	0.48
186	C_6H_{12}	0.47
197	C_6H_{14}	12.27
300	C_7H_{16}	4.18
379	C_8H_{16} and C_8H_{18}*	2.01
450	C_9H_{18} and C_9H_{20}*	1.05
540	$C_{10}H_{20}$ and $C_{10}H_{22}$*	0.15

* These scans may be mixtures

** This has been estimated based on the past experience

Table 4. GC/MS Analysis of PAO Stressed at 371°C for 6 Hours in the Gaseous Phase.

Scan No.	Component	% Concentration
35	CH_4, C_2H_4, and C_2H_6*	13.06
38	C_3H_6 and C_3H_8*	31.06
47,49	C_4H_8 and C_4H_{10}*	24.94
82,89	C_5H_{10} and C_5H_{12}*	12.51
186,198	C_6H_{12} and C_6H_{14}*	6.92
301	C_7H_{16}	4.11
389	C_8H_{18}	2.46
467	C_9H_{20}	1.59
537	$C_{10}H_{22}$	0.80
601	$C_{11}H_{24}$	0.37
661	$C_{12}H_{26}$**	-
717	$C_{13}H_{28}$**	-
770	$C_{14}H_{30}$**	-
819	$C_{15}H_{32}$**	-

* These are mixtures.

** These are present only in trace amounts.

294

polyalphaolefin fluid indicating the presence of double bond as unsaturated hydrocarbons. It is clear that thermal degradation of polyalphaolefin fluid resulted in more unsaturated hydrocarbons than the silahydrocarbon fluid. However, the concentration of unsaturated hydrocarbons was small as indicated by GC/MS data in tables 3 to 6.

The data in table 7 represents thermal degradation data for the experimental synthetic basestock fluids at $316^{\circ}C$ and $371^{\circ}C$ as a function of time. At stress temperature of $371^{\circ}C$, both from the % concentration and viscosity data it is clear that the sialhydrocarbon fluid decomposed at a lower rate than the polyalphaolefin fluid as all of the PAO fluid was decomposed for 72 hours and about 4% of silahydrocarbon was left undecomposed under above conditions. At the degradation temperature of $316^{\circ}C$, about 1% of the fluid degraded when the fluids were stressed for 72 hours. It is noted that as far as degradation is concerned, viscosity is a better indicator indicator than gas chromatographic analysis, however, GC analysis has its usefulness in identifying degradation products. Table 8 represents the thermal degradation data for the two fluids stressed for 6 hours as a function of temperature. Both fluids were found thermally stable upto stress temperature of $343^{\circ}C$, but increased degradation was observed when the fluid was heated at temperatures above $343^{\circ}C$. The silahydrocarbon fluid was found more thermally stable than the polyalphaolefin fluid.

The thermal degradation data of the two fluids and their mixtures stressed at $371^{\circ}C$ is given in table 9. The data point for 75% SHC and 25% PAO could be off due to some experimental error, otherwise the data indicates a linear relationship between the extent of degradation and mixture concentration. As far as the pure fluids are concerned the silahydrocarbon fluid is more thermally stable than the polyalphaolefin fluid, and mixing the two fluids did not provide any advantage as far as the thermal stability was concerned.

As pointed out earlier, the unstressed silahydrocarbon fluid was found to contain four silahydrocarbon compounds labelled as components A, B, C, and D. The data in table 10 indicates the % concentration of the four components remaining in the stressed fluid when the fluid was stressed at $371^{\circ}C$ as a function of time. From the data, it appears that there may be a slight difference in the rate of degradation of the four components, indicating the order of thermal stability as A > B > C > D. The length of the substituent group does effect the thermal stability of the compound, smaller the length of the substituent group more stable the compound. In view of the fact that this interpretation is based on the degradation data of a mixed silahydrocarbon where one of the components D is less than 3% and there is significant uncertainity in the experimental data, so it would be necessary to investigate the single compounds to verify this small difference in stability. A similar observation was made in a previous effort (5), when it was determined that $CH_3Si(n-C_8H_{17})_3$ decomposed about 10% whereas $CH_3Si(n-C_{10}H_{21})_3$ decomposed about 20% when stressed at $371^{\circ}C$ for 6 hours.

3. CONCLUSIONS

Based upon the results presented in the previous section, it is concluded that the silahydrocarbon candidate advanced fluid is more thermally stable than the polyalphaolefin base fluid. Both the fluids did not show any appreciable degradation below the temperature of $343^{\circ}C$, but increased degradation was observed at higher temperatures. Mixing the two fluids did not provide any advantage as far as the thermal stability was concerned. Viscosity is a better indicator of degradation than gas chromatographic analysis, however, the gas chromatography coupled with mass spectrometry is very useful in identifying and quantifying the

Table 5. GC/MS Analysis of SHC Stressed at 371°C for 24 Hours in the Liquid Phase.

Scan No.	Component	No. of C Atoms	% Concentration
49	C_5H_{10} and C_5H_{12}	5	6.60
53	C_6H_{12} and C_6H_{14}	6	5.95
55	C_7H_{14} and C_7H_{16}	7	6.95
57,59	C_8H_{16} and C_8H_{18}	8	11.65
359	$H-Si-CH_3(C_8H_{17})_2$*	17	0.26
371	$(CH_3)_2- Si - (C_8H_{17})_2$	18	4.35
404	$CH_3-Si-C_2H_5 (C_8H_{17})_2$	19	3.31
424	$CH_3-Si-C_3H_7 (C_8H_{17})_2$	20	2.48
433	$(CH_3)_2-Si- C_8H_{17} C_{10}H_{21}$	20	3.90
447	$CH_3-Si-(C_8H_{17})_2 C_4H_9$	21	1.56
463	$CH_3-Si-C_2H_5 C_8H_{17} C_{10}H_{21}$	21	2.74
473	$CH_3-Si-C_5H_{11}(C_8H_{17})_2$	22	1.33
481	$CH_3-Si-C_3H_7 C_8H_{17} C_{10}H_{21}$	22	1.46
489	$(CH_3)_2-Si-(C_{10}H_{21})_2$	22	0.58
498	$CH_3-Si-C_6H_{13} (C_8H_{17})_2$*	23	1.14
502	$CH_3-Si-C_4H_9 C_8H_{17} C_{10}H_{21}$	23	0.76
517	$CH_3-Si-C_2H_5(C_{10}H_{21})_2$*	23	0.25
523	$CH_3-Si-C_7H_{15}(C_8H_{17})_2$	24	1.32
525	$CH_3-Si-C_5H_{11}C_8H_{17}C_{10}H_{21}$	24	1.04
533	$CH_3-Si-C_3H_7 (C_{10}H_{21})_2$	24	0.18
539,553	$CH_3-Si-(C_8H_{17})_3$	25	13.66
572	$CH_3-Si-C_7H_{15}C_8H_{17}C_{10}H_{21}$	26	2.03
587,602	$CH_3-Si-(C_8H_{17})_2C_{10}H_{21}$	27	17.46
618	$CH_3-Si-C_8H_{17}C_9H_{19}C_{10}H_{21}$	28	0.97
644	$CH_3-Si-C_8H_{17}(C_{10}H_{21})_2$	29	6.20
692	$CH_3-Si- (C_{10}H_{21})_3$	31	1.08

* Indicates may be

Table 6. GC/MS Analysis of PAO Stressed at 371°C for 6 Hours in the Liquid
Phase.

Scan No.	Component	% Concentration
49	C_5H_{12}	3.29
52	C_6H_{14}	2.97
55	C_7H_{16}	4.00
59	C_8H_{18}	5.28
71	C_9H_{20}	5.29
90	$C_{10}H_{22}$	1.08
117	$C_{11}H_{24}$*	-
150	$C_{12}H_{26}$*	-
187	$C_{13}H_{28}$*	-
225	$C_{14}H_{30}$*	-
263	$C_{15}H_{32}$*	-
300	$C_{16}H_{34}$*	-
335	$C_{17}H_{36}$	1.49
369	$C_{18}H_{38}$	1.61
402	$C_{19}H_{40}$	0.74
415	$C_{20}H_{42}$	0.37
427	$C_{21}H_{42}$	0.12
433	$C_{20}H_{42}$	0.45
445	$C_{21}H_{44}$	0.11
474	$C_{22}H_{46}$	0.16
617,627,and 640	$C_{30}H_{62}$ and its isomers	73.22

* These components are present only in trace amounts.

Table 7. Thermal Degradation Data for the Experimental Synthetic Basestock Fluids as a Function of Time and Temperature.

Time (Hrs)	Fluid A Silahydrocarbon		Fluid B PAO	
	% Conc.	Visc. (cSt)	% Conc.	Visc. (cSt)
Degradation Temperature 371°C				
0 (unstressed)	99.7	9.9	99.2	17.4
6	78.9	8.6	62.2	4.4
16	82.6	9.2	61.3	5.9
24	37.6	5.8	53.2	5.0
48	10.3	3.3	2.4	1.7
72	4.4	2.8	-*	-*
Degradation Temperature 316°C				
0 (unstressed)	99.7	9.9	99.2	17.4
6	98.2	9.8	99.8	16.5
16	98.8	9.7	98.7	15.5
24	96.6	9.9	98.2	15.8
48	97.6	9.8	99.0	14.2
72	97.9	9.7	98.6	14.6

* All of the test fluid was lost during the degradation process.

Table 8. Thermal Degradation Data for the Experimental Fluids Stressed for
6 Hours as a Function of Temperature.

Temperature (°C)	Fluid A Silahydrocarbon		Fluid B PAO	
	% Conc.	Visc. (cSt)	% Conc.	Visc. (cSt)
Unstressed	99.7	9.9	99.2	17.4
316	98.2	9.8	99.8	16.5
329	99.5	9.5	99.6	13.6
343	94.3	9.4	97.2	10.9
357	97.6	9.0	96.0	8.3
371	78.9	8.7	62.2	4.4

Table 9. Thermal Degradation Data for the Experimental Fluids and Their
Mixtures Stressed at 371°C for 6 Hours.

% Silahydrocarbon	% PAO	% Conc.	Visc. (cSt)	Visc. (cSt)*
0	100	62.2	4.4	17.4
25	75	64.0	5.8	15.1
50	50	75.3	6.2	12.9
75	25	84.3	7.2	11.2
100	0	78.9	8.6	9.9

* These are viscosities of the unstressed fluid measured at 37.8°C.

Table. 10. Thermal Degradation Data for the Four Components of the Silahydrocarbon Fluid Stressed at 371°C for 6 Hours.

Time (Hrs)	% Conc. of Components							
	A	A*	B	B*	C	C*	D	D*
0	28.8	100.0	45.6	100.0	20.1	100.0	2.7	100.0
6	23.3	82.0	36.6	80.2	16.6	82.6	2.5	92.6
16	24.1	84.8	38.2	83.8	17.5	87.0	2.7	100.0
24	13.6	47.9	17.2	37.6	6.7	33.3	0.9	33.3
48	3.2	11.3	3.5	7.6	1.3	6.5	0.1	3.7
72	1.7	6.0	2.1	4.6	0.6	3.0	0.0	0.0

* These values are % Conc. Normalized on the scale of 0 to 100.

300

degradation products. In case of silahydrocarbons, it is felt that a smaller hydrocarbon chain (alkyl group) as a substituent may improve the thermal stability as compared to a longer hydrocarbon chain as substituent.

4. ACKNOWLEDGEMENTS

The author would like to thank the Air Force Systems Command, the Air Force Office of Scientific Research, and the Universal Energy Systems Inc. for providing the opportunity and financial support to conduct the above research at Materials Laboratory, Wright-Patterson Air Force Base, Ohio 45433.

REFERENCES

1. Snyder, C.E. Jr., Gschwender, L.J., and Tamborski, C., "Synthesis and Characterization of Silahydrocarbons - A Class of Thermally Stable Wide-liquid Range Functional Fluids", ASLE Trans. 25, 299-308, 1981.

2. Snyder, C.E. Jr., Tamborski, C., Gschwender, L.J., and Chen, G.J., "Development of High Temperature (-40°C to 288°C) Hydraulic Fluids for Advanced Aerospace Applications", ASLE Preprint No. 80-LC-IC-1.

3. Snyder, C.E. Jr., and Gschwender, L.J., "Development of Nonflammable Hydraulic Fluids for Aerospace Applications Over a -54°C to 315°C Temperature Range", Lubr. Eng. 36, 458, 1980.

4. Tamborski, C. Chen, G.J., Anderson, D.R., and Snyder, C.E. Jr., "Synthesis and Properties of Silahydrocarbons, A Class of Thermally Stable Wide-liquid Range Fluids", I EC Product Research Development, The American Chemical Society, 22, 172, 1983.

5. Gupta, V.K., and Weatherby, D.W., "Thermal Stability Characteristics of Silahydrocarbons", USAF-SCEEE Summer Faculty Research Program 1984, Technical Reports vol. II, # 56-1 to 56-38, 1984.

6. Gupta, V.K., Snyder, C.E. Jr., Gschwender, L.J., and Fultz, G.W., "Thermal Decomposition Investigations of Candidate High Temperature Base Fluids I. Silahydrocarbons", 41st Annual meeting of the Scoiety of Tribologists and Lubrication Engineers held in Toronto, Ontario, Canada, May, 1986 (in press).

7. Gupta, V.K., Snyder, C.E. Jr., Gschwender, L.J., and Chen, G.J., "Thermal Decomposition Investigations of Candidate High Temperature Base Fluid II. Model Silahydrocarbons", 43rd Annual meeting of The Society of Tribologists and Lubrication Engineers (STLE) held in Cleveland, Ohio, May 1988 (in press).

Mathematical Modeling of Two-Phase Flow Thermal Oscillations in Single Channel Upflow System

L. Q. FU
Shanghai Institute of Electric Power
Shanghai, PRC

T. N. VEZIROĞLU
Clean Energy Research Institute
University of Miami
Coral Gables, Florida, USA

ABSTRACT

This paper describes experimental and theoretical investigations of thermal oscillations in forced convection boiling in a single channel upflow system, and particular (1) to examine two-phase flow thermal oscillations having different tube sizes by performing experiments under various conditions; (2) to describe the mathematical modeling of two-phase flow thermal oscillations; (3) to compare the theoretical results with the experimental results.

INTRODUCTION

In recent years, considerable interest has been expressed concerning the phenon- menon of flow instabilities in two-phase flow systems. there are three basic types of oscillations, namely, pressure-drop type, density-wave type and thermal oscillations, which are encountered in boiling flows, each one being associated with a specific mode of operation, depending on mass flow rate, variation of heat input, inlet temperature and system pressure. In the previous studies the pressure-drop type and density-wave type oscillations in single and parallel channel open loop, forced convection upflow system, have already been observed in the laboratory of Clean Energy Research Institute, University of Miami. In the experimental observations, in addition to pressure-drop type and density-wave type oscillations, thermal oscillations were observed. The periods and amplitudes of thermal oscillations apear to be associated with the inlet temperature of the fluid, mass flow rate, heat input, gas-vapour mixture volume in the system, heater tube sizes and material.

The thermal oscillations are accompanied by large fluctuations in the wall temperature with periods of 30-120 seconds. It would cause thermal fatigue resulting from a continual cyclig of the thermal stress set up in the heat transfer surfaces. It is very important to study the phenomenon of the thermal oscillations with consideration of the safety of the operation in tne chemical engineering, power engineering, nuclear engineering, and so on.

Since high pressure and high temperature are used in the boiling systems, the thermal oscillations are the most dangerous one of the three basic type oscilla- tions, because they cause large amplitude of temperature fluctuation in addition to the pressure and mass flow fluctuation.

In this study, a special concern has been made with the thermal oscillations in forced convection boiling in a single channel upflow system with different sizes of heater tube. A mathematical model has been developed to simulate the steady

state characteristics, pressure drop oscillations and thermal oscillations. the
steady-state model is saved as initial conditions. The thermal oscillation model
is based on the pressure-drop type oscillation model. Using the pressure-drop
type oscillation model, the mass flow rate and pressure fluctuation can be obtained
in the system. As the heat transfer coefficient oscillates between wet and dry
conditions and the constant heat generation in the wall, the heat input in the
fluid keeps change. Thus thermal oscillations take place, A thermal oscillation
model is used to simulate the temperature fluctuation in the tube wall and property
fluctuation of the fluid inside the tube.

This model can be used to predict the effects of system geometry, heat input, inlet
temperature, mass flow rate, quality, inlet and exit restrictions, heater wall
thermal capacity, gas-vapour mixture volume in the system, amplitudes and periods
of pressure-drop type and thermal oscillations. Experimental results of single channel electrically heated forced convection upflow system have been used to compare
the model predictions. The mass flow rate versus pressure-drop in the steady-state
conditions and the periods as well as amplitudes of thermal oscillations are in
good agreement with the experimental results.

MODEL FOR STEADY-STATE CHARACTERISTICS

The physical situation to be simulated by the mathematical model is the one in
which the fluid flow in the circular tube, due to a pressure difference between
the inlet and exit of the system, and heat input through the tube wall. Figure 1
is a schematic diagram of the two-phase system, which has been simulated in this
mathematical model. Test fluid freon-11 is supplied from the main tank pressurized
by Nitrogen gas. A thermostatically controlled immersion heater in the main tank
and a cooling unit before the test section provide an inlet temperature range of
-20 ℃ to 90 ℃, with a control accuracy of   1.0 ℃. Following the electrically
heated test section is a recovery section consisting of a condenser and a collector tank. The mixture of saturated liquid and vapour is led through the condenser
coil, which is cooled by refrigerated brine at 0 ℃. The condensed liquid is then
stored in a recovery tank which is maintained at a constant pressure. This arrangement ensures a constant level of container and exit pressure.

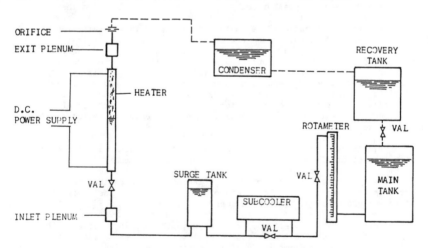

FIGURE 1. Schematic drawing of the experimental system

The test section includes a surge tank, a vetical heater tube, two plenums and a riser followed by exit restriction. The surge tank acts as a capacitance and is an important dynamic component of the loop. The heater tube is heated by controlled D.C. electricity. The exit restriction is a sharp-edged orifice with an inner diameter of 1.6 mm.

Using the experimental system described above, different sets of experiments corresponding to various heat inputs, inlet temperature, different tube sizes, wall thickness and heater tube materials have been conducted while the mass flow rate was varied in small decrements. Stability boundaries, steady-state and oscillatory characteristics were determined for each individual test.

In the theoretical investigation, first of all, a steady-state model [6] has been developed to simulate the relationships of the pressure drop versus mass flow rate for the liquid, two-phase and vapour flow under steady-state conditions. These relationships which are the steady-state solutions of the conservation equations, are also used to determine the initial conditions for the pressure drop and thermal oscillations. This model is based on assumptions of homogeneous two-phase flow and thermodynamic equilibrium of the phase. Compressibility effects in the two-phase region and thermal capacity of the heater wall have been included.

Governing Equations

Using the above assumption and definitions, the conservation laws for mass, momentum and energy can be written as follows:

Continuity:

$$\frac{\partial \rho}{\partial t} + \frac{\partial (\rho u)}{\partial z} = 0 \tag{1}$$

Momentum:

$$\rho \frac{\partial u}{\partial t} + \rho u \frac{\partial u}{\partial z} + \frac{\partial P}{\partial z} = -2 \frac{f}{d} \rho u^2 - \rho g \tag{2}$$

Energy:

$$\rho \frac{\partial h}{\partial t} + \rho u \frac{\partial h}{\partial z} = \phi \tag{3}$$

In addition to the conservation equations, the equation of state is necessary to close the equation set. It can be written as follows:

For subcooled liquid properties:

$$\eta_1 = \eta_1 (T) \tag{4}$$

For saturated liquid and vapour properties:

$$\eta_1 = \eta_1 (P) \tag{5}$$
$$\eta_v = \eta_v (P) \tag{6}$$

In two-phase regime:
Quality:

$$x = \frac{(h - h_{sat})}{h_{lv}} \tag{7}$$

Density:

$$\rho = \frac{\rho_1}{[1 + x (\rho_1/\rho_v - 1)]} \tag{8}$$

Enthalpy:

$$h = xh_v + (1 - x) h_1 \tag{9}$$

Where the subscripts '1' and 'v' refer to liquid and vapour phases respectively.

Finite difference method has been used to prepare a computer code for the solution of the governing equations. The predictions of the steady-state model along with the experimental recordings are illustrated in Fig. 2 and Fig. 3.

Fife curves are plotted on Fig. 2 for different heat input corresponding to 500 W, 400 W, 300 W, 200 W and a constant inlet temperature 20°C. Fig. 3 shows the curves for various inlet temperatures corresponding to -8°C, 0°C, 8°C, 16°C, 24°C, 32°C and a constant heat input 500 W. The theoretical results can be seen in good agreement with the experiments.

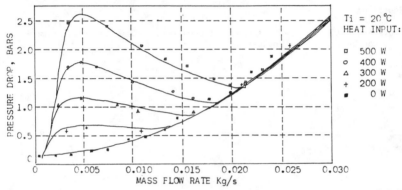

FIGURE 2. COMPARISON OF STEADY-STATE MODEL PREDICTIONS WITH EXPERIMENTAL DATA.

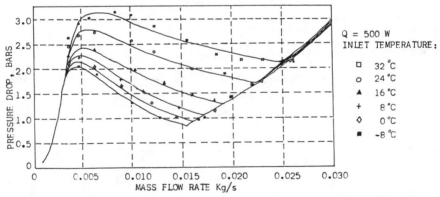

FIGURE 3. COMPARISON OF STEADY-STATE MODEL PREDICTIONS WITH EXPERIMENTAL DATA.

MODEL FOR PRESSURE-DROP TYPE OSCILLATIONS

Since thermal oscillations are triggered by the pressure-drop type oscillations, thus the model for the pressure-drop type oscillations have to be created first. The mechanism of the pressure-drop type oscillations can be driven from the steady-state pressure drop versus mass flow rate relations [1].

It can be seen in Fig. 4, the starting point is assumed in the middle of the negative slop region (two-phase flow region) of the steady-state curve. When the pressure P_s in the surge tank increases a small amount by a static instability, the mass flow rate from main tank into the surge tank G_1 will be decreased. As the pressure in the surge tank increases and the exit pressure of the system P_e remains constant , the pressure drop between the surge tank and exit of the system increases. In the negative slop of the steady-state characteristic curve, the mass flow rate, G_2, from surge tank to the heater decreases with the increasing of the pressure drop. Since G_2 decreases more than G_1, the liquid level in the surge tank will rise and P_s in the surge tank will increases further. The system is then suffered unsteady operation. The operating point will move along the boiling curve from a stable point A to point B, until it reaches a position at which the pressure in the surge tank gets to a maximum. The process cannot stop at point B because of the inbalance between the surge tank inlet and exit flow rates. The flow rate G_2 will increases suddenly and a flow excursion occurs taking the operating point from B to C in the liquid region. At point C, G_2 is greater than G_1, the increasing flow rate through the heater now works in evacuating the surge tank. The pressure P_s in the surge tank will decrease with the decreasing of flow rate G_2. In the meantime, the flow rate G_1 entering the surge tank will increase gradually with the decreasing of pressure P_s. The operating point will move along the curve in the liquid region from point C to point D. At point D, the mass flow rate G_2 is still larger than G_1. The surge tank pressure P_s decreases further more. The flow rate G_2 decreases suddenly and another excusion occurs taking the operating point from D to E. At the point E, G_2 is less than G_1. The pressure in the surge tank increases again and the flow rate in the system G_2 also increases. The process then takes place from point E to point B, which is the peak of a steady-state characteristic curve. Thus the process repeats itself along the limit cycle of BCDE. These are called pressure-drop type oscillations. During the pressure-drop type oscillations, the mass flow rate and the heat transfer coefficient keeps change, thus the heat input into the fluid also change. The oscillation cycle will cross different steady-state characteristic curve which is belong to different heat input, as shown in Fig. 4.

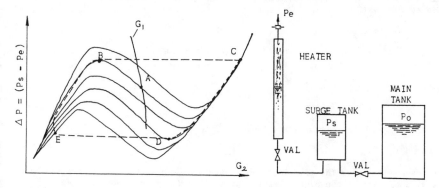

FIGURE 4. Schematic of flow diagram and limit cycle of pressure-drop type oscillations.

307

Governing Equations

The system under consideration is the same as Fig. 1. The equations which describe the surge tank dynamics must be included in the analysis, since the air-vapour mixture inside the surge tank plays an important role in generating and maintaining the oscillations.

For the surge tank, the continuity equation can be written as [3]:

$$\frac{dPs}{dt} = P_{sa}^2 \frac{(G_1 - G_2)}{P_{sao} V_{go} \rho_1} \tag{10}$$

Where Ps is the surge tank pressure, Psa is the unsteady and Psao is the steady pressure of the air and vapour mixture in the surge tank. G_1 and G_2 are the inlet and outlet mass flow rate to and from the surge tank, and ρ_1 is liquid density.

The momentum equation for the mass flow rate from main tank to surge tank can be written as:

$$G_1 = \left[\frac{(P_o - P_s)\rho_1}{R} \right]^{\frac{1}{2}} \tag{11}$$

Where R is the restriction coefficient from the main tank to the surge tank, obtained from experimental data.

The mass flow rate from surge tank to system exit is the solution of the steady-state model, which depends on the heat input Qi, pressure Ps and fluid inlet temperature Ti. The relations can be written as following:

$$G_2 = f(Ps, Qi, Ti) \tag{12}$$

During the oscillations, heat input Qi and pressure Ps change, thus G_1 and G_2 also change. When Ps increases, the decrements of G_1 and G_2 are not equal, so that makes the system unstable.

Methed of Solution

An explicit forward difference technique is used to solve the nonlinear differential equations which describe the system dynamics. The governing equations can be approximated as follows:

From main tank to surge tank:

$$G_1^{j+1} = \left[\frac{(P_o - P_s^{j+1})\rho_1^{j+1}}{R} \right]^{\frac{1}{2}} \tag{13}$$

From surge tank to system exit:

$$G_2^{j+1} = f(P_s^{j+1}, Qi^{j+1}, T_i^{j+1}) \tag{14}$$

The continuity equation of the surge tank:

$$P_s^{j+1} = \Delta t \, (G_1^{j+1} - G_2^{j+1})(P_s^j)^2 / P_{sao} V_{go} \rho_1^{j+1} + P_s^j \tag{15}$$

In the above equations, the time increment must be chosen with care, otherwise the calculation can not be converged. Experimenting with different time increments is necessary to determine the maximum value for which the solution converges and saving computer time, In this numerical scheme, 0.5 seconds are available for the time increment or it can be determined from:

$$\Delta t \leq \frac{\partial z}{u_{max}} \qquad (16)$$

The calculations start with given inlet fluid temperature and pressure from the surge tank. The flow parameters and properties, corresponding to the given inlet mass flow rate and heat input, are calculated using the steady-state program. These results are saved as the initial conditions at the stable operating point which is located at the negative slop portion of the steady-state characteristic curve.

The system is perturbed by a small increment of pressure Ps in the surge tank. Using Eq.(11), the surge tank inlet mass flow rate G_1 is calculated. Using Eq.(12) by calling the steady-state program, the mass flow rate G_2 along the heater is computed. Since the pressure in the surge tank is increased, the flow rate through the heater decreases. The heat transfer coefficient and the heat input into the fluid also change. Using Eq.(13), the pressure Ps in the surge tank is calculated. As the exit mass flow rate G_2 from the surge tank decreases more than the mass flow rate G_1 into the surge tank, the pressure in surge tank increases further during the time step Δt. In the successive time step, the procedure described above is repeated.

As the flow through the heater decreases, the surge tank pressure Ps increases till it reaches a maximum. Thus makes the pressure difference between the surge tank and the exit of the system increasing along the negative slop of the steady-state curve until it get to the top of one of the curve, which depends on the transient heat input into the fluid. Then the calculation point will leap automatically from the top of the curve to the liquid region of the steady-state characteristic curve. The mass flow rate through the heater increases suddenly. In the liquid region, the exit mass flow rate G_2 from surge tank is large than the mass flow rate G_1 into the surge tank. The pressure Ps will decrease till it reaches a minimum. Then another excursion of the calculation point will take place from liquid region to the vapour phase region. This time, the mass flow rate G_1 is larger than G_2. Ps increases again until the calculation point reaches the top of the steady-state curve again. These limit cycles are then sustained.

Fig. 5 shows the limit cycles of the pressure-drop type oscillations, which are simulated by the mathematical model.

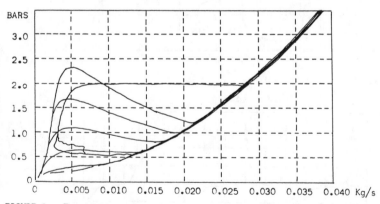

FIGURE 5.   The limit cycles of the pressure-drop type oscillations.

309

Fig.6 shows the theoretical results of the pressure-drop type oscillations along with experimental recordings, for constant electrical heat input (Q_o= 500 W).

Fig.7 and 8 show the mass flow rate and heat input into the fluid fluctuations during the oscillations.

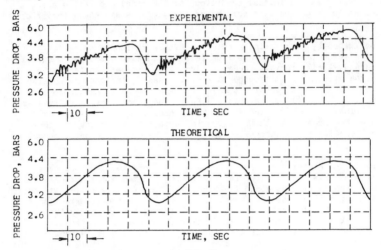

FIGURE 6. Pressure-drop type oscillations.

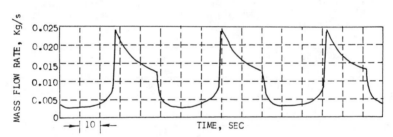

FIGURE 7. Mass flow rate fluctuations.

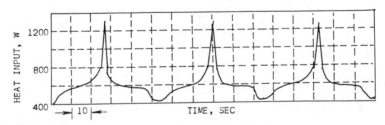

FIGURE 8. Heat input into the fluid fluctuations.

MODEL FOR THERMAL OSCILLATIONS

During the pressure-drop type oscillations, mass flow rate, heat transfer coefficient and heat input into the fluid sustains fluctuations as shown in Fig.7 and 8. However, the heat generated by the D.C. current is constant. When the limit cycle enters the vapour region, the heater wall temperature increases, because the heat transfer coefficient is usually low at the dry region. Whereas the limit cycle enters liquid region, the heater wall temperature decreases as the liquid heat transfer coefficient is usually high. Thus the heater wall temperature maintains fluctuations during the limit cycles. These are the thermal oscillations.

Governing Equations

Heat input into the fluid is given by:

$$\Delta Q_I = \pi d \Delta z \alpha (T_w - T_f) \tag{17}$$

The heater wall temperature can be calculated from the energy balance for the heater, that is:

$$\frac{d(T_w)}{dt} = \frac{(Q_o - Q_I)}{m_h c_h} \tag{18}$$

Method of Solution

In the pressure-drop type oscillations model, fluid parameters and properties are calculated along the system during the oscillations. The fluid temperature inside the heater at any node is known. The heat transfer coefficient is calculated using the experimentally developed correlation as following:

$$\alpha = 1.2\alpha_1 - \frac{0.2\alpha_1}{0.3^n} \left| 0.3 - x \right|^n \tag{19}$$

Where

$$n = 0.1 \ln \frac{1.2\alpha_1 - \alpha_v}{0.2\,\alpha_1} \tag{20}$$

While the oscillations occure, the fluid quality changes. From Eq.(19), it can be seen that the heat transfer coefficient oscillates between wet and dry region. At first, the heat input into the fluid is assumed to be constant which is equal to the heat generated by the D.C. current. Then the heater wall temperature can be calculated. During the oscillations, the heat transfer coefficient and heat input into the fluid change, so the heater wall temperature changes accordingly.

The governing equations are written in finite difference from as following:

Heat input into the fluid at any segment:

$$\Delta Q_{I,i}^{j+1} = \pi d \Delta z \alpha_i^{j+1} (T_{w,i}^{j+1} - T_{f,i}^{j+1}) \tag{21}$$

Heater wall temperature at any node:

$$T_{w,i}^{j+1} = \Delta t \frac{(Q_{o,i} - Q_{I,i}^{j+1})}{m_h c_h} + t_{w,i}^{j} \tag{22}$$

The temperature at any node of the heater during the thermal oscillations can be obtained by solving above equations.

Comparison of the Thermal Oscillation Model Predictions with Experiments

The results of the thermal oscillation model are plotted along with the experimental recordings in Fig.9 and 10. These plottings show the thermal oscillations of the wall temperature near the end of the heater for constant electrical heat input (Q_o= 500 W), with various inlet fluid temperatures, initial mass flow rate and different heater tube sizes. Table 1 and 2 are the summary of the comparison of the theoretical predictions with the experimental results. The theoretical results are in good agreement with the experiments.

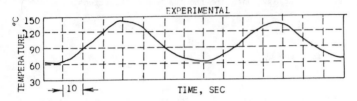

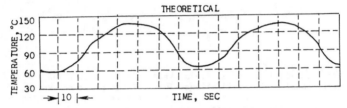

FIGURE 9. Thermal oscillations at the end of the heater. Q_o= 500 W, T_i= 8 $^{\circ}$C, m= 0.00682 Kg/s. Heater: I.D. 0.43", O.D. 0.50". Material: Stainless steel.

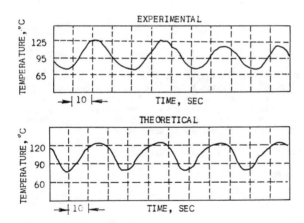

FIGURE 10. Thermal oscillations at the end of the heater. Q_o= 500 W, T_i= 16 $^{\circ}$C, m= o.00904 Kg/s. Heater: I.D. 0.305", O.D. 0.375". Material: Stainless steel.

TABLE 1.   Comparison of the experimental and theoretical results[1]

Inlet temperature	Mass flow rate	Oscillation type & location	Experimental period	amplitude	Theoretical period	amplitude
°C	g/s		s	°C or bar	s	°C or bar
-8	5.40	Temperature at heater end	80	128	85	135
		Pressure at heater inlet	80	0.24	85	0.45
0	6.80	Temperature at heater end	50	85	55	75
		Pressure at heater inlet	50	0.55	55	0.60
8	6.35	Temperature at heater end	56	114	56	115
		Pressure at heater inlet	56	1.26	56	1.35
16	9.04	Temperature at heater end	30	48	30	45
		Pressure at heater inlet	30	1.20	30	1.00

[1] Heater: I.D. 7.75 mm, O.D. 9.53 mm. Material: Stainless steel. Heat input: 500 W.

TABLE 2.   Comparison of the experimental and theoretical results.[2]

Inlet temperature	Mass flow rate	Oscillation type & location	Experimental period	amplitude	Theoretical period	amplitude
°C	g/s		s	°C or bar	s	°C or bar
-8	5.42	Temperature at heater end	115	92	115	95
		Pressure at heater inlet	115	1.1	115	1.1
0	4.97	Temperature at heater end	80	60	77	60
		Pressure at heater inlet	80	1.02	77	1.05
8	6.82	Temperature at heater end	75	75	73	75
		Pressure at heater inlet	75	0.66	73	0.60
16	8.20	Temperature at heater end	45	33	45	33
		Pressure at heater inlet	45	0.54	45	0.50

[2] Heater: I.D. 10.92 mm, O.D. 12.7 mm. Material: Stainless steel. Heat input: 500 W.

CONCLUSIONS

The heater tubes with different outside and inside diameters have been used to conduct the experiments in a single channel upflow forced convection system to study the effects of heat input, inlet temperature, initial mass flow rate, tube sizes, air-vapour mixture in the surge tank, etc. on the behavior of the thermal oscillations. A theoretical model has been used to simulate the pressure-drop type and thermal oscillations effected by these factors. Based on the experimental and the theoretical studies, certain conclusions can be summarized below:

1. The steady-state characteristics, and the amplitudes and periods of the oscillations predicted by the mathematical model are in good agreement with experimental results.
2. The amplitudes of the pressure-drop type and thermal oscillations increase with decreasing inlet temoerature. The periods also increase with decreasing inlet temperature, i.e. increasing inlet subcooling.
3. the periods of the thermal oscillations are the same as the pressure-drop type oscillations, but the phase of pressure-drop type oscillations lag behind the thermal oscillations's.
4. The amplitudes and periods of the oscillations increase with increasing heat power.
5. The amplitudes and periods of the oscillations increase with increasing comprassible air-vapour mixture volume in the surge tank.
6. The periods of the oscillations decrease with increasing exit pressure of the system.
7. The amplitudes and periods of the oscillations increase with decreasing mass flow rate at the initial operating point on the negative slope of the steady state characteristic curve.
8. If the heater tube thermal capacity is low, the amplitudes of thermal oscillations will be high.

NOMENCLATURE

C_h Specific heat of heater wall material, J/(Kg K)
d Inside diameter of the heater tube, m
F_m Two-phase flow friction multiplier, dimensionless
f Friction factor, dimensionless
G_1 Mass flow rate from main tank to surge tank, Kg/s
G_2 Mass flow rate from surge tank to system exit, Kg/s
g Gravitational acceleration, m/s
h_1 Liquid enthalpy, J/Kg
h_v Vapour enthalpy, J/Kg
h_{iv} Latent heat of evaporation, J/Kg
m_h Mass of the heater, Kg
P Pressure, Pa
P_e Exit pressure, Pa
P_o Main tank pressure, Pa
P_s Surge tank pressure, Pa
P_{sa} Unsteady surge tank pressure, Pa
P_{sao} Steady-state surge tank pressure, Pa
Q_i Total heat input rate into the fluid, W
ΔQ_i Local heat input into the fluid, W
Q_o Electrical heat generation rate in the heater wall, W
R Resistance coefficient for the inlet restriction, dimensionless
T_f Fluid temperature, °C
T_o Inlet temperature, °C
T_w Heater wall temperature, °C
t Time, s

Δt Time increment in the numrrical scheme, s
u Velocity, m/s
V_{go} Steady-state volume of the gas in the surge tank, m
x Quality of the liquid-vapour mixture, dimensionless
z Axial distance along the flow path, m
Δz Increment in the distance along flow path for the numerical scheme, m
α Heat transfer coefficient, $W/(m \ K)$
η General fluid property
ρ Density, Kg/m
ϕ Heat input into the fluid per unit inner volume of the heater, W/m

Subscripts

e Exit condition
f Fluid parameter
g Gas
h Heater
i Inlet condition
l Liquid
o Steady-state condition
s Surge tank
v Vapour
w Wall condition

REFERENCES

1. T.N.Veziroglu and S.S.Lee, "Boiling Flow Instabilities", Int. Symp. on research of Concurrant Gas-Liquid Flow, 1968.

2. T.N.Veziroglu, S.S.Lee and S.Kakac, "Fundamentals of Two-Phase Oscillations and Experiments in Single-Channel Systems", Eds. S.Kakac, et al., Two-Phase Flow and Heat Transfer, Vol. 1, Hemisphere Publishing Corp., U.S.A., 1977.

3. T.N.Veziroglu and S.Kakac, "Two-Phase Flow Instabilities", 1983., Final Report to N.S.F. on Project CME 79-20018.

4. S.Kakac, T.N.Veziroglu, L.Q.Fu, et al., "Two-Phase Flow Instabilities in a Vertical Boiling Channel", Int. Symp. on Fundamentals of Gas-Liquid Flows, ASME Winter Annual Meeting, Nov. 1988, Chicago.

5. L.Q.Fu, et al., "Mathematical Modeling of Two-Phase Flow Instabilities in Parallel Channels", Hemisphere Publishing Corp., Washington, 1988.

6. T.N.Veziroglu, S.Kakac and L.Q.Fu, "Two-Phase Flow Thermal Instabilities in a Single Channel System", First Annual Report: N.S.F. Project CBT 86-12282, Clean Energy Research Institute, Coral Gables, Florida, U.S.A., 1988.

The Effects of Unequal Heating Power and Inlet Restrictions on Density Wave Instability of a Parallel Channel System

YANG-QIANG RUAN, LONG-ZHOU FU, and XUE-JUN CHEN
Department of Energy and Power Engineering
Xi'an Jiaotong University
Xi'an, PRC

ABSTRACT

In this paper, we study the density wave instability of parallel multi-channel system with nonidentical operating conditions for every channel. The analysis is based on the multivariable frequency domain method. According to this theoretical study, a new version of the computer program MADWO has been developed. It can be conveniently used to evaluate stability margin and investigate parametric effects of a parallel multi-channel system with unequal heating power and inlet restrictions between any two channels. The predicted stability margins show well agreement with the previous experimental data and better than that obtained by the traditional single channel and single variable method. It was also found that increase in inlet restrictions difference between any two channels may decrease the system stability when other paramters are held to be unchanged.

1. INTRODUCTION

The operating conditions of every channel are not identical in actual parallel multi-channel systems, such as in steam generators or in nuclear reactor cores . As several experimental studies have shown, in some cases the nonidentical operating conditions may significantly affect the density wave instability margin of the system because of thermal, hydrodynamic or nuclear coupling interaction. Since most previous analyses for parallel multi-channel systems relied on the assumption of the single representative channel or the hottest channel, the effects due to the nonidentical operating conditions for every channel are usually not included in the stability margin evaluation.

In this paper, we present a study on the density wave instability of a parallel multi-channel system under the nonidentical operating conditions for every channel, in particular the system with different heating power and inlet restrictions. The analysis is based on the multivariable frequency domain method (MFDM) (Ruan, 1986). A new mathematical model has been developed which accounts for the nonuniform distribution of heating power, flow splitting, and the coupling interaction between the parallel channels and its external loop. The predicted results show well agreement with the previous experimental results and they are better than the results obtained by the traditional single channel and single variable method (SCSVM).

2. PERTURBATION WAVES PROPAGATION AND FEEDBACK PROCESS IN A PARALLEL MULTI-CHANNEL SYSTEM

The previous paper (Ruan, 1986) have presented the foundamental aspects in developing a multivariable frequency domain model of a parallel multi-channel system for analyzing density wave oscillations. In this paper, we just describe some new parts of the model which relate to the present subject.

It is well known that the delayed and feedback processes of perturbation wave propagation play an important role in the mechanism of density wave instability phenomena. The delayed time are necessary for (1) the liquid enthalpy wave to propagate in the subcooled region and (2) for the mixture density wave to propagate in the two-phase region.

Assume an incompressible flow in the single-phase region, the enthalpy wave propagation equation can be written as

$$\frac{\partial h}{\partial t} + V\frac{\partial h}{\partial z} = \frac{q'}{\rho_1 A_c} \tag{1}$$

where the velocity $V(z,t)$ is equal to the inlet velocity

$$V(z,t) = V_{in}(t) \tag{2}$$

and $h(z,t)$ is the enthalpy, A_c the flow area, ρ_1 the density, and $q'(z,t)$ is the heating power per unit length. In order to allow for an arbitrary axial power distribution, the actual heat generation curve was replaced by a staircase function (see Fig. 1), which can be written as the sum of a series of unit functions

$$
\begin{aligned}
q'(z,t) &= q(t) \cdot q(z) \\
&= q(t) \cdot \sum_{n=1}^{N} q_n \cdot \left[U(z-z_n) - U(z-z_n) \right]
\end{aligned} \tag{3}
$$

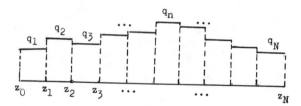

Fig. 1 A staircase function of heating power

In the two-phase region, the density wave equation can be derived from the void equation (Zuber, 1967) or from the mass conservation equations for liquid and vapor phases (Ruan, 1986):

$$\frac{\partial \rho_m}{\partial t} + c_k\frac{\partial \rho_m}{\partial z} = -\rho'\Omega' \tag{4}$$

318

where the mixture density and the velocity of the density wave are defined as

$$\rho_m = \alpha \rho_g + (1 - \alpha) \rho_f \tag{5}$$

$$\rho' = (1 - C_o) \rho_f + \rho_m C_o \tag{6}$$

$$C_k = C_o j + V_{gj} + \alpha \frac{\partial V_{gj}}{\partial \alpha} + j \frac{\partial C_o}{\partial \alpha} \tag{7}$$

and the characteristic frequency, Ω' is given by

$$\Omega' = C_o \frac{V_{fg} \, q'}{h_{fg} \, A_c} \tag{8}$$

In equations (7) j is the volumatric flux of the mixture, V_{gj} the drift velocity, C_o the void distribution parameter, α the void fraction.

In a parallel multi-channel system, the instability mechanism become more complicated because of the nonidentical thermal boundary (e.g. different heating power) and the complex hydrodynamic boundary (e.g. different inlet restrictions and coupling interaction between the channels) in the system. The nonidentical operating conditions of every channel and the coupling interaction between the parallel channels and its external loop may produce a dynamic process of inlet flow rate redistribution among the channels, which may affect the perturbation wave propagation and feedback processes in every individual channel. Therefore, in a parellel multi-channel system the stability phenomena appear as the behavior of the whole system rather than as that of a single boiling channel.

If we assume that the mixing in the inlet plenum and exit plenum is perfect, the pressure drop imposed on every channel are equal each to each, and then the total flow rate, W_t, entering the inlet plenum is distributed between individual channels without any mass storage. Therefore, the flow rate redistribution problem will be reduced to find the solution of the following equations

$$\Delta P_m = \Delta P_{ext} \tag{9}$$

$$W_t = \sum_{m=1}^{M} W_m \tag{10}$$

$$m = 1, 2, \cdots, M$$

where M is the number of channels, W_t is the total flow rate, W_m the flow rate of the m'th channel, ΔP_{ext} and ΔP_m are respectively the pressure drop of external loop and the m'th channel.

3. PRESSURE DROP PERTURBATION PROPAGATION

Consider the m'th channel between the inlet plenum (IP) and the exit plenum (EP) of a parallel multi-channel system such as the one shown in Fig.2. The channel is divided into: inlet region (Im) single-phase region (S_m), two-phase region (T_m), and exit region (E_m). In the figure, P denotes pump, C the condenser, PH the preheater, PR the condenser pressure regulator, TP the inlet subcooling regulator, and the subscript λ denotes the boiling boundary.

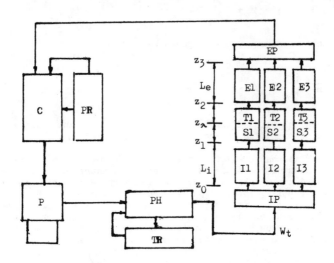

Fig. 2 Schematic diagram of, a parallel
three-channel system (D'Arcy, 1969)

The pressure drop across the m'th channel may be written as two parts

$$\Delta P_m = \Delta P_{1m} + \Delta P_{2m} \tag{11}$$

where ΔP_{1m} is the pressure drop in the single-phase region, i.e., from the inlet plenum to the boiling boundary, ΔP_{2m} the pressure drop in the two-phase region, i.e. from the boiling boundary to the exit plenum. In the single phase region, the pressure drop can be obtained by integrated the momentum equation from the inlet to the boiling boundary

$$\Delta P_1 = \frac{1}{2} K_i \rho_1 V_i^2 + L_i \left(\rho_1 \frac{\partial V_i}{\partial t} + g \rho_f + \frac{f}{2D_i} \rho_1 V_i^2 \right)$$

$$+ \int_{z_1}^{z_\lambda} \rho_1 \left(\frac{\partial V_i}{\partial t} + g + \frac{f}{2D_n} V_i^2 \right) dz \tag{12}$$

where K_i is the inlet restriction, f the friction coefficient of subcooled liquid, ρ_l the density, L_i the length of the inlet region, D_i and D_h are the hydraulic diameters of the inlet region and the heated region, respectively.

In the two-phase region, the pressure drop is

$$\Delta P_2 = \int_{z_\lambda}^{z_3} \left[\rho_m \left(\frac{\partial V_m}{\partial t} + V_m \frac{\partial V_m}{\partial z} + g_z \right) + \frac{f\phi_{lo}^2}{2D_h\rho_f} G_m^2 \right.$$

$$\left. + \frac{\partial}{\partial z} \left(\frac{\rho_f - \rho_m}{\rho_m - \rho_g} \frac{\rho_f \rho_g}{\rho_m} V_{gj}'^2 \right) \right] dz$$

$$+ \left(\frac{K_e}{2\rho_f} + \frac{fL_e}{2D_e\rho_f} \right) \phi_e^2 G_e^2 \tag{13}$$

where ρ_m is the mixture density, V_m the mass velocity of the mixture, G_m the mass flux, D_e the hydraulic diameter of exit region, ϕ_{lo}^2 and ϕ_e^2 are respectively the two-phase friction loss multiplier in the two-phase region and in the exit region, which take the form

$$\phi^2 = A(G,P) + B(G,P) X^a + C(G,P) X^b \tag{14}$$

In Eq.(14) X is the quality, the coefficients A, B, C, and the constants a and b depend on the correlation used.

Consider the system shown in the Fig.2, the pressure drop of external loop from the exit plenum to the inlet plenum is written as two parts

$$\Delta P_{ext} = \Delta P_{pump} + \sum_{\nu=1}^{L} \frac{1}{2} K_\nu \frac{G_t^2}{\rho_l} \tag{15}$$

where ΔP_{pump} represents the pressure rise of pump, K_ν is the lumped pressure drop coefficients which account for the local and friction losses in the ν'th segment of external loop.

Applying the linearization and Laplace-transformation techniques to the Eqs. (1)-(15), we can obtained the transfer functions of pressure drop perturbation versus flow rate perturbation. They can be written as

$$\delta\Delta P_{1m} = G_{1m}(s, K_{m1}, K_{m2}, \ldots, K_{mJ}) \delta W_m - \delta R_{1m} \tag{16}$$

$$\delta\Delta P_{2m} = G_{2m}(s, K_{m1}, K_{m2}, \ldots, K_{mJ}) \delta W_m - \delta R_{2m} \tag{17}$$

$$\delta \Delta P_{ext} = G (s, K_{t1}, K_{t2}, \ldots, K_{tL}) \delta W_t + \delta R_t \qquad (18)$$

$$m = 1, 2, \ldots, M$$

where $\delta \Delta P_{1m}$, $\delta \Delta P_{2m}$ and $\delta \Delta P_{ext}$ are respectively the pressure drop   in the   single phase region, the two-phase region, and in the external loop, $\{K_{m1}, K_{m2}, \ldots, K_{mJ}\}$ are the channel parameters in the m'th channel, $\{K_{t1}, K_{t2}, \ldots, K_{tL}\}$ are the external loop parameters, δR_{1m}, δR_{2m} and δR_t denote the lumped external input perturbations entering the system in the correspondent regions. Perturbating the Eg.(10) yields

$$\delta W_t = \sum_{m=1}^{M} \delta W_m \qquad (19)$$

where δW_t is the perturbation of the total flow rate.

4.   SYSTEM TRANSFER MATRIX MODEL AND STABILITY CRITERION

The Eqs. (16)-(19) can be manipulated into the matrix form (Ruan, 1986b)

$$\underline{\delta W} = Q(s, \ldots) \underline{\delta \Delta Pe} \qquad (20)$$

$$\underline{\delta \Delta Pe} = \underline{\delta R} - F(s, \ldots) \underline{\delta W} \qquad (21)$$

where $\underline{\delta W}$ is the flow rate vector, $\underline{\delta \Delta Pe}$ the pressure drop vector, $\underline{\delta R}$ the lumped external perturbation entering the system, $Q(s, \ldots)$ and $F(s, \ldots)$ are the system transfer matrices of MXM dimensions

$$F(s, \ldots) = (f_{mn}) \qquad (22)$$

$$Q(s, \ldots) = (q_{mn}) \qquad (23)$$

and the elements are

$$f_{mn} = \delta_{mn} \cdot G_{1m}(s, K_{m1}, K_{m2}, \ldots, K_{mJ}) \qquad (24)$$

$$q_{mn} = \delta_{mn} \cdot G_{2m}(s, K_{m1}, K_{m2}, \ldots, K_{mJ})$$
$$+ C_{mn} \cdot G_t(s, K_{t1}, K_{t2}, \ldots, K_{tL}) \qquad (25)$$

where C_{mn} are contants which relate to the channel flow rates in the steady state. The Eqs. (20) and (21) represent a standard feedback configulation of multi-input and multi-output system. Therefore, the stability of the system can be analized by the multivariable control theory.

According to the generized Nyquist stability theory (MacFarllane, 1980), a general stability criterion for density wave osicillations of parallel multi-channel system has been obtained (Ruan, 1986)

$$\sum_k N_k = P_o \qquad (26)$$

This criterian can be stated in the following way: the multi-channel system which can be presented by the Eqs.(20) and (21) will be stable, if and only if, the net sum of anticlockwise encirclements of the critical point (-1, jo) by the set of characteristic loci of the matrix $Q(s, \ldots) F(s, \ldots)$ is equal to the total number of the right-half plane poles for the open-loop system. Characteristic loci are also called generized Nyquist diagrams, which can be effectively computed and drawed by some softwares.

In our case, P_o can be proved to be zero (Ruan, 1986b). Therefore, the relative stability parameter K_g can be defined such that its absolute value is equal to the distance between the critical point and the nearest cross point at which the locus goes through the real axis. If the loci encircle the critical point, the system is unstable and K_g takes negative sign; otherwise, the system is stable and K_g takes positive sign. The greater the algebraic value of K_g, the more stable the system. Some of generized Nyquist diagrams are shown in Fig.3-6, the system parameters are given on the Table 1.

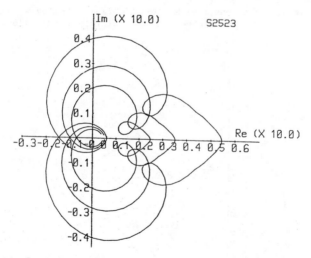

Fig.3   The Generized Nyquist Diagrams of a Three-channel System with Unequal Heating Power

323

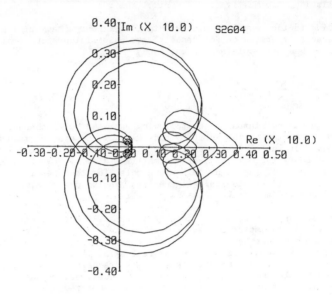

Fig.4   The Generized Nyquist Diagrams of a Three-
          channel System with Unequal Inlet Restrictions

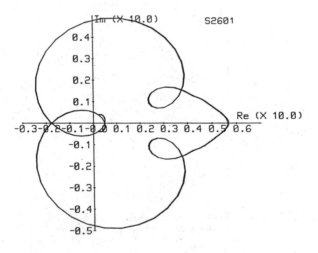

Fig.5   The Generized Nyquist Diagrams of a Three-channel
          System with Identical Channel Parameters

324

Table 1   System Parameters for Numerical Examples

Channel number	$M = 3$
Length of inlet region	$L_i = 2.4638$ m
Length of heated region	$L_h = 3.429$ m
Length of exit region	$L_e = 1.2446$ m
Flow area of inlet region	$A_{ci} = 6.2819893 \times 10^{-5}$ m^2
Flow area of heated region	$A_{ch} = 1.5390873 \times 10^{-5}$ m^2
Flow area of exit region	$A_{ce} = 6.2819893 \times 10^{-5}$ m^2
Response time constant of heater	$T_c = 0.8$ sec
System pressure	$P_s = 68.947$ bar
Inlet subcooling	$T_{sub} = 0.0 \quad 72.0$ $^\circ$C
Heating power	$Q = 40.0 \sim 115.0$ KW/channel
Inlet mass flux	$G = 300.0 \sim 450.0$ Kg/m^2.sec
Inlet restrictions	$K_{in} = 2.0 \sim 50.0$
Exit restrictions	$K_e = 0.2 \sim 30.0$
Nondimensional pump slope	$K_p = 0.0 \sim 20.0$
Lumped loss coefficient	$K_t = 0.2 \sim 50.0$

5.   RESULTS AND DISCUSSION

Figure 6 show a stability boundary map on a N_{sub}--N_{pch} plane, which gives a comparision of the predicted stability boundary lines obtained by different models with experimental data. The system parameters are given on the Table 1. As the figure shown, the predicted stability boundary line by multivariable frequency domain method (MFPM) are quite agreement with the experimental data in the medium and high inlet subcooling conditions. The line predicted by the model which accounts for unequal heating power is more close to the line of the experimental data than that with equal heating power. However, they are both better than the results obtained by single channel single variable method (SCSVM) and the simple criterion (Saha, 1976). In the figure, N_{sub} and N_{pch} are respectively the subcooling and phase-change numbers (Ishii, 1970), which are computed by the channel average parameters in the multi-channel cases.

Figure 7 gives a comparison of stability margins between the cases of equal and unequal heating power under the condition with or without coupling.

In Fig.8 the effects of unequal inlet restriction on stability margin are shown for a parallel three-channel system (see Table 1). It can be seen from the figure that the increase in the difference of the inlet restriction decrease the system stability when other parameters are held constant.

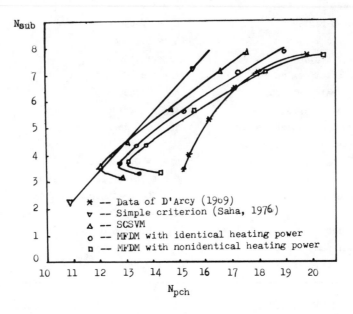

Fig.6   Stability Plane and Comparisions with the Results by
        Other Model and Experimental Data

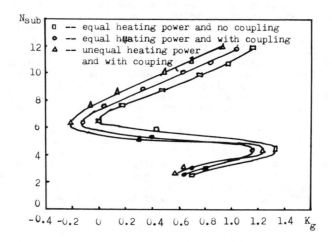

Fig.7   Comparision of Stability Margins Between the
        Cases of Equal and Unequal Channel Heating Power

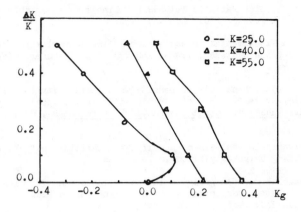

Fig.8   Effects of Unequal Inlet Restrictions
(K_c=10.0, N_{sub}=5.376, N_{pch}=15.822)

6.   CONCLUSIONS

In this paper, an investigation is presented on the density wave instability of a parallel multi-channel system with nonidentical operating conditions. The analysis is based on the multivariable frequency domain method
(MFDM). According to this study, a new version of the computer program MADWO
has been developed. The new mathematical model accounts for the nonuniform
distribution of heating power, flow splitting, and the coupling interaction
between the parallel channels and its external loop. Numerical examples show
that the predicted results are agreement with the previous experimental data
and better than the results obtained by the traditional single channel and
single variable method. We found that, although the total heating power are
held to be unchanged, the heating power difference between channels may
significantly affect the stability of parallel multi-channel systems, in
particular the ones with a stiff pump characteristic line.It was also found
that the increase in the difference of the inlet restrictions decreases the
system stability when other parameters are held constant.

7. REFERENCES

D'Arcy, D.F., "An Experimental Investigation of Boiling Channel Flow Stability",
Symp. On Two-phase Flow Dynamics, EUR-4288e, Vol. 2, pp.1173-1224, 1969.

Ishii, M. and Zuber, N., "Thermally Induced Flow Instabilities in Two-Phase
Mixtures," 4th International Heat Transfer Conference, Paris, Paper No. B5-11
(1970)

MacFarlane, A.G.J., Complex Variable Methods for Linear Multivariable Feedback Systems, Taylor & Francis Ltd., London, 1980.

Saha, P., et al, "An Experimental Investigation of the Thermally Induced Flow
Oscillations in Two-phase Systems", Trans. ASME, J. Heat Transfer, 616-622
(1976)

Ruan, Y.Q., and Fu, L.Z., "Analysis of Density Wave Instability in Multi-
channel Boiling Flow System", Proc. of the 4th Miami Int. Symposium (December
15-17, 1986) on Multiphase Transport and Particulate Phenomena, Hemisphere,
Vol.3, pp.69-79, 1988.

Ruan, Y.Q., "Multivariable Analysis of Density Wave Oscillations", M.S.
Thesis, Xi'an Jiaotong University, 1986[b], (In Chinese)

Zuber, N., and Staub, F.W., "An Analytic Investigation of the Transient
Response of the Volumetric Corcentration on a Boiling Forced-Flow System",
Nucl. Sci. Eng., 268-278 (1967).

An Investigation of Two-Phase Flow Pressure Drop Type Instability in the Vertical Upflow Tube

BAO-HUA XU, XUE-JUN CHEN, and TIN-KUAN CHEN
Xi'an Jiaotong University
Xi'an, PRC

NEJAT VEZIROĞLU and SADIK KAKAÇ
Clean Energy Research Institute
University of Miami
Coral Gables, Florida, USA

ABSTRACT

An experimental and theoretical study of two-phase flow pressure drop type instability in vertical upflow tube has been made at medium-high pressure water and uniform heated conditions. Using Lumped method to formulate the pressure drop type instability system governing equation into the two-order differential equation. According to the theory of the equation stability, the pressure drop type instability only occurs in the region of the negative slope in the curve of steady-state system pressure drop vs. mass flow rate. The equation is solved by means of the four-order of Rung-Kutta method and the effects of the various parameters on the pressure drop type instability are investigated, the result is in general agree with the experiment.

INTRODUCTION

Water-vapor two-phase flow and heat transfer has been widely used in many important areas. The instability phenomena of two-phase flow is important not only influence the safety operation of equipment, but also restriction the operating parameters. In recent years, as problems of two-phase flow extensively investigated, a lot of experimental and theoretical work about two-phase flow instability has been done.

After Ledinegg (1) successful analyzing static two-phase flow instability, systematic investigation about two-phase flow instability started. Later, Stenning and Veziroglu (2) distinguished three types of two-phase flow dynamic instability: pressure drop type instability, density wave type instability and thermal type instability. They give well explanation of the mechanism of pressure drop type instability. The pressure drop type instability is compound dynamic instability generated by the operation of static excursive effects in cooperation with the effects of compressible volume inertia and system delayed feed-backs. The pressure drop type instability will occur in a system with a sufficient upstream compressible volume, and when the system pressure drop decreases with an increase of mass flow rate (3). According to the chracteristics of pressure drop type instability, the assumption of quasi-steady state was obtained. Basis on these achievement, Ozawa (4) using Lumped-method investigated phenomena of pressure drop type instability. The equation was solved by the method of isoclines. But the analytical result of oscillation period was much less than the experimental ones.

Though the analysis about two-phase flow pressure drop type instability has been achived great development, but not all the aspects associated with the flow instability are fully understood, especially the bility using correct model to study the mechanism of dynamic characteristic is still limited. This paper carried out the experimental and theoretical work about water-vapor two-phase flow pressure drop type instability. By formulating and analyzing of the model, the pressure drop type instability only occurs in the region of negative slope in the curve of steady-state system pressure drop vs. mass flow rate. The equation was solved by means of four-order Rung-Kutta method and effect of various parameters on pressure drop type instability were investigated. The analyzing result is in general agree with the experimental ones.

EXPERIMENTAL SYSTEM AND TEST PROCEDURE

The experiment was carried out in a high pressure water-vapor system. The experimental apparatus and working loop are shown in figure 1. From the ion exchanger, the distilled water via water container and filter gets into high pressure water pump. Untill reached the required value of the system pressure, one part of water comes back to the water container by through the system bypass tube, the another part of water passes through the system orifice flowmeter into the preheater section. The total heating power of the preheater is 400 kw. Water is heated by preheater and after got our required experimental inlet temperature then flows into the inlet surge tank which provided the experimental inlet compressible volume by filling up nitrogen from the top of the surge tank. Later, one part of water flows out via the test bypass section, another part of water flows into the test section. The schematic diagram of the test section is shown in figure 2. The test section is a vertical 1Cr18Ni9Ti stainless steel tube with 16 mm O.D and 12 mm I.D. It is heated directly by A. C power and the total heating power is 120kw. Two sheathed thermocouples and pressure conduct chambers are installed in the inlet and outlet of heater section to measure the water temperature and pressure. Using NiCr-NiSi thermocouples which are symmetrically distributed along the heater section to measure the wall temperature of heater tube. An orifice flowmeter is installed in the inlet of test section to record the variation of the inlet mass flow rate during oscillation. Water heated by heater section and brought to a pre-determined value flows through the outlet orifice then joins with the water flowed through the test bypass section, and then passes through the outlet surge tank into the condenser. The cool water finally comes back to the water container.

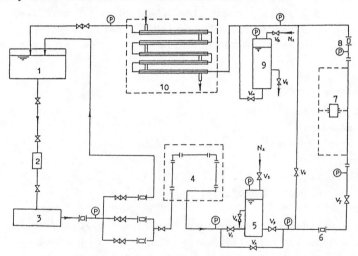

Fig 1 Schematic diagram of experimental system
1 Water container 2 Filter 3 Water pump
4 Preheater 5 Inlet surge tank 6 Inlet orifice
7 Heater 8 Outlet orifice 9 Outlet surge tank
10 Condenser

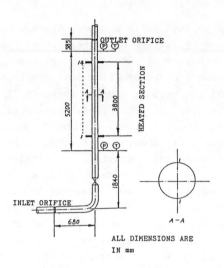

Fig 2 Schematical diagram of the test
section

During the experiments, the system pressure, inlet mass flow rate and inlet water temperature was keeped constant. The test section was heated gradually and decided whether or not oscillation appearance by observed the inlet mass flow rate. In this experiments, the bypass valve V6 and inlet valve of outlet surge tank V10 was always closed.

EXPERIMENTAL RESULT

Range of test parameters:

System pressure:	30-70	bar
Inlet mass flow rate:	600-1100	kg/m^2s
Heat flux:	0-700	kw/m^2
Inlet subcooling:	13-100	^{o}C
Diameter of the outlet orifice:	4-6	mm
Test tube:	16x2	mm

From the experiments, the pressure drop type instability only occured in the region of negative slope of the curve of steady-state system pressure drop vs. mass flow rate. The oscillation period is relatively long. The pressure drop type instability always accompanied with the density wave type instability. The wall temperature fluctuated outphase with the inlet mass flow rate and in certain conditions, the amplitude of wall temperature is

very high. The typical trace of inlet mass flow rate and wall
temperature vs. time during pressure drop type oscillation is shown
in figure 3,4, and 5. The corresponding steady-state curve of figure
3 of system pressure drop vs. mass flow rate which was calculated
by homogenious model method is shown in figure 6.

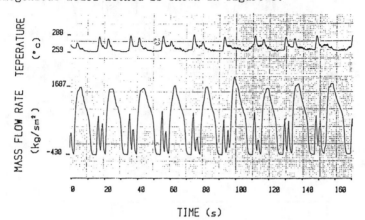

Fig 3

$\beta = 0$ .333   P= 30 (bar)   T= 145.9 (°c)
G= 870.639 (kg/sm²)   Q= 390.772 (kv/m²)

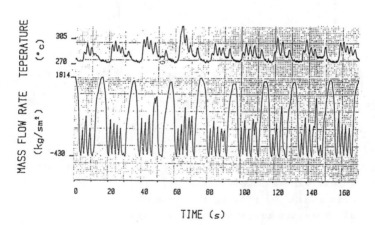

Fig 4

$\beta = 0$ .333   P= 30 (bar)   T= 145.9 (°c)
G= 1081.963 (kg/sm²)   Q= 475.312 (kv/m²)

333

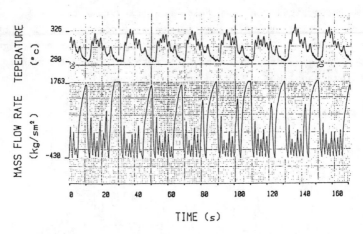

Fig 5

β=0 .333   P= 30 (bar)   T= 145.9 ($^\circ$c)
G= 1011.032 (kg/sm^2)   Q= 558.161 (kw/m^2)

MASS FLOW RATE (kg/sm^2)

Fig 6

β=0 .333   P= 30 (bar)   T= 145.9 ($^\circ$c)
Q= 390.772 (kw/m^2)

THEORETICAL ANALYSIS

1   ESTABLISHMENT OF THE MODEL AND ANALYSIS

In figure 7 an analytical model of the experimental system is shown. Making following assumptions:

(1)   Inlet mass flow rate W_O is constant.

(2)   Outlet system pressure Pe is constant.

(3)   Nitrogen in the surge tank is an ideal gas.

According to the above assumptions and the notations in figure 7, the following equations of the system can be written:

334

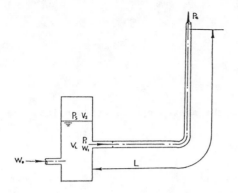

Fig 7   Schematical diagram of pressure drop instability

Continuity equation of the surge tank:

$$W_1 - W_0 = -\rho_f dV_L/dt \qquad (1)$$

Momentum equation of the surge tank:

$$P_s - P_1 = (H/A_s) dW_1/dt \qquad (2)$$

State equation of the surge tank:

$$P_s V_s = const \qquad (3)$$

Volume equation of the surge tank:

$$V_L + V_s = const \qquad (4)$$

Momentum equation of the system:

$$P_1 - P_e = f(W_1) + (L/A) dW_1/dt \qquad (5)$$

Substitued equation (3) and (4) into equation (1), We obtained:

$$W_1 = W_0 - (\rho_f V_s/P_s) dP_s/dt \qquad (6)$$

Now we introduce the following parameters:

$$I_1 = H/A_s$$

$$I_2 = L/A$$

$$C_s = \rho_f V_s / P_s$$

So the equations of the system can be written into the following form:

$$P_s - P_1 = I_1 dW_1/dt$$

$$P_1 - P_e = I_2 dW_1/dt + f(W_1) \qquad (7)$$

$$W_1 - W_0 = -C_s dP_s/dt$$

Eliminating P_s and P_1 from the equation (7), we get:

$$(I_1 + I_2) d^2 W_1/dt^2 + (df(W_1)/dW_1)(dW_1/dt) + (W_1 - W_0)/C_s = 0 \qquad (8)$$

Now, we obtain two-order constant differential equation which described the inlet mass flow rate vs. time during pressure drop type instability. It will be disscused for two cases in the following section.

(1) The curve of steady-state system pressure drop vs. mass flow rate has no negative slope region.

In this case, we assume:

$$df(W_1)/dW_1 = C_1 W_1 + C_2 \qquad (9)$$

Substitute equation (9) into equation (8), we obtain:

$$(I_1 + I_2) d^2 W_1/dt^2 + (C_1 W_1 + C_2) dW_1/dt + (W_1 - W_0)/C_s = 0 \qquad (10)$$

Let

$$W^* = (W_1 - W_0)/W_0$$

$$t^* = t/(C_s(I_1 + I_2))^{\frac{1}{2}}$$

$$E_1 = C_1 W_0 (C_s/(I_1 + I_2))^{\frac{1}{2}}$$

$$E_2 = 1 + C_2/(C_1 W_0)$$

So the equation (10) can be written into the following form:

$$d^2W^*/dt^{*2}+E_1(W^*+E_2)dW^*/dt^*+W^*=0 \tag{11}$$

For convenience of writing, let:

$$y=W^*$$

$$t=t^*$$

We obtain:

$$d^2y/dt^2+E_1(y+E_2)dy/dt+y=0 \tag{12}$$

It's equivalent with the following equations:

$$dx/dt=-E_1E_2x-y-E_1xy \tag{13}$$

$$dy/dt=x$$

The roots of the eigon value of one-order approximate equation are:

$$\lambda=(-E_1E_2\pm(E_1^2E_2^2-4)^{\frac{1}{2}})/2 \tag{14}$$

With the conditions of the curve having upface valley and monotonicly increasing, we obtain:

$$c_1>0 \tag{15}$$

$$c_1W_0+c_2>0$$

Then we obtain:

$$E_1>0 \tag{16}$$

$$E_2>0$$

The real part of the eigon value of one-order approximate equation (12) is negative. According to the Liapunov's theory of equation stability, the equation (12) is asymptotic stable. So the system is stable when the curve of steady-state system pressure drop vs. mass flow rate has no negative slope region and the pressure drop type instability will not occur.

(2) The curve of steady-state system pressure drop vs. mass flow rate has a negative region.

In this case, we assume:

$$df(W_1)/dW_1 = C_1(W_1-W_a)(W_1-W_b) \tag{17}$$

Where: W_a: the point of the function $f(W_1)$ has a maximum value
W_b: the point of the function $f(W_1)$ has a minimum value
C_1: a positive constant

Substitute equation (17) into equation (8), we obtain:

$$(I_1+I_2)d^2W_1/dt^2 + C_1(W_1-W_a)(W_1-W_b)dW_1/dt + (W_1-W_0)/C_s = 0 \tag{18}$$

Let:

$$W^* = (W_1-W_0)/W_0$$

$$t^* = t/(C_s(I_1+I_2))^{\frac{1}{2}}$$

$$W_a^* = (W_0-W_a)/W_0$$

$$W_b^* = (W_b-W_0)/W_0$$

$$E_1 = (C_s/(I_1+I_2))^{\frac{1}{2}}C_1W_0^2W_a^*W_b^*$$

$$E_2 = (W_a^*-W_b^*)/(2(W_a^*W_b^*)^{\frac{1}{2}})$$

$$y = W^*/(W_a^*W_b^*)^{\frac{1}{2}}$$

For convenience of writing, let:

$$t = t^*$$

So the equation (18) can be written as following form:

$$d^2y/dt^2 + E_1(y^2+2E_2y-1)dy/dt + y = 0 \tag{19}$$

338

It is equivalent with the following equations:

$$dx/dt = E_1 x - y - E_1(2E_2 y + y^2)x \qquad (20)$$

$$dy/dt = x$$

The roots of the eigon value of one-order approximate equation are:

$$\lambda = (E_1 \pm (E_1^2 - 4)^{\frac{1}{2}})/2$$

When $W_a < W_o < W_b$, then $E_1 > 0$, the real part of the eigon value of one-order approximate equation (19) is positive. According to Liapunov's theory of equation stability, the equation (19) is unstable.

When $W_o > W_b$ or $W_o < W_a$, then $E_1 < 0$, the real part of the eigon value of one-order approximate equation (19) is negative. According to Liapunov's theory of equation stability, the equation (19) is asymptotic stable.

2   Comparison between the numerical calculation and the experimental results

The equation (19) was solved by four-order Rung-Kutta method. The numerical result is shown in figure 8 and the experiment result is shown in figure 3. Comparison two figures, the analyzing result is in general agree with the experimental ones.

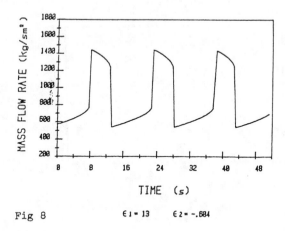

Fig 8 $\epsilon 1 = 13$ $\epsilon 2 = -.684$

3   The effect of the various parameters on the pressure drop type
instability

(1) The effect of the inlet compressible volume on the pressure
drop type instability

When the other parameters keep constant, the effect of the
inlet compressible volume on the pressure drop type instability can
be analyzed by discussing the two dimensionless parameters E_1 and
E_2. The dimensionless parameter E_1 vs. inlet compressible volume is
shown in figure 9. The analyzing result is shown in figure 10 and
figure 11. The period of pressure drop type instability increases
with inlet compressible volume.

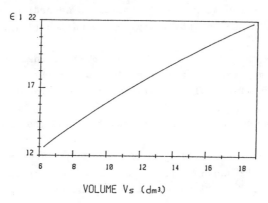

VOLUME Vs (dm³)

Fig 9   β=0 .333   P= 30 (bar)   T= 145.9 (°c)
D= 390.772 (kw/m³)   E₂=-0.604

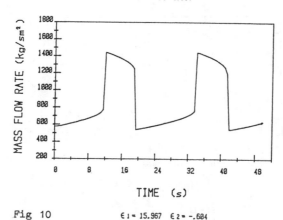

TIME (s)

Fig 10   E₁= 15.967   E₂= -.604

340

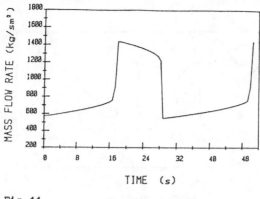

Fig 11 $\quad \epsilon_1 = 19.49 \quad \epsilon_2 = -.604$

(2) The effect of the inlet mass flow rate on the pressure drop type instability

The dimensionless parameters E_1 and E_2 vs. inlet mass flow rate is shown in figure 12. The analyzing results are shown in figure 13, 14 and 15. The period of pressure drop type instability nonmonotonicly increase with inlet mass flow rate. When $E_2>0$ the period of pressure drop type instability increases with inlet mass flow rate, when $E_2<0$ the period of pressure drop type instability decreases with inlet mass flow rate, when $E_2=0$ the period has a minimum value. As increasing in mass flow rate, the part of crest in one wave length increases with inlet mass flow rate during pressure drop type instability.

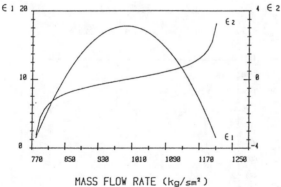

Fig 12 $\quad \beta=0.333 \quad P=30 \text{ (bar)} \quad T=145.9 \text{ (°c)}$
$Q=390.772 \text{ (kw/m}^2\text{)}$

341

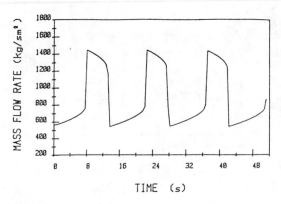

Fig 13 $\epsilon_1 = 14.591$ $\epsilon_2 = -.466$

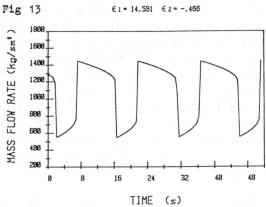

Fig 14 $\epsilon_1 = 13.889$ $\epsilon_2 = .528$

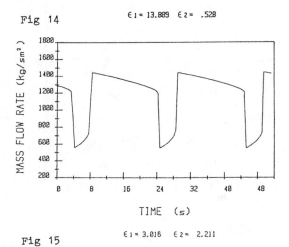

Fig 15 $\epsilon_1 = 3.016$ $\epsilon_2 = 2.211$

CONCLUTION

(1) Pressure drop type instability only occurs in the region of negative slope of the curve of steady-state system pressure drop vs. mass flow rate. The period of pressure drop type instability is relatively long. The wall temperature fluctuates outphase with the inlet mass flow rate and in certain conditions, the amplitude of wall temperature is very high.

(2) The pressure drop type instability always accompanied with the density wave type instability.

(3) The period of pressure drop type instability increases with the inlet compressible volume, but the correlation of the period and inlet mass flow rate is nonmonotonic.

(4) The part of crest in one wave length increases with inlet mass flow rate during pressure drop type instability.

(5) The analytical result is in general agree with the experimental ones.

ACKNOWLEDGEMENT

This work is supported by the Natural Science Foundation of China

NOMENCLATURE

A:	area	m^2
L:	length	m
H:	height	m
V:	volume	m^3
ρ:	density	kg/m^3
P:	pressure	N/m^2
W:	mass flow rate	kg/s
t:	time	s
T:	inlet temperature	$^\circ C$
Q:	heat flux	kw/m^2
G:	mass flow velocity	$kg/m^2 s$
β:	ratio of the diameter of outlet orifice	

Subscribe

e:	exit
f:	fluid
s:	surge tank

REFERENCES

(1) Ledinegg. M. " Instability of flow during natural and forced circulation " Die Wärme. 61. 891. 1938

(2) Stenning. A. H. and Veziroglu. T. N. " Flow oscillation modes in forced convection boiling " Proc. 1965 Heat Transfer and Fluid Mechanics Institute Stanford University Press 301 1965

(3) T. N. Veziroglu and S. Kakac " Two-phase flow instability " Final Report NSF Project CME 79-20018

(4) Ozawa " Flow instabilities in boiling channels " Bulletin of JSME Vol 22 No 170 1979

Experimental Research on Pressure Drop Type Oscillation in Medium-High Pressure System

QIAN WANG and XUE-JUN CHEN
Engineering Thermophysics Research Institute
Xi'an Jiaotong University
Xi'an, PRC

ABSTRACT

This paper described the experimental results of pressure drop type oscillations in medium-high pressure system. The experimental results show that the pressure drop type oscillations occur only when the negative slops of steady state curves of the total pressure drop versus mass flow rate are high enough. The inlet subcooling is one of the most important factors for pressure drop type oscillations. The frequency of the pressure drop type oscillations is about 0.05-0.1 Hz. The oscillation of mass flow rate is always accompannied by wall temperature oscillation.

INTRODUCTION

Two-phase flow instability is one of the most important problems in two-phase flow system. Once the system instability occured, the flow rate and the pressure drop oscillated, and the wall temperature could increase over the value limited by the wall material. The safe operating regime of two-phase flow system can be determined by the instability threshold values of such system parameter as flow rate, pressure, wall temperature and exit mixture quality.

The pressure drop type oscillation is one of the main instability modes in two-phase flow system, which occurs in regime of decreasing the total pressure drop with increasing flow rate. The oscillation periods are about 20s governed by the volume and compressibility of the vapor in the system. The pressure drop type oscillation have been investigated by Stenning, [1], Veziroglu, [2], Ozawa, [3] and Yildirim, [4]. But in the literature, there are rare the results for pressure drop type oscillation in high pressure system.

In this paper, the pressure drop type oscillations have been investigated in medium-high pressure system. The experiments have been carried out for various system pressure, inlet subcoolings, mass flow rates and heat inputs in water as working fluid.

APPARATUS AND TECHNIQUES

The experimental loop is shown in Fig.1. The test section is single vertical tube with 12mm inner diameter. The surge tank acts as a capacitance, and is an important dynamic component of the loop during pressure drop type oscillation. The surge tank is connected to a pressurized nitrogen tank to provide control for the liquid level in it.

An orifice is put at exit of test section, different diameter orifices can be changed for investigations. The different orifice diameter affects the threshold of pressure drop oscillation quite differently.

Both average and transient flow rate and total pressure drop have been measured. The wall temperature have been measured by 28 thermocouples connected on the tube wall. The experimental data have been obtained by data collecting system connected with IBM computer.

The experiments have been carried out at wide region of parameter, as follows:

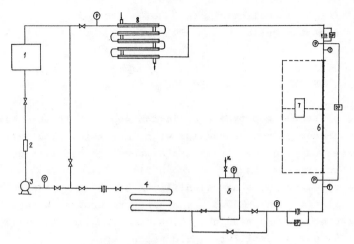

1-water tank   2-filter   3-pump   4-preheater
5-surge tank   6-test section   7-electric heater
8-condenser   ⋈ -valve   ⫘-orifice

Fig.1   Experimental Loop

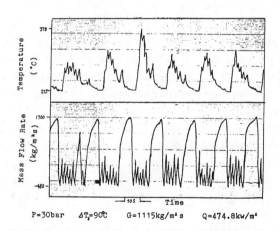

P=30bar ΔT_s=90℃   G=1115kg/m²s   Q=474.8kw/m²

Fig2 Oscilation of Mass Flow Rate and
Wall Temperature

347

```
Pressure  P:              30-70       bar
Mass flow rate G:         600-1300    kg/m² s
Heat input Q:             0-600       kw/m²
Inlet subcooling ΔT_sub:  10-90       °C
Diameter ratio β :        0.33, 0.417, 0.5
```

DISCUSSION OF EXPERIMENTAL RESULTS

The pressure drop type oscillation appeared in the region
of total pressure drop decreasing with increasing flow rate.
The typical curve of flow rate and temperature oscillation
obtained in experiments are shown in Fig.2.

The diameter ratio has significant effect on pressure drop
type oscillation. When β=0.33, the pressure drop type oscillation sustained with period about 20s. β increasing to 0.417,
pressure drop type oscillation did not occur.

Decreasing diameter ratio causes pressure drop in two-
phase regime increasing greatly. The curves of pressure drop
versus flow rate are steepened, see Fig.3. So with the same
increasing vapor quality, mass flow rate varies greatly. the
mass flow rate fluctuation feedback to pressure drop and vapor
quality, the oscillations sustain.

The region of pressure drop type oscillation decreased with
decreasing inlet subcooling. When inlet subcooling Δt_sub=90 °C
the typical pressure drop oscillation curves were observed.
When operating point near the peak of pressure drop versus
mass flow rate curve, the pressure drop oscillation superimposed with density wave oscillations. At the same time, the
wall temperature could increase to 600°C or even more. Inlet
subcooling decreasing to Δt_sub=60 °C, the pressure drop oscillations did not superimposed with density wave oscillations, the
periods of oscillation were about 15s. Fig.4. When inlet subcooling equal 30°C, the pressure drop oscillations did not
observed.

The region of pressure drop oscillation also decreased

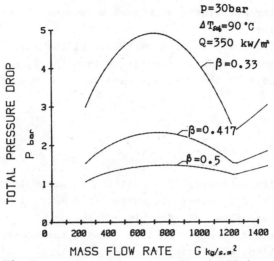

Fig.3   Comparison of The Steady-State
Characteristics With Different Diameter Ratio

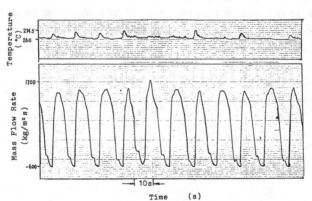

p=30bar ΔT_{sub}=60 °C   G=1115kg/m²s   Q=272.2kw/m².

Fig.4   Oscillation of Mass Flow Rate and Wall Temperature

349

with increasing system pressure. When system pressure equal 70 bar, the pressure drop type oscillations disappeared.

During the oscillation, the amplitude of the mass flow rate oscillation is quite large, and the periods varied from 13 to 20s, increasing with increasing heat inputs and decreasing system pressure.

During the mass flow rate oscillation, the wall temperature alsc has a oscillation, its amplitude is quite large. The wall temperature oscillation have a time delay with mass flow rate oscillation.

CONCLUSIONS

1. The pressure drop type oscillation occur only when the negative slops of steady-state curves of the total pressure drop versus mass flow rate are big enough.

2. The small diameter ratio β created the pressure drop type oscillation, which did not occur with large diameter ratio.

3. The pressure drop type oscillation superimposed with density wave oscillation when operating point near the peak of the pressure drop versus flow rate curve. At this time the amplitude of wall temperature oscillation was auite large. That is very dangerous for operating system.

4. The inlet subcooling is one of the most important factors for pressure drop pressure drop type oscillations. The pressure drop type oscillation disappeared as inlet subcooling decreasing.

5. In our experimental system, the periods of pressure drop type oscillations were in the range from 13 to 20s. The oscillation of wall temperature delayed with oscillation of mass flow rate.

REFERENCES

[1]. Stenning,A.H., and Veziroglu,T.N., Flow Oscillation
Modes in Forced-Convection Boiling, Proc. Heat Transfer and
Fluid Mechanics Institute, 1965 Vol.1, pp 301-316.

[2]. Veziroglu,T.N., Lee,S.S. and Kakac,S., Fundamentals
of Two-Phase Flow Oscillations and Experiments in Singgle
Channel System, Two-Phase Flow and Heat Transfer, Vol.1,
pp 423-466, 1977.

[3]. Ozawa,M., Nakanishi,S., Mizuta, Y. and Tarui, H.,
Flow Instabilities in Bioling Channels, Bulletin of JSME,
Vol.22, No.170, pp 1113-1116,1979.

[4]. Yildirim, O.R., Yuncu,H. and Kakac,S., The Analysis
of Two-Phase Flow Instabilities in a Horizontal Single Channel,
Thermal Sciences Proceeding of 16th Southeastern Seminar,
1982 Miami, pp 761-787.

LOCA AND OTHER
ACCIDENT PHENOMENA

Nonequilibrium Effects on the Two-Phase Flow Critical Phenomenon

S. M. SAMI*
Department of Mechanical Engineering
University of Sherbrooke
Sherbrooke, Quebec
Canada J1K 2R1

Abstract

The choking criterion for nonhomogeneous nonequilibrium two-phase flow has been obtained by solving the two-fluid conservation equations. The method of characteristics is employed to predict the critical flow conditions. Numerical results revealed that the proposed model reliably predicted the critical two-phase flow phenomenon and compared well with existing experimental data.

1. INTRODUCTION

In recent years considerable efforts have been made on both experimental and analytical levels to predict the two-phase flow critical conditions. Early modelling of critical two-phase flow has been based on an isentropic expansion with thermal equilibrium and equal velocities of the two-phase [1-2a] Thermal equilibrium models with empirical correlations for the different velocities of the two-phases were reported by Fauske [3], Zivi [4], Moody [5], and Cruver [6]. Nonequilibrium effects between phase in critical two-phase flow were studied by Starkman et al. [7] and Henry [8,9] and were reviewed in details by Furness [10]. Other comprehensive reviews of two-phase critical flow have been cited in the literature by Hsu [11], Henry et al. [12], Smith [13], and Weisman-Tentner [14]. On the other hand, Ardron and Furness [15] discussed some of these analytical models and compared them with various experimental data reported in the literature on critical two-phase flow. It appears from their comparative study that the majority of analytical models are in agreement with experimental data only over a restricted range of conditions.

Two-phase thermal-hydraulic system analysis codes such as RELAP 4 [16], RELAP 5 [17], RETRAN [18], SOPHT [19] and FIREBIRD [20] include a simple momentum or critical flow, sonic flow model, Moody critical flow, Henry-Fauske critical flow model and finally homogeneous equilibrium critical flow model to account for the critical flow phenomena [21].

Recently, a hydrodynamic model has been presented by Trapp and Ransom [22]. It is established after the two-fluid equations, relative phasic acceleration, nonequilibrium and derivative-dependent exchange of mass. Possible criticisms of their model involve the calculation of nonequilibrium mass and heat transfer terms, thermal resistance to heat flux and virtual mass coefficient, and the thermal equilibrium derivatives of the equation of state.

Analytical Model

The choking criterion presented in this model for nonhomogeneous, nonequilibrium two-phase flow is obtained by solving the two-fluid model conservation equations. The method of characteristics is employed to predict the

*Present address: Mechanical Engineering/School of Engineering, University of Moncton, Moncton, N.B., Canada, E1A 3E9

critical flow conditions. Critical flow is established after the magnitude of the characteristic slopes (velocities). Critical flow conditions are reached when the smallest characteristic slope becomes equal to zero.

The one dimensioned two-fluid, nonhomogeneous nonequilibrium, conservation equations can be written in the following forms after neglecting effects of surface tension, turbulent kinetic energy associated with phase change, and viscous energy dissipation [22,23]:

Mass continuity equations

$$\frac{\partial}{\partial t}(\alpha_g \rho_g) + \frac{\partial}{\partial x}(\alpha_g \rho_g v_g) = \Gamma_g \tag{1}$$

$$\frac{\partial}{\partial t}(\alpha_f \rho_f) + \frac{\partial}{\partial x}(\alpha_f \rho_f v_f) = \Gamma_f \tag{2}$$

$$\Gamma_f = -\Gamma_g \tag{3}$$

Momentum equations

$$\alpha_g \rho_g \left[\frac{\partial v_g}{\partial t} + v_g \left(\frac{\partial v_g}{\partial x}\right)\right] + \alpha_g \left(\frac{\partial p}{\partial x}\right) + \Gamma_g (v_{gI} - v_g)$$

$$+ C \alpha_g \alpha_f \rho \left[\frac{\partial v_g}{\partial t} + v_f \left(\frac{\partial v_g}{\partial x}\right) - \frac{\partial v_f}{\partial t} - v_g \left(\frac{\partial v_f}{\partial x}\right)\right]$$

$$+ \alpha_g \rho_g v_g^2 \frac{1}{A}\frac{dA}{dx} + Fi_{stat} = 0 \tag{4}$$

$$\alpha_f \rho_f \left[\frac{\partial v_f}{\partial t} + v_f \left(\frac{\partial v_f}{\partial x}\right)\right] + \alpha_f \left(\frac{\partial p}{\partial x}\right) + \Gamma_f (v_{fI} - v_f)$$

$$+ C \alpha_f \alpha_g \rho \left[\frac{\partial v_f}{\partial t} + v_g \left(\frac{\partial v_f}{\partial x}\right) - \frac{\partial v_g}{\partial t} - v_f \left(\frac{\partial v_g}{\partial x}\right)\right]$$

$$+ \alpha_f \rho_f v_f \frac{1}{A}\frac{dA}{dx} + Fi_{stat} = 0 \tag{5}$$

Energy equations

$$\alpha_g \left[(\rho_g \frac{\partial h_g}{\partial p})_{sat} -1\right] (\frac{\partial p}{\partial t} + v_g \frac{\partial p}{\partial x}) + \alpha_g \rho_g (\frac{\partial h_g}{\partial t} + v_g \frac{\partial h_g}{\partial x})$$

$$- Q_g + \Gamma_g \Delta h_g = 0 \tag{6}$$

$$\alpha_f \left[(\rho_f \frac{\partial h_f}{\partial p})_{sat} -1\right] (\frac{\partial p}{\partial t} + v_f \frac{\partial p}{\partial x}) + \alpha_f \rho_f (\frac{\partial h_f}{\partial t} + v_f \frac{\partial g_f}{\partial x})$$

$$+ Q_f + \Gamma_f \Delta h_f = 0 \tag{7}$$

where Q_g and Q_f are the interfacial heat transfer rate for gas and liquid respectively, v_{gI} and v_{fI} are the average interphase velocity for momentum transfer caused by mass transfer, and $C \, \alpha_g \, \alpha_f \, \rho$ is the coefficient of virtual mass.

It is assumed in this study that the interface velocities $v_{gI} = v_{fI} = v_I = 0.5 \, (v_g + v_f)$. This was recommended by Trapp-Ransom [22] and Wallis [23]. Q_g and Q_f are interrelated by the energy equation for the interface [14]:

$$\Gamma_g \, h_{fg} + Q_g + Q_f = 0 \qquad (8)$$

Expressions have been developed to determine the nonequilibrium mass and heat exchanges in terms of the system dependent parameters derivatives. In addition, comprehensive flow regime maps have been employed in the calculation of interfacial heat and momentum transfer rates.

The preceeding set of the conservation equations and their constitutive relations can be rewritten in the following form:

$$A(U) \left[\frac{\partial U}{\partial t}\right] + B(U) \left[\frac{\partial U}{\partial x}\right] + C(U) = 0 \qquad (9)$$

The characteristic velocities of the system of equations in equation (9) are obtained after the roots of the following characteristic equation;

$$\det \left[A\lambda - B \right] = 0 \qquad (10)$$

The real and imaginary parts of these roots (λ_n) give the velocity of propagation of small perturbations and the rate of growth or decay of propagation respectively. Critical conditions exist at;

$$\lambda_i \geq 0 \qquad (i \leqslant n, \; n = 6) \qquad (11)$$

3. RESULTS AND DISCUSSIONS

In this section, we describe the numerical solution of the proposed model as well as its validation. The preceeding mathematical formulations have been programmed in order to obtain the characteristic velocities of the system. The logical chart of this computer program is shown in Figure 1. Calculations started with the initialization of dependent variables and the definition of the time step. Then series of subroutines were called to calculate subcooled length and solve conservation equations. Once the vapor and liquid velocities were determined, the characteristic equation was solved and the six roots were compared to determine the critical node thermalhydraulic conditions such as: pressure, critical mass flow rate, quality, enthalpy, density and temperature.

Experimental data obtained at sharp-edged pipes (0.25") diameter [24] have been simulated by this model, compared with Malnes model [25] and plotted in Figure 2. It is quite clear from the results shown in this figure that cold subcooled water data were fairly predicted by our model as well as Malne's model. However, at lower subcooling, where the nonequilibrium effects are significant, the Malnes model poorly estimated the subcooled and saturated water data, while our model was in excellent agreement with the critical flow rate data. This is partly because our model takes into account the nonequilibrium effects between phases.

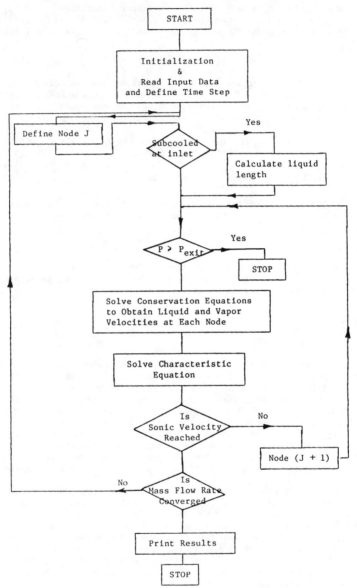

Figure 1: Logical flowchart for the present model

358

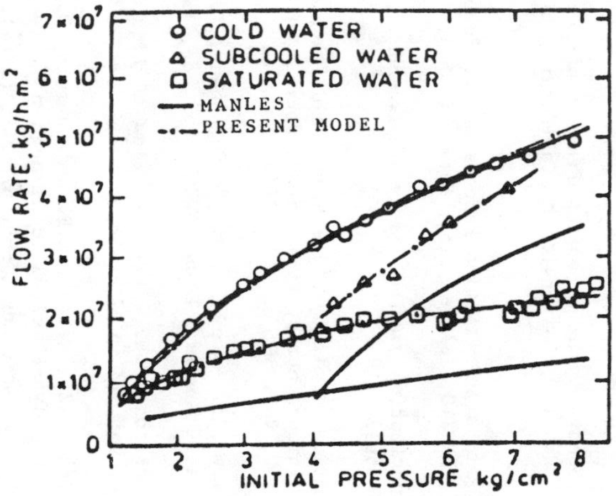

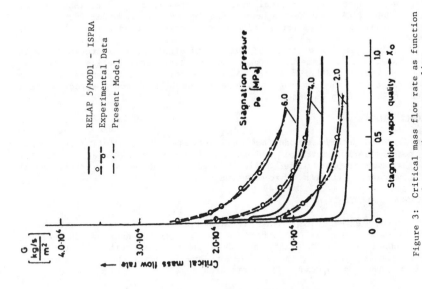

Figure 2: Sharp-edged pipes (0.25") diameter data compared with our model

Figure 3: Critical mass flow rate as function of stagnation vapor quality

RELAP 5 predictions have been compared with LOBI break nozzle calibration tests carried out at Westinghouse Canada [24]. These tests were performed for upstream pressures of 2.0 MPa to 8.0 MPa and for subcooled and saturated conditions with vapor qualities between 0.0 and 1.0.

Figure 3 compares the critical mass flow rates calculated by INEL version of RELAP 5/MOD1 [26] with Westinghouse experiments and our model for a nozzle with 1.0 A break (A = cross sectional area of nozzle). Good agreement between our model and LOBI data can be observed for all stagnation pressures. On the other hand, INEL version of RELAP 5/MOD1 showed significant discrepancies with the LOBI experiments. This is in part due to the underestimation of the critical velocity by the RELAP 5 model. Also, the comparison presented in Figure 2 demonstrated that the improved homogeneous equilibrium sound velocity model employed in RELAP 5 is inadequate.

4. CONCLUSIONS

An improved numerical model has been developed for predicting nonhomogeneous, nonequilibrium critical two-phase flow conditions. Numerical results showed that the proposed model reliably predicted the existing experimental data and compared well with other existing model.

5. ACKNOWLEDGMENTS

The author extends his thanks to Mrs. P. Jutras for typing this text. This research work has been supported by The Natural Sciences and Engineering Research Council of Canada Grant, No. U0444 and the University of Sherbrooke. Also, the author wishes to thank F. Mighri and T. Duong for their contributions during the development of this research work.

REFERENCES

[1]   Dancely, D.J., "A Study of the Expansion Process of Low Quality Steam Through a De Laval Nozzle", University of California, Livermore Report UCRL 6230 (1962).

[2]   Malnes, D., "Slip Ratios and Friction Factors in the Bubble Flow Regime in Vertical Tubes", KR-110, IFA, Kjeller, Norway (1966).

[3]   Fauske, H.K., "Contribution to the Theory of Two-Phase, One-Component Critical Flow", ANL-6633, (1962).

[4]   Zivi, S.M., "Estimation of Steady-State Steam Void Fraction by Means of the Principale of Minimum Entropy Production", Journal of Heat Transfer, Trans. ASME, Series C, Vol. 86, p. 247, (1964).

[5]   Moody, F.S., "Maximum Flow Rate of a Single Component, Two-Phase Mixture", Journal of Heat Transfer, Trans. ASME, Vol. 87, p. 134, (1965).

[6]   Cruver, J.E. and Moulton, R.W., "Critical Flow of Liquid-Vapor Mixtures", AICHE Journal, Vol. 13, n° 1, (1967).

[7]   Starkman, S., Schrock, V.Z., Neusen, K.F. and Maneely, D.J., "Expansion of a Very Low Quality Two-Phase Fluid Through a Convergent-Divergent Nozzle", Trans. ASME, J. Basic Eng., 86D, p. 247, (1964).

[8] Henry, R.E. and Fauske, H.K., "The Two-Phase Critical Flow of One-Component Mixtures in Nozzles, Orifices an Short Tubes", ASME Paper 70-WA/HT-5 (1971).

[9] Henry, R.Z., "A Study of One- and Two-Component Two-Phase Critical Flows at Low Qualities", ANL-7430, (1968).

[10] Furness, R.A., "A Review of Theoretical and Experimental Studies on the Critical Discharge Flow Rates of Initially Saturated Water", CEGB Report RD/B/N2997 (1976).

[11] Hsu, Y.Y., "Review of Critical Flow, Propagation of Pressure Pulse and Source Velocity", NASA Report, NASATND-6814 (1972).

[12] Henry, R.E., Grohmes, M.A. and Fauske, H.K., "Pressure Drop and Compressible Flow of Cryogenic Vapor Mixture Heat Transfer at Low Temperatures (Ed. Frost W.) Plenum Press, New York (1975).

[13] Smith, R.V., "Critical Flow for Cryogenic Fluids", National Bureau of Standards, NBS Tech. Note 633, (1973).

[14] Weisman, J. and Tentner, A., "Models for Estimation of Critical Flow in Two-Phase Systems", Progress in Nuclear Energy, Vol. 2, p. 183, (1978).

[15] Ardron, K.H. and Furness, R.A., "A Study of the Critical Flow Model Used in Reactor Blowdown Analysis", Nucl. Eng. and Des., 39, p. 257, (1976).

[16] Moore, K.V. and Retting, W.H., "RELAP 4: A Computer Program for Transient Thermal-Hydraulic Analysis", ANCR-1127, Aerojet Nuclear Company, (1973).

[17] Ransom, V.H. et al, "RELAP-MOD1, Code Manual, NUREG/CR 1826-EFF-2070, (1982).

[18] Moore, K.V. et al., "RETRAN-4 Program for One Dimensional Transient Thermalhydraulic Analysis of Complex Fluid Flow Systems", EPRI-CCM-5, Vol. 1-4, Electric Power Research Institute, (1978).

[19] Chang, Y.F., "A Thermalhydraulic System Simulation Model for the Reactor, Boiler and Heat Transport System (SOPHT)", CNS-37-2, Ontario Hydro, (1977).

[20] Lin, M.R. et al., "FIREBIRD III - Program Description", Atomic Energy of Canada Limited, AECL 7533, (1979).

[21] Sami, S.M., "Comparative Study on the Modelling of Code SOPHT and Other Transient Thermal-Hydraulic Codes", GAN-12, Montréal, (1982).

[22] Trapp, J.A. and Ransom, V.H., "A Choked Flow Calculation Criterion for Nonhomogeneous, Nonequilibrium, Two-Phase Flows", Int. J. Multiphase Flow", Vol. 8, n° 6, p. 669, (1982).

[23] Wallis, G.P., "One Dimensional Two-Phase Flow", McGraw-Hill, New York, (1969).

[24] Fauske, H.K., "The discharge of Saturated Water Through Tubes", Chem. Eng. Progr. Symp., 61, p. 210, (1965).

[25] Malnes, D., "Critical Two-Phase Flow Based on Non-equilibrium Effects", ASME Winter Annual Meeting, Houston, Texas, Nov. 30 - Dec. 5, (1975).

[26] Stadtke, H., Kolar, W. and Worth, B., "Assessment of RELAP 5 Chocking Model", Proceedings of 3rd International Topical Meeting on Reactor Thermal Hydraulics, Newport Rhode Island, U.S.A., Oct. (1985).

362

Transient Liquid Film Thickness during Drainage on a Wall

ALI NOURI-BORUJERDI
Department of Mechanical Engineering
Sharif University of Technology
Tehran, Iran

ABSTRACT

A theoretical study is presented for the deposition of the liquid film on a surface under an accelerated drainage. The results demonstrate that the film thickness approaches a constant value after some distance travelled by the liquid over the surface. The results are consistence with the previous experimental data. Furthermore, flowing of the liquid through a flow channel deposits more liquid on the wall when the channel diameter decreases.

INTRODUCTION

In the context of reactor safety, various types of phenomena have to be considered in the analysis of few probability accidents for both light water reactor (LWR) and the liquid Metal Fast breeder reactor (LMFBR). One such phenomenon that is possible is the rapid vaporization of the coolant and the possible formation of shock waves during physical context between the hot core materials and the more volatile coolant.

For example, voiding of the cores sodium coolant (LMFBR) causing a rise in reactor power and subsequent fuel vaporization and expansion, the sodium as a work enhancing fluid or as a quenching liquid depending upon the relative mass of sodium coolant intermixed with the two-phase fuel and the characteristic length scale for heat transfer between the constituents. If the entrained mass is small and the characteristic size small, the sodium can vaporized and become the working fluid enhancing the disruptive expansion work.

During the expansion transient, it is expected that the solid structure also acted as a heat sink to the fuel vaporization. The surface area available for solid heat transfer will be covered by a coolant film during expansion of the coolant. The thickness of the film is a time dependent and decreases as the acceleration of the expanding fluid increases. The heat transfer to the solid structure will be to the film not to the solid structure. On the other hand, the vaporization of the coolant left on the wall enhance the heat transfer between the hot fuel and the coolant (because of increasing of

363

the surface area of the coolant).

Drainage of any liquid from a vessel or withdrawal of a flat plate from a
bath of liquid is usually accompanied by establishment of a liquid film on
the solid surface. The transient behavior of the liquid film has been studied both theoretically and experimentally by many investigators.
Jeffreys |1| derived a time dependent function for the film thickness by
neglecting inertia, surface tension effects, and a constant speed for drai-
nage . Van Rossum |9| verified that there was consistency between the Jef-
frey's model and his experimental work. Long and Tallmadge |5| also propo-
sed a theoretical model to predict the film thickness under the influence of
gravity. They considered a vertical plate immersed in a quiescent bath of a
wetting newtonian liquid. The solid plate was then withdrawal from the liquid bath at constant speed for a certain distance.
Ozgu et al |7| conducted some experimental investigations to measure the
local water thickness for some points on the wall after passing water in a
circular channel .

The results show that the film thickness decreases with time and finally it
reaches about .025 of the channel diameter . Khilar et al |3| also experim-
entally investigated the transient profile of the drainage film of a viscoelastic liquid. They postulated that in comparison with the newtonian liquid,
the viscoelastic liquid initially produced a thinner film and drained more
slowly. They also found that the viscoelastic coating film had a more uniform thickness during the initial period of drainage.
It should be noticed that all of the past works have been done to predict
the film thickness at low velocity or under the gravity effect. Therefore,
this paper is investigating the time history of the film thickness at the
high liquid acceleration.

MATHEMATICAL MODEL FOR LIQUID FILM THICKNESS

In this model a slug of liquid with an initial depth h_o and under an effective pressure difference ΔP in a vertical circular tube is investigated (Fig.
1). In order to simplify the equations of mass and momentum, the following
assumptions are made:
- The liquid slug velocity is uniform across the control volume except for
 the liquid film next to the wall.
- The friction factor of the transient flow is the same as that of the steady flow.

Because of the wall effect, the velocity of the liquid slug next to the channel wall can change in both r and z directions (Fig. 1).
For the first approximation, assume a second order polynomial to describe
the velocity of the liquid film next to the wall as:

$$V = C_1 r^2 + C_2 r + C_3 \tag{1}$$

with the following boundary conditions

$$V(R) = 0 \tag{2a}$$

$$\frac{dV}{dt}(R-\delta) = 0 \tag{2b}$$

$$V(R-\delta) = V_z \tag{2c}$$

Where V_z is the velocity of the liquid slug outside of the liquid film next
to the wall.
Introducing the above boundary conditions into Eq.(1), the result is:

$$\frac{V}{V_z} = 1 - (\frac{R-\delta-r}{\delta})^2 \tag{3}$$

where $\delta = \delta(z)$.

Since the sum of the mass of the control volume and liquid film left on the
wall is constant then,

$$\rho A h_0 = \rho A h + \int_0^Z \int_{R-\delta}^R \rho(2\pi r dr dz) \tag{4}$$

$$h_0 = h + 4 \int_0^Z [\frac{\delta}{D} - (\frac{\delta}{D})^2] \, dz \tag{5}$$

where D is the channel diameter and h is the depth of the liquid slug after
a travelled distance Z.

The momentume equation for the non-inertial control volume also requiers
that

$$\Sigma F_s + \iiint_{C.V.} g\rho dV - \iiint_{C.V.} \rho \frac{dV_z}{dt} \, dV$$

$$= \iint_{C.S.} \vec{V}_{rel} (\rho \vec{V}_{rel} \cdot d\vec{A}) + \frac{\partial}{\partial t} \iiint_{C.V.} \vec{V}_{rel} (\rho dV) \tag{6}$$

where V_{rel} is the velocity of the thickness relative to the control volume and $\sum F_s$ are the surface forces. A mass balance for the control volume also requiers that

$$\frac{\partial}{\partial t} \iiint_{C.V.} \rho dV + \iint_{C.S.} \rho \vec{V}_{rel} \cdot d\vec{A} = 0 \qquad (7)$$

If $\bar{V}_{rel}$ is the average of the liquid film velocity relative to the control volume, then the above equation becomes

$$\bar{V}_{rel} = - \frac{D^2}{4(\delta D - \delta^2)} (\frac{dh}{dt}) \qquad (8)$$

Introducing Eq.(8), the first integral on the right side of Eq.(6) changes to

$$\iint_{C.S.} \vec{V}_{rel} (\rho \vec{V}_{rel} \cdot dA) = - \frac{\pi \rho D^4}{16(\delta D - \delta^2)} (\frac{dh}{dt})^2 \qquad (9)$$

The second integral on the right side of Eq.(6) is also zero because the velocity of the liquid sluge does not change for a coordinate system XYZ attached to the control volume.

$$\frac{\partial}{\partial t} \iiint_{C.V.} \vec{V}_{rel} (\rho dV) \simeq 0 \qquad (10)$$

Introducing Eqs.(9) and (10) into Eq.(6) the result becomes:

$$\frac{dV_z}{dt} = \frac{\Delta P}{\rho h} + g - \frac{4\tau}{\rho D} + \frac{D^2}{4h(D\delta - \delta^2)} (\frac{dh}{dt})^2 \qquad (11)$$

where τ is the shear force over the wetted area and ΔP is the pressure difference across the control volume. In the absence of the surface tension effect, the mechanical energy for the control volume requiers:

$$\frac{\partial}{\partial t} \iiint_{C.V.} (\frac{V^2}{2} - gz) \rho dV - \dot{W} - \iiint_{C.V.} \phi dV = 0 \qquad (12)$$

where ϕ is the energy dissipation by viscous effect. Introducing Eq.(3), the dissipated energy yields

$$\phi = 2\mu (\frac{\partial V}{\partial r})^2 = \frac{8\mu}{\delta^4} (R - \delta - r)^2 V_z^2 \qquad (13)$$

Introducing Eqs. (3) and (13) into Eq.(12) the result becomes:

$$\left(\frac{dh}{dt}\right)^2 + \frac{2(D\delta-\delta^2)}{V_z D^2}(V_z^2 - 2gz)\frac{dh}{dt} - \frac{16\tau h(D\delta-\delta^2)}{\rho D^3}$$

$$+ \frac{64\mu h V_z}{3\rho D^4}(3D\delta-2D^2-\delta^2) = 0 \tag{14}$$

with initial conditions

$$V_z = \frac{dz}{dt}\ (0) = 0 \tag{15a}$$

$$h(0) = h_0 \tag{15b}$$

$$\delta(0) = 0 \tag{15c}$$

On the assumption of

$$\tau = \frac{1}{2} f\rho\ V_z^2$$

and the funning friction factor is

$$f = \frac{16}{Re} \qquad\qquad R_e < 2100$$

$$f = .079\ R_e^{-.25} \qquad\qquad 2100 < R_e < 10^5$$

$$f = .046\ R_e^{-.2} \qquad\qquad R_e > 10^5$$

The simultaneous solution of the Eqs.(5), (11) and (14) for different imposed pressure differences across the control volume is shown in Fig.2. The result shows that the film thickness will decrease when the effective pressure increases. The film thickness also approaches to a constant value after a distance about thirty times the channel diameter.
Figure 3 shows the effect of the channel diameter on the film thickness. The figure illustrates for a same initial liquid depth, the film thickness will decrease when the diameter of the flow channel increases.

It should be noted that the predicated film thickness is the same order of the experimental results by Ozgu |7| at low pressure and small channel diameter. In order to show the percentage change in (h) due to the liquid film on the wall, the parameter h/h_0 is plotted against z/h_0 in Fig.4 for different diameter. The figure illustrates that the liquid depth will decrease when the channel diameter decreases. For instance, after a travelled distance

about three times the initial liquid depth (52.5 cm), the change of liquid
depth for D=2 inches is about 5 times larger than that for D=8 inches. That
is, the change of the liquid depth due to the film thickness is significant
when the flow channel decreases.

CONCLUSIONS

The liquid film thickness on a vertical cylinder was obtained by the mass,
momentum and energy equations integral methods. It has been found to predict
the liquid holdup with good accuracy when compared with the experimental results. The theory can be used to predict the film thickness and the liquid
pickup for a flat-plate withdrawal of a liquid bath or falling a liquid sheet on a vertical wall under different conditions.
It was found that the accelerated liquid slug through a channel has more
influence on the liquid film thickness for the small channel diameter.
Furthermore, the liquid film reaches a reasonably constant value after a certain distance.

NOMENCLATURE

A   Cross section area of the flow channel.

C   Constant.

D   Diameter of the flow channel.

f   Friction factor coefficient.

F   Force.

g   Gravity acceleration.

h   Liquid depth.

R   Radius of the flow channel.

R_e Rynolds Number, $R_e = \dfrac{V D \rho}{\mu}$

t   Time.

V   Velocity.

$\dot{W}$ Work due to the surface tension effect.

Z   Travelled distance by slug of liquid.

ρ Fluid density.

$\dot{\phi}$ Dissipation of energy due to the viscous effects.

μ Viscosity.

Subscripts.

O   Initial Value.

r   Radial direction.

rel   Relative.

Z   Axial direction.

REFERENCES

1. Jefferys, H,. Proc. Cambridge Phil. Soc. 26. 204 (1930).

2. Jenekhe, Samson A., and Schuldt, Spencer B., "Coating Flow of Non-Newtonian Fluids on a Flat Rotatong Disk," Ind. Eng. Chem. Fundam.,Vol. 23. No. 4,PP. 432 - 436 (1984).

3. Khilar, K. C., C.B. Weinberger, and J.A. Tallmadge, "Postwithdrawal Drainage of Viscoelastic Speran Solution," Ind. Eng. Chem. Fundam., 266 - 274 (1980).

4. Lee, Chie Y., and Tallmadge, John A., "Meniscus Shapes in Withdrawal of Flat Sheets".

5. Long. K.C. and J.A., Tallmadge, "A postwithdrawal Expression for Drainage on Flat Plates," Ind. Eng. Chem. Fundam., Vol. 10, No.4, 648 (1971).

6. Matsumoto, Shiro, and Takashima, Yolchi, "Film Thickness of a Bingham Liquid on a Rotating Disk." Ind. Eng. Chem. Fundam. Vol. 21, No.3, PP 198 - 202, (1982).

7. Ozgu, M.R. and J.C. Chen. "Local Film Thickness During Transient Voiding of a Liquid-Filled Channel," J. of Heat Transfer, Transaction of ASME, 159 - 165 (May 1976).

8. Spiers. R.P., Subbaraman, C.V., and Wilkinson, W.L., "Free Coating of a Newtonian Liquid onto a Vertical Surface," Chem. Engineering Science,Vol. 29, PP 389 - 336 (1974).

9. Van Rossum, J.J., "Viscous Lifting and Drainage of Liquid," Appl.Sci. Research. A7, 121 - 144 (1958).

10. Vijayraghvan, K. and Gupta. Jal P., "Thickness of the Film on a Vertically Rotating Disk Partially Immersed in a Newtonian Liquid", Ind. Eng. Chem. Fundam., Vol. 21, No. 4, PP. 333 - 336 (1982).

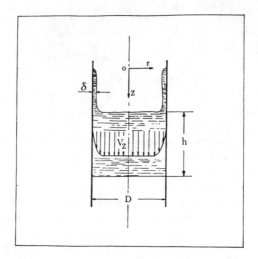

Fig. 1 Schematic Diagram of the Liquid Film Thickness on the Channel Wall

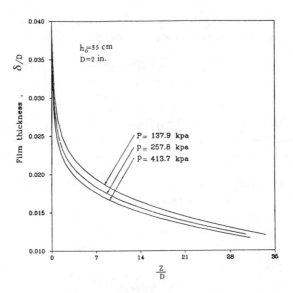

Fig. 2 Variations of the Nondimensional Liquid Film Thickness versus Nondimensional Travelled Distance by Water Slug for Different Effective
Pressures.

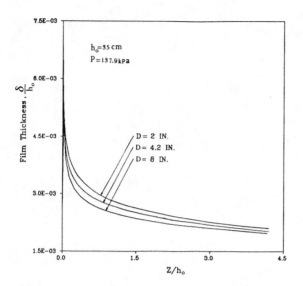

Fig. 3 Variations of the Nondimensional Liquid Film Thickness versus Nondimensional Travelled Distance by Water Slug for Different Channel
Diameters.

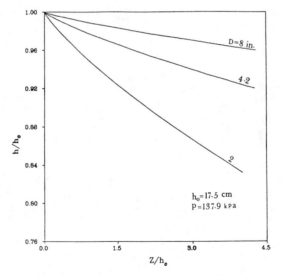

Fig. 4 Variations of the Nondimensional Water Depth versus Nondimensional
Distance by Water Slug for Different Channel Diameters.

371

Comparison between the Critical Heat Fluxes for a Finned Tube and for a Plain Tube in a Cross Flow

G. MEYER, E. S. GADDIS, and A. VOGELPOHL
Institut für Thermische Verfahrenstechnik
Technische Universität Clausthal
3392 Clausthal-Zellerfeld, FRG

Abstract

This paper describes an experimental investigation of the critical heat flux for a finned tube made of brass. The finned tube was placed horizontally in a channel and was cooled from the outside by a cross flow of freon 113 perpendicular to earth gravity. Heating the tube from the inside was by means of a synthetic oil. The undisturbed coolant velocity and the coolant subcooling were varied. Critical heat flux measurements for a plain tube of external diameter equal to the root diameter of the finned tube (25 mm) are also presented. Comparison is made between the results for the finned tube and for the plain tube.

1. INTRODUCTION

Vaporizing equipment, particulary those used in refrigerating plants, are mostly provided with finned tubes. The medium to be cooled flows in most cases inside the tubes, while the refrigerant boils on the external surface of the finned tubes. Many investigators have examined the boiling of liquids on finned tubes (e.g. [1-5]). Most of these investigations concentrated mainly on determining the external heat transfer coefficient for the finned tube. The maximum thermal load per unit tube length is an important parameter required by the designer of vaporizing units equipped with finned tubes. However, little attention has been paid to its determination. One of the most important investigations related to this subject was made by Bondurant and Westwater [1], who determined experimentally the maximum thermal load per unit tube length for a number of finned and plain tubes under pool boiling conditions. In the investigation presented in this paper, the critical heat flux and the maximum thermal load per unit tube length were determined for a finned tube placed in a forced flow across the tube; the coolant examined was subcooled and had different degrees of subcooling. Comparison is made between the experimental results for the finned tube and those for a plain tube with an external diameter equal to the root diameter of the finned tube.

2. EXPERIMENTAL EQUIPMENT

The experimental investigation has been carried out in the same facility described briefly in [6]. A detailed description of the equipment is found in [7]. The finned tube was made of brass

and had a tip diameter, a root diameter and an internal diameter
of 32 mm, 25 mm and 20 mm respectively. Other dimensions of the
finned tube are shown in Fig. 1. The finned tube was placed
horizontally in a channel of a square cross section 250 mm x 250
mm. Heating the finned tube from the inside was by means of a
synthetic oil. Freon 113 as a coolant flowed across the finned
tube perpendicular to earth gravity (see Fig. 1). The undisturbed
coolant velocity was varied from 0.12 m/s to 1.2 m/s and the
coolant subcooling was 1 K, 10 K and 16 K. The coolant pressure
at the plane of the finned tube was 1.75 bar. The plain tube,
required for the comparison with the finned tube, had an external
diameter of 25 mm and an internal diameter of 19 mm. The plain
tube was provided with jacketed thermocouples (not shown in Fig.
1- see [8]) to measure the wall surface temperature. These
thermocouples were led outside the tube through a protection tube
of 6 mm external diameter concentric with the main tube.The
finned tube was not provided with such thermocouples. The coolant
pressure and the examined ranges of the undisturbed coolant
velocity and coolant subcooling were the same for the finned tube
and the plain tube. In an earlier investigation [8], the present
authors have shown that the critical heat flux depends on the
thermal resistance of the tube wall and the thermal resistance of
the heating fluid inside the tube. For that reason, the same
material (brass) and nearly the same wall thickness (2.5 mm and
3.0 mm respectively) were used for the finned tube and for the
plain tube. The velocities of the oil inside the finned tube and
the plain tube were chosen such that the inside heat transfer
coefficients for the finned tube and for the plain tube were
nearly the same.

3.   EXPERIMENTAL PROCEDURE

     At a particular undisturbed coolant velocity and a part-
icular subcooling the temperature of the oil has been increased
in small steps while keeping the volumetric flow rate of the oil
constant. The heat flow rate from the oil to the refrigerant
(freon 113) was calculated at each step from the volumetric flow
rate of the oil, the oil thermal properties and the oil
temperature drop along the examined tube length. Correction was
made to take into account the small heat losses at the tube ends.
This procedure was carried on until remarkable reduction in the
heat flow to the refrigerant was noticed. The maximum thermal
load was determined from the plot of thermal load against oil
temperature.

4.   DISCUSSION OF EXPERIMENTAL RESULTS

     Figure 2 shows the critical heat flux for the plain tube as
a function of the undisturbed coolant velocity with the coolant
subcooling as a parameter. The critical heat flux decreases with
increasing undisturbed coolant velocity at the three examined
degrees of subcooling up to a critical undisturbed coolant
velocity somewhere between 0.2 m/s and 0.25 m/s. The critical
heat flux increases then with the undisturbed coolant velocity;
the rate of increase is higher at higher subcooling. At low

374

TABLE I: RATIO OF MAXIMUM THERMAL LOADS FOR THE FINNED TUBE AND
THE PLAIN TUBE

subcooling	undisturbed coolant velocity (m/s)				
(K)	0.12	0.25	0.36	0.80	1.20
1	2.27	2.24	2.02	2.06	2.04
10	1.95	2.15	1.99	1.90	1.85
16	1.81	2.00	1.85	1.90	1.76

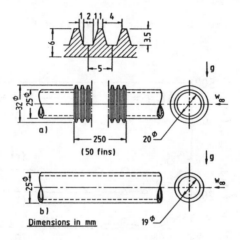

Fig. 1.- Test sections,
a) finned tube,
b) plain tube.

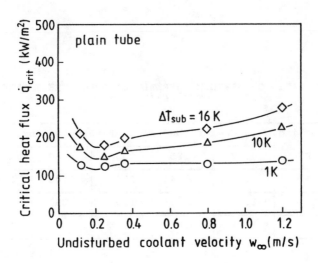

Fig. 2.- Critical heat flux for the plain tube.

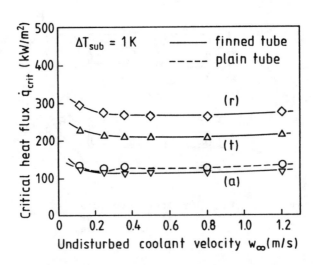

Fig. 3.- Comparison between the critical heat flux for the finned tube and for the plain tube at 1 K subcooling.

subcooling of 1 K, the critical heat flux for the plain tube is virtually constant for undisturbed coolant velocities in the range between 0.4 m/s and 1.2 m/s. An explanation for the minimum value of the critical heat flux at the examined flow geometry and a theoretical model to predict the critical undisturbed coolant velocity at which the critical heat flux is a minimum, is given in [6].

The critical heat flux for the finned tube can be calculated based on one of the following three reference areas:
1. the surface area of a plain tube with an external diameter equal to the root diameter of the finned tube,
2. the surface area of a plain tube with an external diameter equal to the tip diameter of the finned tube or
3. the actual surface area of the finned tube.
Critical heat fluxes based on the above different areas are marked in Figs. 3 to 5 by the letters (r), (t) and (a) respectively.

Figures 3 to 5 show the dependence of the critical heat flux for the finned tube (full lines) on the undisturbed coolant velocity at different degrees of subcooling 1 K, 10 K and 16 K respectively. The critical heat flux values for the plain tube are also shown for comparison in Figs. 3 to 5 as dotted lines.

The critical heat flux for the finned tube has, similar to the plain tube, a minimum. However, the critical undisturbed coolant velocity, at which the critical heat flux is a minimum, is somewhat higher for the finned tube and the rate of increase of the critical heat flux just after exceeding the critical undisturbed coolant velocity is lower for the finned tube compared with the plain tube. The higher critical undisturbed coolant velocity for the finned tube might be due to the generation of larger vapour bubbles. During bubble growth on the surface of a finned tube, the bubbles receive heat not only from the cylinderical surface between the fins but also from the sides of the fins. The larger bubbles generated between the fins of a finned tube have a higher rise velocity compared with the smaller bubbles generated on the surface of a plain tube. It has been shown in [6] that the critical undisturbed coolant velocity for the plain tube at the examined flow geometry is equal in magnitude to the bubble rise velocity. Comparing the numerical values of the critical heat flux in Figs. 3 to 5 for the finned tube based on the fictitious surface areas (r) or (t) with those of the plain tube gives an indication of the gain in the maximum thermal load due to the presence of the fins. Carrying out the comparison with the critical heat flux for the finned tube based on the actual surface area (a) shows that the critical heat flux at low subcooling (1 K, Fig. 3) is nearly the same for the finned tube and for the plain tube. However, as the subcooling increases (Figs. 4 and 5) the critical heat flux for the finned tube based on the actual surface area is somewhat lower than that for the plain tube. Visual observations have shown that at high subcooling particularly at high undisturbed coolant velocity the tips of the fins at the forward half of the tube facing the main flow were free from bubble nucleation (see Fig. 6); the heat flow

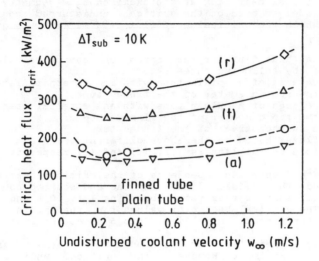

Fig. 4.- Comparison between the critical heat flux for the finned tube and for the plain tube at 10 K subcooling.

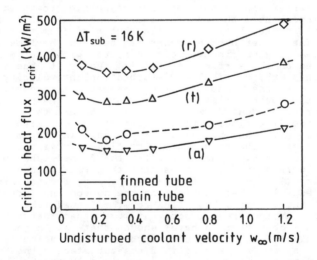

Fig. 5.- Comparison between the critical heat flux for the finned tube and for the plain tube at 16 K subcooling.

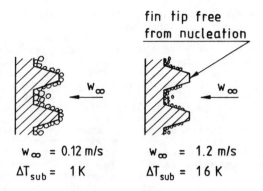

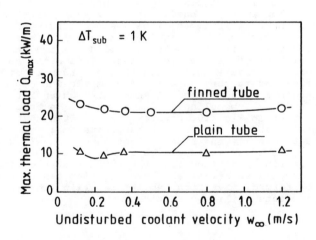

Fig. 6.- Bubble nucleation on the surface of the fins near the maximum thermal load.

Fig. 7.- Comparison between the maximum thermal load per unit tube length for the finned tube and for the plain tube at 1 K subcooling.

to the coolant from that area is thus only by forced convection. Another factor, which might be responsible for the reduction in the critical heat flux based on the actual surface area of the finned tube, is the reduced fin efficiency at high thermal loads due to higher temperature drops along the fins.

Figures 7 to 9 show the dependence of the maximum thermal load per unit tube length on the undisturbed coolant velocity and coolant subcooling for the finned tube and for the plain tube. Table 1 shows the ratio of the maximum thermal load for the finned tube to the maximum thermal load for the plain tube at the examined undisturbed coolant velocities and coolant subcoolings. It is seen in this table that this ratio decreases from about 2.3 at low undisturbed coolant velocity and low coolant subcooling to about 1.8 at high undisturbed coolant velocity and high coolant subcooling. The ratio of the maximum thermal loads for the finned tube and the plain tube passes at the higher degrees of subcooling (10 K, 16 K) through a maximum near the critical undisturbed coolant velocity for the plain tube. This is mainly due to the higher reduction in the value of the maximum thermal load per unit tube length for the plain tube at the critical undisturbed coolant velocity and the higher degrees of subcooling.

Figures 10 to 13 show the thermal load for the finned tube and for the plain tube as a function of oil inlet temperature for the four different combinations of the lowest and the highest examined values of undisturbed coolant velocity and coolant subcooling. These curves have the character of the boiling curve. Figure 10 (w_∞=0.12 and ΔT_{sub}=1 K) shows that the finned tube has, even at low oil inlet temperatures at which the plain tube operates at nucleate boiling region, higher thermal loads than the plain tube. Thus, the finned tube has a better thermal performance even at partial load. At higher subcooling and for the same low value of the undisturbed coolant velocity (see Fig. 11) the plain tube yields in the nucleate boiling region thermal loads comparable with those of the finned tube. This might be due to the suppression of nucleate boiling on the forward part of the fins facing the flow and to the reduced fin efficiency, which have been mentioned previously. Fig. 12 shows at high undisturbed coolant velocity and low subcooling (1.2 m/s and 1 K) a behaviour nearly similar to that in Fig. 10. However, the limited number of experimental points shown in Fig. 13 suggests that at high undisturbed coolant velocity and high coolant subcooling (1.2 m/s and 16 K) the plain tube may yield a higher thermal load in the nucleate boiling region. Heat flow from the surface of the finned tube to the coolant at such relatively low oil inlet temperatures may be mainly by forced convection, while the plain tube may be operating under nucleate boiling conditions.

5. CONCLUSION

The following may be concluded:

1. The maximum thermal load for the finned tube is much higher

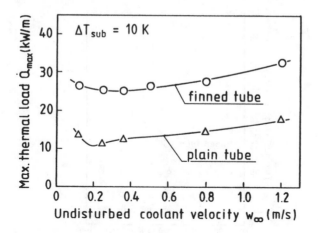

Fig. 8.- Comparison between the maximum thermal load per unit
tube length for the finned tube and for the plain tube
at 10 K subcooling.

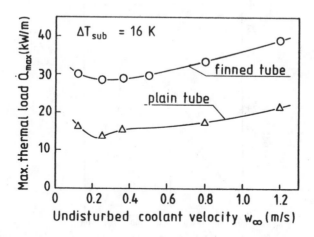

Fig. 9.- Comparison between the maximum thermal load per unit
tube length for the finned tube and for the plain tube
at 16 K subcooling.

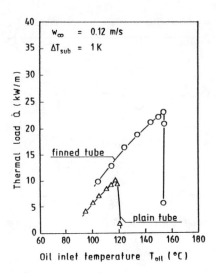

Fig. 10.- Dependence of the thermal load per unit tube length on the inlet temperature of the heating oil for the finned tube and for the plain tube at 0.12 m/s undisturbed coolant velocity and 1 K subcooling.

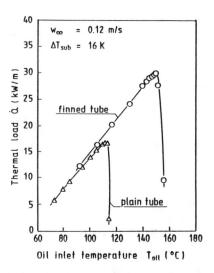

Fig. 11.- Dependence of the thermal load per unit tube length on the inlet temperature of the heating oil for the finned tube and for the plain tube at 0.12 m/s undisturbed coolant velocity and 16 K subcooling.

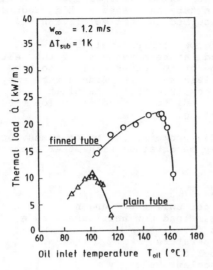

Fig. 12.- Dependence of the thermal load per unit tube length on the inlet temperature of the heating oil for the finned tube and for the plain tube at 1.2 m/s undisturbed coolant velocity and 1 K subcooling.

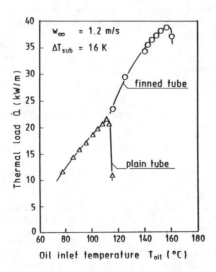

Fig. 13.- Dependence of the thermal load per unit tube length on the inlet temperature of the heating oil for the finned tube and for the plain tube at 1.2 m/s undisturbed coolant velocity and 16 K subcooling.

383

than that for the plain tube.

2. The ratio of the maximum thermal loads for the finned tube and the plain tube decreases at high degrees of subcooling and high undisturbed coolant velocities.

3. At low undisturbed coolant velocities and low degrees of subcooling and at temperatures of the heating fluid lower than the temperature corresponding to the maximum thermal load for the plain tube, the finned tube may yield a higher thermal load compared with the plain tube. The reverse may occur at high undisturbed coolant velocities and high degrees of subcooling.

4. The examined fin geometry is by no means optimized. Better performance may be obtained with other geometries.

6. NOMENCLATURE

g	earth gravity
$\dot{Q}$	thermal load per unit tube length
$\dot{Q}_{max}$	maximum thermal load per unit tube length
$\dot{q}_{crit}$	critical heat flux
T_{oil}	oil inlet temperature
ΔT_{sub}	coolant subcooling
w_{∞}	undisturbed coolant velocity

Letters in the Figures

a   critical heat flux based on actual surface area of the finned tube

r   critical heat flux based on the outside surface area of a plain tube of external diameter equal to the root diameter of the finned tube

t   critical heat flux based on the outside surface area of a plain tube of external diameter equal to the tip diameter of the finned tube

Acknowledgement

The authors thank the "Deutsche Forschungsgemeinschaft" for the financial support.

REFERENCES

1. Bondurant, D. L. and J. W. Westwater, Performance of transverse fins for boiling heat transfer, Chem. Engng Progr. Symp. Ser. 67, 30-37 (1971).
2. Slipcevic, B. and F. Zimmermann, Wärmeübergangskoeffizienten beim Blasensieden von Kältemitteln an einzelnen Rippenrohren, Klima Kälte Heizung 10, 1157-1162 (1982).
3. Slipcevic, B. , Sieden von Halogen-Kältemitteln an einzelnen Rippenrohren, Maschinenmarkt 89, 2090-2093 (1983).
4. Hahne, E. and J. Müller, Boiling on a finned tube and a finned tube bundle, Int. J. Heat Mass Transfer 26, 849-859 (1983).

5. Windisch, R. and E. Hahne, Heat transfer for boiling on finned tube bundles, Int. Comm. Heat Mass Transfer 12, 355-368 (1985).
6. Meyer, G. , E. S. Gaddis and A. Vogelpohl, Critical heat flux on a cylinder of large diameter in a cross flow, Proc. 8th Int. Heat Transfer Conf.- San Francisco 5, 2125-2130 (1986).
7. Meyer, G., Kritische Wärmestromdichte querangeströmter Rohre, Diss. Technische Universität Clausthal (1986).
8. Meyer, G. , E. S. Gaddis and A. Vogelpohl, Dependence of the critical heat flux over a cylinder in a cross flow on the thermal resistances of tube wall and heating medium, Wärme- und Stoffübergang 22, 309-313 (1988).

Freon-12 Upflow Experiments on Pressure Loss, Heat Transfer Coefficient, and Boiling Crisis in a Tube Downstream of a Contraction

D. R. H. BEATTIE and K. R. LAWTHER
Australian Nuclear Science and Technology Organization
Private Mail Bag 1
Menai, 2234, Australia

ABSTRACT

Experiments are described for upflow of Freon-12 at 1 MPa in a tube 1 cm ID with a heated length of 0.9m downstream of a contraction from a 2.7 cm ID tube. Mass fluxes ranged from 1150 to 8000 kg m^{-2} s^{-1}, and inlet qualities from -0.3 to +0.8. Pressure loss behaviour, including that for the contraction, are correlated with semi-theoretical regime-dependent equations. Wall temperature measurements demonstrated hystereses in pre-crisis heat transfer, in post-crisis heat transfer, and in critical heat flux. Explanations are given for these phenomena. Boiling crisis characteristics may be related to those for wall shear stress.

1. INTRODUCTION

Two-phase flow friction factor characteristics can undergo abrupt transitions at certain flow conditions, with the transition usually occurring with no change in wall shear stress [1]. Thus, transition between 'friction regimes' can be obtained from equations giving friction factors for each regime. Qualities at regime transitions usually decrease with increasing mass flux, and, on the basis of friction factor equations, can even occur at negative qualities at sufficiently high mass fluxes. This is unrealistic, and implies that the friction factor equations must be replaced by alternatives at high mass fluxes.

An extensive analysis of published pressure loss data for a range of two-phase flow conditions [1] provides partial support for this inference. However no one data set was able to provide direct confirmation, partly because of the relatively limited mass flux range of most data sets.

In view of an observed dependence of critical heat flux (CHF) characteristics on friction factor characteristics [2, 3], the above inference of a change in friction characteristics on reaching a particular mass flux (for a given flow system) implies corresponding changes in CHF characteristics. From an examination of CHF data, Smolin et al [4] independently conclude that such a change does occur. Interestingly, on limited and inconclusive evidence, they relate this change to pressure loss behaviour.

Boiling Freon-12 experiments, covering a wider-than-usual range of mass fluxes, were performed to test the inference of changes in friction factor and CHF characteristics on reaching sufficiently high mass fluxes. These experiments are discussed here.

2. EXPERIMENTAL

The experiments were performed on the Freon-12 loop described in [5]. This loop provides subcooled liquid to a heated test section. To achieve the purposes of the tests described here, the 'test section' consisted of two independently heated tubes in series. The upstream tube effectively acted as a preheater for the downstream tube, on which the tests were performed. As upper limits of the inlet quality to the downstream tube were controlled by CHF in the upstream tube, high qualities could only be achieved with a long, large diameter upstream tube. A contraction to a smaller diameter tube permitted higher downstream mass fluxes while maintaining high flow qualities when required.

The upstream tube was 26.6 mm ID, with a heated length of 2139mm and distance between two wall pressure tappings of 2215 mm. The downstream tube was 10.5 mm ID with a heated length of 914 mm and distance between two pressure tappings of 991 mm. The two pressure tappings adjacent to the unheated contraction were 294 mm apart.

Chromel-alumel thermocouples were placed at the end of the heated lengths of each tube and at four additional axial positions in the smaller tube. A chart recorder monitored the responses of the thermocouples on the tube as well as the voltage applied to it. CHF was determined from the thermocouple responses, and, in the case of the downstream tube, also from Wheatstone bridge detection of temperature-induced changes of electrical resistance.

The exit pressure was kept at 1.1 MPa. Mass fluxes ranged from 1160 to 7900 kg m^{-2} s^{-1} for two-phase tests, with single phase tests extending to 19,000 kg m^{-2} s^{-1}. Inlet qualities for the small tube ranged from -0.29 to 0.82 at 1160 kg m^{-2} s^{-1}, and from -0.29 to 0.21 at 7900 kg m^{-2} s^{-1}. Exit qualities extended to 1.15 at the lower mass fluxes and to 0.5 at the high mass fluxes.

Fluid temperatures were measured upstream and downstream of the test arrangement. Several preliminary tests were performed with single phase exit conditions to confirm that enthalpy rises determined from these agreed closely (within a few percent) with rises calculated from measured power inputs to the test section. These tests were performed for each section heated separately and for both sections heated. Heat balance tests of this type are usually confined to liquid only conditions and hence to low power inputs. In the experiments described here, the attainment of superheated vapour at low mass fluxes allowed the confirmation tests to also be performed at high power conditions.

Tests under two-phase flow conditions were performed at seven mass fluxes (approximately 1160, 1690, 2230, 3350, 4450, 6600, and 7900 kg m^{-2} s^{-1}). For each of these, the inlet quality was varied systematically in steps of about 0.05. For each of these, downstream power was systematically varied from zero to a level set by requiring exit temperatures to be less than 280°C. Pressure drops were recorded with increasing power; temperatures were recorded for both increasing and decreasing power levels.

An attempt to complement the tests with γ-ray determinations of average and local voidage values was abandoned because effects of power, applied to the test sections, on γ detection equipment could not be readily eliminated in the time available for the experiments.

3. RESULTS

3.1. Friction Factor Behaviour

Frictional pressure drops for the small tube were estimated by subtracting gravitational and acceleration pressure drops, calculated using a homogeneous flow model, with the calculation for the gravitational component further assuming the mean void fraction is that corresponding to the mean test section quality. Errors due to these assumptions are small. Local friction factors were obtained by assuming the average frictional pressure gradient coincides with that midway between the two pressure tappings. This result holds if the frictional pressure gradient is constant, or varies linearly with quality. The approximation is reasonable for most of the tests performed, particularly those with only a small quality increment over the heated length. The approximation is poor if large discontinuities in frictional pressure gradient exist between the pressure tappings, as may, for example, occur at boiling crisis locations. For this reason, the following discussion is confined to data where the approximation is reasonable, i.e. to data obtained a) under adiabatic conditions; b) at the boiling crisis onset; and c) for heat fluxes well above CHF (where the test section is essentially all under dryout conditions).

As shown in fig. 1, single phase friction factor data are consistent with the Colebrook equation for a surface roughness value ϵ expected for drawn tubing. As with most data from other sources, it can be seen that this equation extends to low quality two-phase flow conditions for appropriately defined physical properties in the friction factor $f = 2\tau_w / (\rho v^2)$ and Reynolds number Re $= \rho v D/\mu$ definitions. The definitions are those for a bubbly sublayer flow [1].

Theory [6] suggests the equation should also hold for post-CHF conditions if gas physical properties, corresponding to a gas-only sublayer, are used. The post-crisis data in fig. 1 conform to this except that most lie on the smooth tube equation, rather than the equation for the appropriate roughness value. Similar trends in other data were previously interpreted [6] as indicating 'pseudo-dryout' instead of true dryout. In pseudo-dryout, superheated liquid remains in the microcrevices of the roughness elements, reducing the apparent roughness for the flow. This point will be returned to in the discussion of heat transfer behaviour, where an explanation for the phenomenon will be offered. With regard to the post-crisis data of fig. 1, true dryout friction factors, i.e. those corresponding to the actual duct roughness, were confined to high mass fluxes. For reasons given later, more true dryout friction factors may have resulted had pressure drops been measured between reducing, as well as increasing, steps in applied power.

Fig. 1 also shows some data, at CHF conditions, for which single phase concepts extend to cover two-phase conditions, but with coefficients of the correlating equation differing from those for the single phase equation. Many such correlating equations have previously been found for various data sets [1]. As with others, coefficients of the present correlation are consistent with constraints suggested in [1] *

Fig. 2 shows data for which friction factor varies with Weber number instead of Reynolds number. The correlating equations are of the form derived in [7] from the concept that attached wall bubbles act as roughness elements and are of a size determined by a balance of surface tension and shear stress forces. Although originally conceived for nucleate boiling, the form of equation often correlates adiabatic data where low level flashing might be expected, as might occur downstream of an obstruction, or when pressure drops are not small compared with the system pressure. Both these effects would contribute to the existence of attached wall bubbles for the adiabatic data of fig. 2. As with similar equations used to fit other data, the coefficients of the correlating equations of fig. 2 are consistent with constraints suggested in [7].

Fig. 3 shows data where friction factors correlate with a 'film function' FF. The form of correlating equation was first suggested in [8] and was derived for an annular flow structure in which the liquid film at the wall falls entirely within the viscous sublayer of the gas/droplet core flow. Thicker film data conform to trends in figs. 1 and 2.

Figs. 1-3 demonstrate that data can be grouped into different friction regimes, each characterised by a correlation of the friction factor data. The nature of the correlating parameters indicates the near-wall flow structure (gas-only, bubbly, attached wall bubbles, or very thin annular film). Note, however, a dependence on Reynolds number does not necessarily imply an absence of attached wall bubbles. As with some single phase flows with artificially roughened walls, attached wall bubble 'roughness' might not alter friction factors.

In considering the boundaries of each of these friction regimes, it is clear that, as anticipated, friction behaviour is different at high and low mass fluxes.

Figs. 1-3 do not directly show variations of friction factors with quality. Fig. 4 presents representative friction factor data, and correlations of figs. 1-3, as a function of quality. Only the correlating equations for low quality adiabatic data and post crisis data, both of which are extensions of the single phase Colebrook equation, cover all the mass fluxes of the present experiments.

3.2 Contraction Pressure Drops

The single phase contraction pressure drop data yielded loss coefficients $\Delta P_c \rho / G^2$ which are independent of Reynolds number and close to 0.66, a value predicted from the method of [9]. As discussed in [10] an independence of Reynolds number implies the contraction pressure losses can be predicted using the 'homogeneous' flow prediction, i.e.

$$\Delta P_c = 0.66 \, G^2 \, \{1 + x \, (\rho_\ell/\rho_g - 1)\}$$

for a coefficient of 0.66, even if the flow is highly non-homogeneous.

Although dominated by the contraction pressure drop, the measured pressure drop has components from the gravitational head and a contribution to acceleration from flashing. Predictions for these, using a homogeneous flow model, have been added to predictions from the above contraction equation, and the totals compared with measured pressure drops in fig. 5. The agreement is satisfactory.

3.3 Heat Transfer Data

Representative heat transfer data are shown in fig. 6. These demonstrate unexpected hystereses in CHF; subcrisis conditions; and post-crisis conditions. (However, the two curves of fig. 6b result from slight drifts in mass flux and subcooling during the experiment. They do not represent a true hysteresis. Such drifts occurred at high qualities). The CHF is well defined in most cases, but becomes less so at high qualities (e.g. fig. 6b).

---

* i.e. the correlating equation is consistent with integer choices for m and n in
$1 / \sqrt{f} = (\sqrt{2})^{m+4} [\log Re \sqrt{f} - (0.11m + 0.50n + 1.10)]$.

An explanation of the small precrisis temperature drop often observed (e.g. figs. 6a, c and d) has been given in [11]. Briefly, evaporating microlayers of attached wall bubbles are usually replenished by liquid from outside the wall bubble layer. However, the thinning of the liquid film during approach to dryout depletes this liquid source, so bubble microlayers disappear. Vapour within the bubble heats through contact with the wall, resulting in equilibrium bubble sizes too small to be removed by the flow. Bubbles then collapse and form a stable, thin froth layer of very small bubbles. Heat transfer is now by conduction through this layer and is superior to that previously occurring - hence the wall temperature decrease - because the froth layer is very thin. Heat transfer is derived to be of the form $\phi \propto (T_w-T_s)^2$. Once formed, the froth layer may persist, producing hystereses in heat transfer. The reverse component of the subcrisis heat transfer hysteresis of fig. 6c is of the above $(T_w-T_s)^2$ form.

Other subcrisis heat transfer behaviour in these experiments was generally of the form $\phi \propto (T_w-T_s)^n$, with n often being ~ 3; sometimes slightly greater than unity (as in fig. 6c for increasing heat flux and fig. 6d for decreasing flux); occasionally unity (e.g. fig. 6d, for increasing flux); and occasionally less than unity (e.g. fig. 6a and fig. 6b). A value of n near three is widely regarded as indicating strong nucleate boiling, and a value slightly greater than unity is consistent with a heat transfer model [12] including components from nucleate boiling and from conduction through liquid between wall bubbles. Hence, for n > 1 and/or a pre-crisis wall temperature drop, a nucleate boiling process is indicated. This is consistent with the conclusion, in section 3.2, in which nucleate boiling is inferred from the dependence of friction factor on Weber number.

Current models have not predicted exponents of unity or less, although, by analogy with single phase behaviour, a unity value has elsewhere [13] been interpreted as implying convective heat transfer with no boiling. However, the current data with slopes of unity or less are often associated with pre-crisis wall temperature drops and/or friction factors depending on Weber number, both of which imply nucleate boiling.

Post-crisis hystereses, as shown in figs. 6a and 6c, were often evident. (As noted, the apparent hysteresis in fig 6b is actually a result of drifts in flowrate and inlet subcooling.) Two aspects of these hystereses are worthy of note. The first concerns predicted post-crisis heat transfer behaviour, shown in figs. 6 as dotted lines. (The prediction method is discussed later.) Concepts of the model appear plausible since predictions agree with, or are parallel to, data where no hystereses occur, and to data from the reverse component of hysteresis cycles. However, post-crisis data from initial parts of hysteresis cycles show a weaker dependence on heat flux than predicted from the dryout model, indicating incompatibility with the dryout model. This suggests that these post-crisis data do not represent dryout as conventionally considered and as built into the dryout model.

The second aspect concerns post-crisis friction factors. As noted previously, these are consistent with theoretical predictions but, as shown in fig.1, some data suggest an apparent reduction in wall surface roughness. This behaviour occurred only during the initial part of post-crisis heat transfer cycles. Friction factors consistent with the true roughness were obtained in the absence of a hysteresis (as in fig. 6d). (Unfortunately, pressure drops were measured only between steps of increasing heat fluxes. Thus, although the above suggests a corresponding hysteresis in post-crisis friction factors, the experimental program did not enable such a hysteresis to be verified).

As discussed in [6], similar friction and heat transfer behaviour are also evident in data from other laboratories. This behaviour points to the existence of a 'pseudo-dryout' condition [6] which may occur after the critical heat flux condition has been reached. As opposed to true dryout, the hypothesised 'pseudo-dryout' condition has static superheated liquid remaining within the microcrevices of the roughened surface. Under such conditions, the flow would experience a reduced wall roughness, and wall temperatures would be lower than for true dryout. Once true dryout was achieved, it would persist, consistent with the hysteresis.

Although the concept of superheated liquid being retained in microcrevices of the roughened surface seems implausible, the concept is consistent with observed post-crisis friction and heat transfer behaviour. Moreover, a possible partial explanation for the existence of superheated liquid is built into the model for precrisis wall temperature drops given earlier. As discussed, this was explained as arising from a restructuring of nucleating bubbles to form a fine stable froth layer caused by the thinning of the film during approach to dryout. Sizes of the froth bubbles can be obtained from standard equations built into the analysis [11]. Calculations show that these bubbles can be extremely small; indeed, much smaller than the roughness elements of commercial tubing, thus allowing the froth to exist within surface microcrevices following the boiling crisis. Moreover, surface tension forces acting on these small bubbles lead to high internal vapour pressure in the froth. The liquid in the froth may thus remain as liquid up to saturation temperatures corresponding to pressures within the froth bubbles, i.e., apparently superheated with regard to the pressure in the liquid component of the froth and in the fluid external to the froth structure.

internal vapour pressure in the froth. The liquid in the froth may thus remain as liquid up to saturation temperatures corresponding to pressures within the froth bubbles, i.e., apparently superheated with regard to the pressure in the liquid component of the froth and in the fluid external to the froth structure.

The post-crisis heat transfer predictions are based on an adaptation of the single phase Dittus-Boelter equation to two-phase flow conditions, i.e.

$$\frac{\phi D}{(T_w - T_v)k_g} = 0.023 \left(\frac{\rho_g \upsilon D}{\mu_g}\right)^{0.8} Pr_g^{0.4.}$$

This is reasonable since theory and data support on extension of the single phase friction factor equation to post-crisis conditions through correct definitions of parameters in dimensionless groups. Similar adaptations of definitions in single phase heat transfer equations should then similarly lead to valid expressions for post-crisis conditions. However, this reasoning suggests that the reference temperature is the mean vapour temperature T_v and not saturation temperature T_s. The two can be related via a thermal analogue of the hydromechanical drift-flux equation:-

$$T_w - T_s = C_o(T_w - T_v) + \Delta$$

The first term on the RHS takes account of the fact that liquid droplets, at saturation temperature, concentrate in the cooler central region, and the second term is a local non-equilibrium term, analogous to the drift velocity in two-phase flow, representing local differences in vapour and liquid temperatures. From the thermal analogy, the distribution parameter is identical to that for the drift-flux equation, and can be calculated as $1 + 2.6\sqrt{f}$, where f is the dry wall friction factor obtained from the post-crisis form of Colebrook's equation (i.e. regime 4a of fig. 1). The model is discussed in more detail in [6].

Current models for thermal non-equilibrium neglect phase and temperature distribution effects (i.e. place $C_0 = 1$), and apply various methods for predicting the local non-equilibrium term Δ. The present analysis has neglected Δ, and allowed for distribution effects. Predictions for the present data are good at high mass fluxes (as in fig. 6(d)), and acceptable at low mass fluxes (as in figs. 6(a), (b), (c)). This suggests that local non-equilibrium, represented by Δ, becomes significant only at low mass fluxes.

A previous analysis [6] of published post-crisis heat transfer data suggests that the existence of pseudo-dryout can lead to non-reproducibility:- CHF may precipitate either true dryout or pseudo-dryout. Clearly, conservative predictions should consider the higher wall temperatures that will occur with true dryout. This suggests that empirical correlations for post-crisis heat transfer be used with caution since data bases for these correlations may be biased towards pseudo-dryout conditions. This could lead to underprediction of wall temperatures for true dryout.

3.4 Critical Heat Flux

As anticipated, exit CHF characteristics at low mass fluxes differed from those at high mass fluxes, with the transition occurring at about 3000 kg m^{-2} s^{-1}. It will be recalled that friction characteristics altered at a similar mass flux. Representative data for high and low mass fluxes are given in fig. 7. In each case, the characteristics can be represented by a series of inter-connected straight lines. As indicated in fig. 7, each line can be associated with specific trends in heat transfer behaviour during approach to CHF. Changes in slope perhaps represent changes within the mechanism inducing the crisis.

The hysteresis in CHF, mentioned in section 3.3, was more evident at low mass fluxes. At these fluxes, the hysteresis is confined to the middle region of the three-region characteristic (see fig. 7(a)) and return-path CHF data fall on an extrapolation of the line describing the high quality data. An interpretation is that pre-crisis hydrodynamic conditions, as set by sublayer flow pattern, for example, are different for each of the three regions of the CHF characteristic, but, on rewetting following CHF, only the low quality and high quality regions return to similar pre-crisis conditions. The intermediate quality region perhaps reverts to conditions previously confined to high qualities.

Other CHF data covering both increasing and decreasing heat flux, e.g. from [14, 15], have no similar hystereses. An exception is found in the study described in [16]. However, in that case, hystereses were small, producing CHF changes of the order of 10% (as with the present hystereses at high mass fluxes), and return path CHFs of [16] were higher than initial CHFs whereas the present return path CHFs are lower.

391

As with the other studies extending to high mass fluxes, e.g. [15], an intermediate quality region, where CHF increases with quality, occurs at high mass fluxes. Also, exit crises of this type were sometimes preceded by upstream crises. In the present study, CHFs of this type appear to be associated with friction factors corresponding to regime 3 in fig. 1.

An inter-relation between CHF and pressure loss characteristics has been proposed in [2, 3]. Juxtaposition of the present CHF and friction factor dependencies on quality support this concept. An example is given in fig. 7. This figure demonstrates that various specific qualities characterising different regions of the CHF curve coincide with those delineating different regions of the friction factor plot. However, the number of friction regimes of the present study [9], together with the comparatively limited extent of some of these and of some regions of CHF plots, prevent comprehensive examination of any relationship between the present CHF and friction factor data.

4. CONCLUDING DISCUSSION

Pressure loss, heat transfer, and critical heat flux data have been obtained for a wide range of mass fluxes. The data indicate that thermo-hydraulic characteristics at high mass fluxes differ from those at low mass fluxes, thus confirming indications from a previous examination of data sets covering less extensive ranges of mass fluxes [1].

Pressure loss data were examined in terms of theoretical concepts so as to yield a number of 'friction regimes', each with specific theoretically-based, or semi-theoretically based, friction factor equations. The forms of these equations provide insight into (among other factors) the near-wall flow structure and so are potentially useful in interpreting the less well understood heat transfer phenomena.

Heat transfer data yielded a number of unexpected hystereses; in particular, significant hystereses of a type not previously reported occurred with CHF in some cases. Moreover, hystereses in post-crisis heat transfer, explained here by a 'pseudo-dryout' phenomenon, suggest empirical correlations for wall temperature predictions may underpredict true dryout temperatures if pseudo-dryout data are used in developing correlations.

As with pressure loss data, an attempt was made to interpret heat transfer data via phenomenalogical modelling. However, interpretations were limited by the current lack of understanding of boiling heat transfer processes. Conclusions drawn from those data where interpretations were possible are consistent with those drawn from the pressure loss data. A theoretical model to explain a pre-crisis wall temperature drop, frequently observed in the present experiments, also provides an explanation of a pseudo-dryout phenomenon.

The present experiments utilised an upstream contraction to provide the high velocity two-phase flows in these experiments. Pressure losses across this agree with predicted values. However it is conceivable that the contraction affected the downstream behaviour, but this was not investigated here. Also, the unusual range of flow conditions used severely limits the availability of similar published data for comparison. Nevertheless, several published data e.g.[15, 17, 18], are for conditions which overlap to some extent those of the present experiments. Although not conclusive, comparisons with these data indicate that the upstream contraction did not alter CHF behaviour for subcooled inlet conditions, nor alter the thermo-hydraulic characteristics for the low quality and post-crisis regimes (friction regimes 2 and 4 in fig. 1). However, there is limited evidence that behaviour may have been affected outside these conditions.

5. NOTATION

Cp	Specific heat at constant pressure
C_0	Distribution parameter
D	Diameter
f	Friction factor, $2\tau_w / (\rho v^2)$
FF	Film function defined in fig. 3
G	Mass flux
k	Thermal conductivity
ΔP_c	Contraction pressure drop
Pr	Prandtl no., $\mu Cp/k$
Re	Reynolds no., $\rho vD/\mu$
T	Temperature
x	Flow quality
x_0	x at pressure tapping immediately downstream of contraction
We	Weber no., $\rho_h v^2 D/\sigma$
ϵ	Surface roughness
β	Homogeneous void fraction
ϕ	Heat flux
μ	Viscosity
ρ	Density
σ	Surface tension
τ	Shear stress

Subscripts

do	Dryout
h	Homogeneous
g	Gas
ℓ	Liquid
s	Saturation
v	Mean vapour value
w	Wall

REFERENCES

1. Beattie, D.R.H., 'Some aspects of two-phase flow drag reduction', Paper D 3, Proc. 2nd Int Conf. Drag Reduction, Cambridge, BHRA Fluid Engineering, 1977.

2. Beattie, D.R.H., and Lawther, K.R., 'Relationship between wall shear stress and the heat transfer crisis phenomenon with vapour/liquid flows', Paper FB23, Proc 6th Int. Heat Transfer Conf., Toronto, Canada, 1978.

3. Beattie, D.R.H., 'Hydrodynamic regime change effects on critical heat flux characteristics', Paper C9, European Two-Phase Flow Group Meeting, Eindhoven, 1981.

4. Smolin, V.N., Shpanskii, S.V., Esikov, V.I., and Sedova, T.K., 'Method of calculating burnout in tubular fuel rods when cooled by water and a water-steam mixture', Thermal Engineering 24 (12) 30-35, 1977.

5. Ilic, V., 'The AAEC freon rig ACTOR and initial boiling crisis (burnout) results' AAEC/TM 632, 1972.

6. Beattie, D.R.H., 'Two-phase fluid mechanics and heat transfer in the dry wall region', Proc. 2nd Australasian Conf. on Heat and Mass Transfer, Sydney, 1977.

7. Beattie, D.R.H., 'An evaluation of two bubble-detachment models for two-phase flow', page 1343, Proc ANS-ASME Int. Meeting on Nuclear Reactor Thermal Hydraulics, Saratoga, NUREG/CP6014, 1980.

8. Beattie, D.R.H. and Sugawara, S., 'An application of mixing length theory to two-phase flow thermohydraulics in ATR', PNC report N 941 85-10, 1985.

9. Weisbach, J., 'Die Experimental Hydraulik', J.S. Engelhardt, Frieberg, 1855.

10. Beattie, D.R.H., 'Two-phase pressure losses for flow through pipeline fittings: A note on the correct application of a model based on near-wall flow pattern'. Submitted to Nuclear Engineering and Design, 1988.

11. Beattie, D.R.H., and Lawther, K.R., 'An examination of the wall temperature drop phenomenon during approach to flow boiling crisis', page 2215 Proc 8th Int. Heat Transfer Conf., San Francisco, 1986.

12. Beattie, D.R.H., 'Heat transfer for convective boiling annular flows', section 8-2 of PhD Thesis 'An extension of single phase flow turbulent pipe flow concepts to two-phase flow', U.NSW, 1983.

13. Aounallah, Y., Kenning, D.R.B., Whalley, P.B., and Hewitt, G.F., 'Boiling heat transfer in annular flow', Paper FB3, Proc. 7th Int. Heat Transfer Conf., Munich, 1982.

14. Bennett, A.W., Hewitt, G.F., Kearsey, H.A., and Keeys, R.K.F., 'Heat transfer to steam-water mixtures flowing in uniformly heated tubes in which the critical heat flux has been exceeded', AERE report R5373, 1967.

15. Groeneveld, D.C., 'The thermal behaviour of a heated surface at and beyond dryout', AECL report 4309, 1972.

16. Era, A., Gaspari, G.P., Hassid, A., Milani, A., and Zavattarelli, R., 'Heat transfer data in the liquid deficient region for steam-water mixtures at 70 kg/cm^2 flowing in tubular and annular conduits', CISE report R-184, 1966.

17. Friedel, L., 'Modellgesetz fuer den Reibungsdruckverlust in der Zweiphasen-stroemung'. VDI-Forschungsheft 572, 1975.

18. Stevens, G.F., Elliott, D.F., and Wood, R.W., 'An experimental investigation into forced convection burnout in freon, with reference to burnout in water: Uniformly heated round tubes with vertical upflow'. AEEW report R321, 1964.

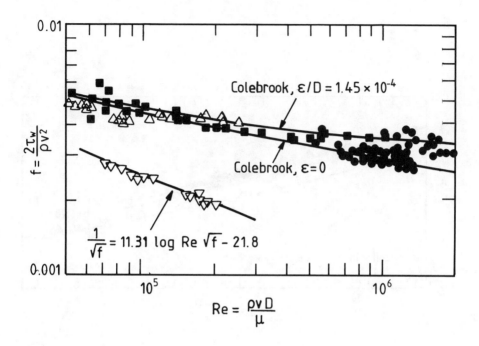

$$f = \frac{2\tau_w}{\rho v^2}$$

Colebrook, $\varepsilon/D = 1.45 \times 10^{-4}$

Colebrook, $\varepsilon = 0$

$$\frac{1}{\sqrt{f}} = 11.31 \log Re \sqrt{f} - 21.8$$

$$Re = \frac{\rho v D}{\mu}$$

Friction Regime/notation		ρ	μ	Conditions
1	■	ρ_ℓ	μ_ℓ	x < 0
2	△	$\rho_\ell(1-\beta) + \rho_g\beta$	$\mu_\ell(1+2.5\beta)$	x < 0.6
3	▽	$\rho_\ell(1-\beta) + \rho_g\beta$	$\mu_\ell(1+2.5\beta)$	CHF x < 0.15, $G > 3000$ kg m^{-2} s^{-1}
4a,b	●	ρ_g	μ_g	$\phi \geqslant$ CHF

Fig.1. Variation of friction factor with Reynolds number. (The friction regime numbers identify regions of fig.4)

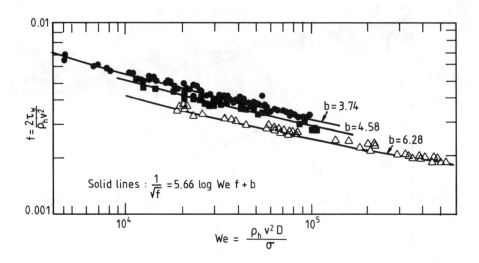

Friction regime/notation		b	Conditions
5	•	3.74	$\phi = 0$, $x > 0.06$, $G < 3000$ kg m^{-2} s^{-1}
6	■	4.58	CHF, $G < 2000$ kg m^{-2} s^{-1}
7	△	6.28	CHF, $x > 0.15$, $G > 2000$ kg m^{-2} s^{-1}

Fig.2. Variation of friction factor with Weber number. (The friction regime numbers identify regions of fig.4).

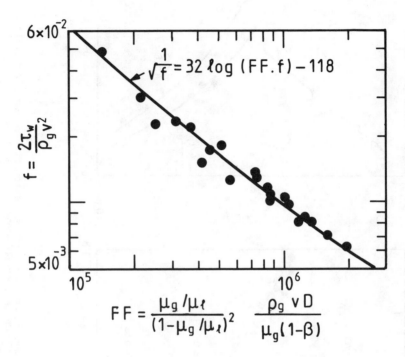

Fig.3. Variation of friction factor with the film function FF. Conditions are $\phi = 0$, $x > 0.06$, $G > 3000$ kg m^{-2} s^{-1}. Denoted as friction regime 8 for the purposes of identifying a region of fig 4b.

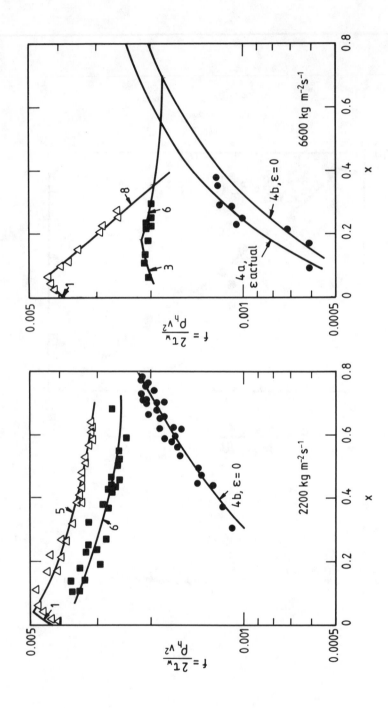

Fig. 4. Representative low mass flux and high mass flux variations of friction factor with quality. $\triangle$ - $\phi = 0$; ■ - at CHF; and ● - $\phi \gg$ CHF. Numbers refer to friction regimes defined in figs. 1-3.

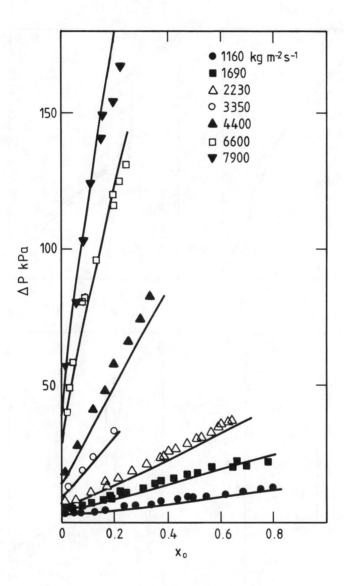

Fig. 5. Comparison of predicted and measured pressure drops across the contraction.

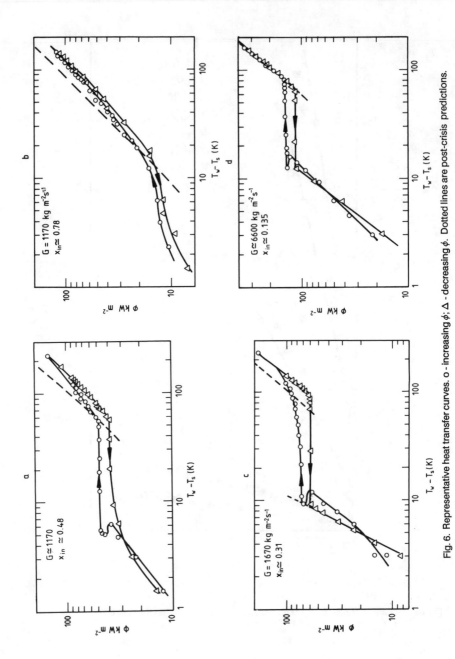

Fig. 6. Representative heat transfer curves. o - increasing φ; Δ - decreasing φ. Dotted lines are post-crisis predictions.

400

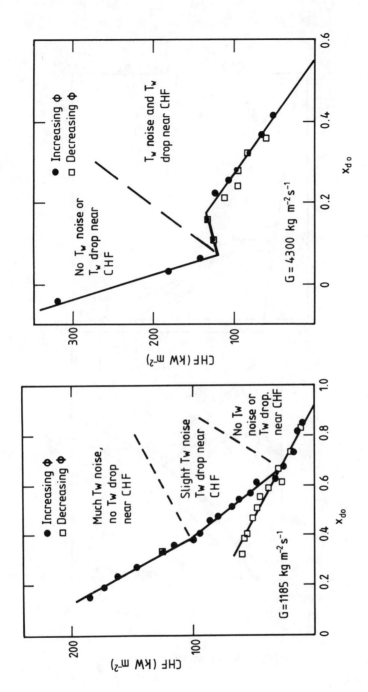

Fig. 7. Representative critical heat flux data for low and high mass fluxes.

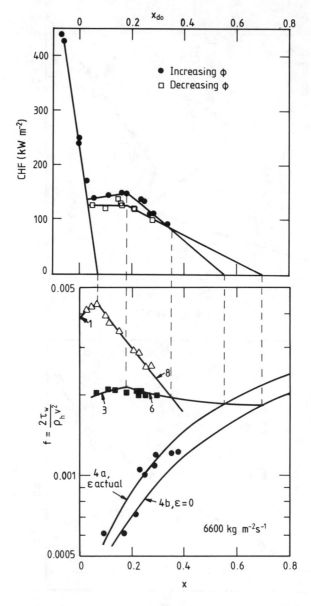

Fig. 8. Comparison of CHF and friction characteristics at G=6600 kg m^{-2} s^{-1}

An Analysis of Critical Heat Flux in Interconnected Subchannels

C. W. SNOEK
Atomic Energy of Canada Limited
Chalk River Nuclear Laboratories
Chalk River, Ontario
Canada K0J 1J0

Abstract

The paper describes an experiment that was designed to get detailed information on the dryout heat flux in subchannels for a range of thermalhydraulic conditions. Several subchannel configurations, formed by placing an insert inside a vertical tube, were studied. Both tube and insert were heated with direct electric current. Different inserts were used to form the following geometries:

a) Three – equal subchannels
b) Three – equal subchannels, but misaligned lengthwise 60 degrees every 50 cm.

The CHF values obtained in the subchannels were determined and compared with predictions from the "CHF Look–Up Table of Groeneveld et al". These comparisons are presented in graphical form as "Discrepancy Factors". The Discrepancy Factor trends of the Three–Equal–Subchannel experiment were analysed both for subchannel and gap. It was found that the Discrepancy Factor is affected by both local quality and mass flux. The analysis resulted in an improved gap and subchannel CHF prediction method.

The CHF prediction method developed for the equal–subchannel configuration was then applied to the 60–degrees–misaligned geometry, to determine the effects of upstream mixing, shape and proximity to another heated wall on CHF. The model predicted the 60–degrees–misaligned subchannel data quite well, indicating that the improved mixing at the misalignment plane does not propagate beyond the next spacer plane.

1. INTRODUCTION

The dryout prediction methods employed by thermalhydraulic codes for the simulation of nuclear reactor circuits are largely based on tube CHF data. The first step taken to reach the objective of this study was to establish the difference between dryout in tubes and subchannels and, if possible, recommend correction factors to the present prediction methodology to improve the prediction accuracy.

The first objective of the experimental program was to get detailed information on the dryout heat flux for different subchannels. To accomplish this, inserts defining the subchannel shapes were placed inside a tube. Both tube and insert were heated with direct electric current. The outer tube was insulated to reduce the heat loss. The studied subchannel shapes are shown in Figure 1. The inside diameter of the outer tube was 1.19 cm (the hydraulic equivalent diameter measured 0.512 cm).

About 67% of the total power was dissipated in the outer tube, to ensure that dryout would first occur in a location where it could be detected with surface-mounted thermocouples.

All experiments were performed at a nominal test-section outlet pressure of 9.6 MPa.

2. EXPERIMENTAL EQUIPMENT AND PROCEDURE

A high-pressure water loop was used for the experiments. The loop operated with inlet pressures up of to 10.3 MPa and a maximum liquid flow of 2 kg/s. The loop could also be operated with a two-phase inlet to the test section with a steam flow of up to 0.35 kg/s.

The first experiments were performed with a three-equal-subchannel test section. The geometrical arrangement is shown in Figure 1. The maximum mass flux that could be achieved was 7 Mg/m^{2}s.

A further experiment was performed to study the effect of flow mixing between subchannels on the CHF. In this case, the insert was centrally located but divided in three sections. Each section was misaligned by 60 degrees with respect to the adjacent section.

Thermocouples were attached on the outside of the outer tube, directly opposite the gaps and subchannels. Several planes of thermocouples were attached at different axial distances from the test section outlet. Dryout was always observed first at the end of the heated length.

The test section was powered by a direct current power supply capable of delivering 190 kW of power to the test section. About 67% of the available power was dissipated in the outer tube, the remainder was consumed by the insert. The maximum heat flux achieved on the outer tube was in the order of 2400 kW/m^2.

The experimental procedure was as follows. First, the loop was adjusted to operate at the particular conditions of interest. The power to the test section was then raised until dryout was observed on the temperature charts. After dryout had been established, the power was backed off to about 95% of the dryout heat flux. Then the power was slowly ramped up while the loop and temperature data were scanned and stored every seven seconds. The power ramping was continued until the power exceeded dryout by approximately 100% or a preset temperature limit was reached.

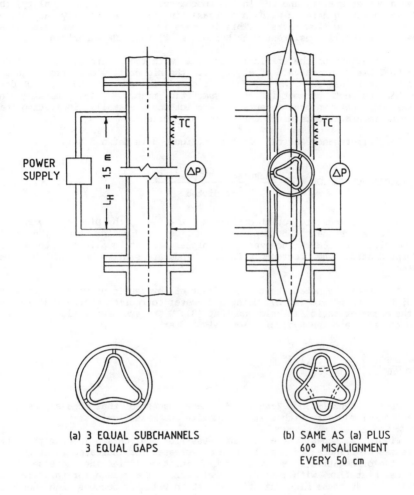

(a) 3 EQUAL SUBCHANNELS
3 EQUAL GAPS

(b) SAME AS (a) PLUS
60° MISALIGNMENT
EVERY 50 cm

Fig. 1 – Geometry of the Test Section and Subchannels.

405

3. EXPERIMENTAL RESULTS

Three-Equal-Subchannel Experiment

The results of the dryout experiments with the Three-Equal-Subchannel test section are shown in Figure 2. This figure shows the relationship between outlet quality and CHF in the subchannel. For constant quality, the CHF increases generally with a decrease in flow, especially at higher qualities and low flow rates. This increase is due to lower entrainment at lower flows and is consistent with Borodin's (1) tube CHF experiments.

A comparison of the Critical Heat Flux with the prediction from the CHF Look-Up Table of Groeneveld et al (2) for the higher flow data from Figure 2 is shown in Figure 3. The CHF-Table is based on over 15 000 tube CHF data and has correct parametric and asymptotic trends. In general, the differences between measurements and predictions are small, indicating that the experimentally determined CHF trends are correct.

The "Discrepancy Factor" plotted in Figure 3 is defined as:

$$\text{DISCREPANCY FACTOR} = \frac{CHF_{measured} - CHF_{predicted}}{CHF_{predicted}} \qquad \qquad ...(1)$$

At low to moderate outlet qualities ($X < 0.4$), the Discrepancy Factors for both subchannel and gap are negative, indicating that the table overpredicts the CHF. However, at higher outlet qualities, the CHF is underpredicted. The Discrepancy Factor trends are similar between subchannel and gap.

It should be noted that the effect of hydraulic diameter on the CHF prediction was considered by using a diameter correction where the exponent of the diameter ratio is held constant (3). The subchannel flow areas were used to calculate the hydraulic equivalent diameter.

$$\frac{CHF\ (D)}{CHF\ (D_o)} = (\frac{D_o}{D})^{1/3} \qquad \qquad ...(2)$$

Mueller-Menzel and Zeggel (4) have suggested that the exponent in Equation (2) may vary with critical quality and mass flux.

Figure 4 shows the effect of mass flux on CHF. In general, CHF increases for increased mass flux and decreases for increased inlet quality (positive or negative). However, at high inlet qualities the increase of the critical heat flux with mass flow levels off. The reason for this trend in CHF is that at these high qualities the steam velocity becomes high enough to increase the liquid entrainment. As the mass flux increases, the increased entrainment has the effect of rapidly depleting the liquid layer at the wall, thereby inducing early dryout. This effect was found in most experiments and is consistent with the CHF-Table.

406

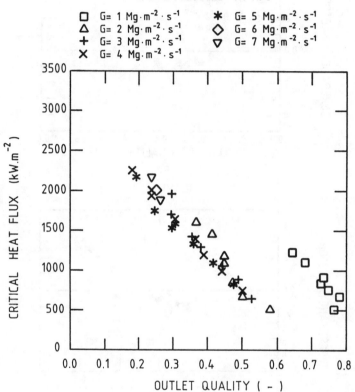

THREE-EQUAL-SUBCHANNEL EXPERIMENT
(SUBCHANNEL)

NOMINAL PRESSURE= 9.6 MPa

☐ G= 1 Mg·m^{-2}·s^{-1} ✳ G= 5 Mg·m^{-2}·s^{-1}
△ G= 2 Mg·m^{-2}·s^{-1} ◇ G= 6 Mg·m^{-2}·s^{-1}
+ G= 3 Mg·m^{-2}·s^{-1} ▽ G= 7 Mg·m^{-2}·s^{-1}
✕ G= 4 Mg·m^{-2}·s^{-1}

Fig. 2 - Effect of Outlet Quality on CHF.

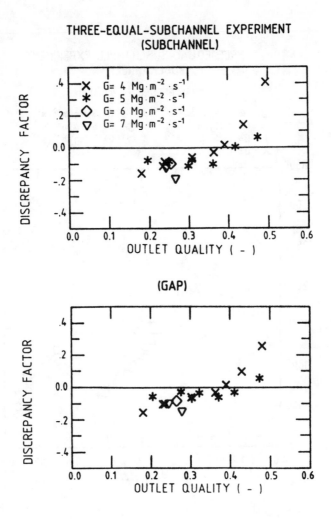

Fig. 3 — Effect of Outlet Quality on the Discrepancy Factor.

THREE-EQUAL-SUBCHANNEL EXPERIMENT

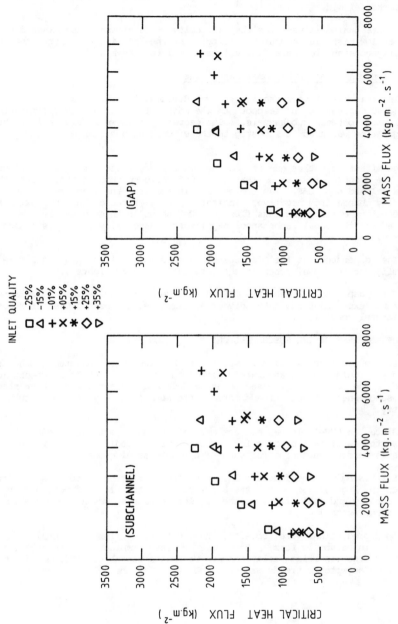

Fig. 4 – Effect of Mass Flux on CHF.

At lower flowrates, the subchannel experienced dryout before the gap. At higher flowrates, however, the gap thermocouples indicated dryout before the subchannel did.

Figure 4 also illustrates the difficulty of determining the exact dryout power at low mass flux and high quality. At the 35% inlet quality, the CHF is fairly constant for mass flux values from 1 to 2 Mg/m^2s.

60-Degrees-Misaligned Subchannel Experiment

In this experiment the insert was divided in three sections, each rotated 60 degrees with respect to the adjacent section, thus simulating misalignment between subchannels. The geometry is shown in Figure 1. This configuration was used to investigate the effects of upstream misalignment on CHF.

The effect of outlet quality on the Discrepancy Factor is shown in Figure 5. The figure indicates that the CHF-Table method overpredicts the subchannel and gap CHF. As the quality increases, the Discrepancy Factors gradually change from negative, indicating an overprediction, to positive. The effect of quality and mass flux on CHF shown in these data is weaker, compared to the equal subchannel experiment, especially in the subchannel geometry.

The dependence of the Discrepancy Factors on quality shown in Figure 5 is similar to the equal subchannel experiment, shown in Figure 3.

The gap and subchannel CHF data from the 60-degrees misaligned subchannel experiment indicated dryout at almost equal heat fluxes for most experimental conditions.

4. DEVELOPMENT OF THE SUBCHANNEL CHF PREDICTION METHOD

In the previous section it was shown that the interconnected subchannel CHF data could not be accurately predicted using the CHF-Table prediction method. The CHF-Table is based on tube data and our experiments have shown that subchannel dryout trends are different from dryout in a tubular geometry.

Since the CHF-Table has correct parametric and asymptotic trends, it was decided to use it as the basis for a more accurate prediction by developing a CHF-Table correction factor for each of the subchannel geometries.

Surprisingly, the functional form of the correction factor is complex. As seen in Figures 3 and 5, the Discrepancy Factor is a function of both local quality and mass flux. (The outlet pressure was not a variable in these experiments.)

Good predictions were obtained when both mass flux and quality were expressed in parabolic form. However, on close examination, the asymptotic trends of the resulting relationship were unacceptable. A good fit with acceptable asymptotic trends was obtained by using the following expression:

410

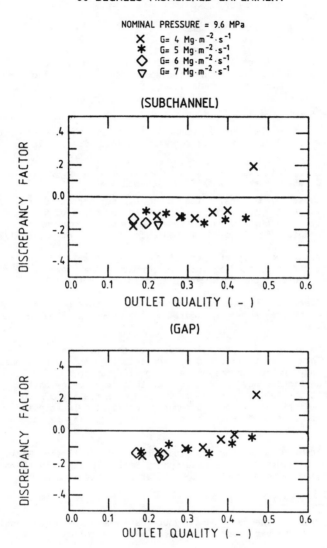

Fig. 5 – Effect of Outlet quality on the Discrepancy Factor.

$$DR = (P1 + P2 * X + P3 * X^{P4}) * (1.0 - P5 * e^{(-P6*G ** P7)}) \quad ...(3)$$

Where DR = Discrepancy Ratio, defined as the ratio of experimental
and predicted CHF,
   G = Mass Flux,
P1-P7 = Constant parameters, and
   X = Dryout Quality.

The particulars of the derived functions are presented in the next sections. The definition of the Discrepancy Ratio is a slight modification from that of the Discrepancy Factor, i.e. DR = 1 + Discrepancy Factor.

Subchannel Discrepancy Ratio

The data from the equal subchannel experiment were used to optimise the coefficients of the Subchannel Discrepancy Ratio. The resulting values are shown below:

$$DR = (0.705+0.742*X-0.725*X^{5.72})*(1.0-0.224*e^{(-0.039*G**3.48)}) \quad ...(4)$$

With these parameters, the original subchannel dryout data were predicted with an average error -0.01% and an RMS error of 6.4%. These error values compare favourably with those commonly found with other dryout prediction models. Equation (4) is valid for positive dryout qualities.

The resulting dryout predictions are shown in Figure 6. The experimental CHF is plotted against the predicted CHF. As can be seen in the figure, most (90%) of the data were predicted within +/- 10%.

The effect of dryout quality and mass flux on the subchannel CHF prediction error are shown in Figure 7. No unusual trends can be detected from this figure.

The parametric and asymptotic trends of the Subchannel Discrepancy Ratio Correlation are shown in Figure 8. From this figure it can be observed that the Discrepancy Ratio increases with quality up to X = 0.7. At higher qualities the Discrepancy Ratio decreases to a value below unity. It should be noted that the CHF-Table predicts CHF = 0 at a quality of 1.0. Since the CHF-Table prediction and the Discrepancy Ratio are multiplied, the exact value of the Discrepancy Ratio close to unity quality is not of practical importance.

The effect of mass flux on the subchannel Discrepancy Ratio is also illustrated in Figure 8. The Discrepancy Ratio increases with mass flux up to 5 Mg/m^{2}s and becomes constant thereafter for constant quality.

Gap Discrepancy Ratio

The equal subchannel gap dryout data were used to optimise the coefficients of the gap Discrepancy Ratio with the following result:

412

THREE-EQUAL-SUBCHANNEL EXPERIMENT

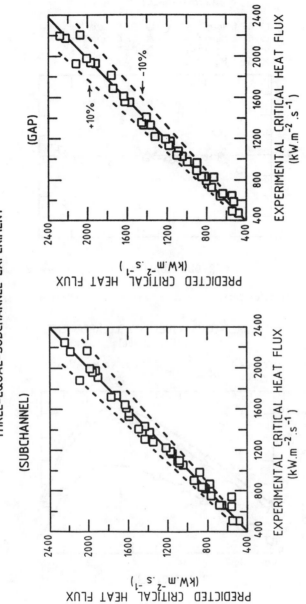

Fig. 6 - Comparison of Experimental and Predicted CHF.

413

THREE-EQUAL-SUBCHANNEL EXPERIMENT
(SUBCHANNEL)

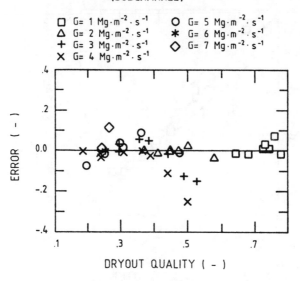

Fig. 7 – Effect of Quality and Mass Flux on the Prediction Error.

THREE-EQUAL-SUBCHANNEL EXPERIMENT
(SUBCHANNEL)

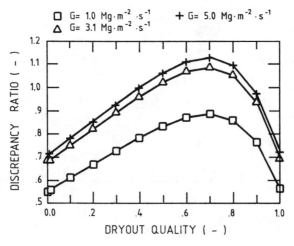

Fig. 8 – Parametric Trends of the Discrepancy Factor.

$$DR = (0.746+0.635*X-0.901*X^{5.72})*(1.0-0.192*e^{(-0.015*G**3.90)}) \quad ...(5)$$

The values are similar to the Subchannel Discrepancy Ratio coefficients. When these coefficients were used to predict the original data, the average and RMS errors were found to be -0.009% and 5.4% respectively. As with Equation (4), Equation (5) is valid for dryout qualities above zero.

The majority (92%) of the data were predicted within +/- 10%, as shown in Figure 6. As in the subchannel case, the error trends for the gap data do not show any bias with mass flux and quality. The parametric and asymptotic trends of the correlation with mass flux and quality are also satisfactory.

5. APPLICATION TO THE MISALIGNED-SUBCHANNEL GEOMETRY

The CHF prediction methods for the equal-subchannel experiment gap and subchannel geometries described in the previous section have also been applied to the 60-degrees-misaligned experiment to investigate the effects of subchannel mixing due to misalignment on CHF.

60-Degrees Misaligned Subchannel Experiment

The CHF prediction method was used to predict both subchannel and gap CHF data of the 60-degrees misaligned experiment. The results are shown in Figure 9. The comparison between experimental and predicted CHF indicates that most of the data are predicted within 10%. Figure 10 shows the effects of mass flux and quality on the misaligned-subchannel data prediction error. The mass flux and quality do not appear to bias the prediction errors.

On average, the subchannel and gap CHF was overpredicted by 3% and 1% respectively. The subchannel mixing, which takes place 50 cm upstream, does not appear to influence the dryout heat flux at the downstream end of the test-section.

6. SUMMARY AND RECOMMENDATION FOR FURTHER WORK

(1) The interconnected subchannel data indicates that the mass flux has a strong effect on CHF in subchannel-shaped geometries at low quality. At higher quality, the effect of the mass flux diminishes.

(2) For similar inlet conditions, the equal-subchannel and 60-degrees-misaligned subchannel experiments gap and subchannel data indicated dryout at almost equal heat flux.

(3) Using the equal-subchannel data, a formulation was developed to correct the difference in the CHF-Table predicted and experimental CHF values. The method uses different constants for the gap and subchannel geometries.

(4) This subchannel-CHF prediction method predicted the 60-degrees-misaligned gap and subchannel data well, indicating that upstream mixing did not affect the downstream CHF.

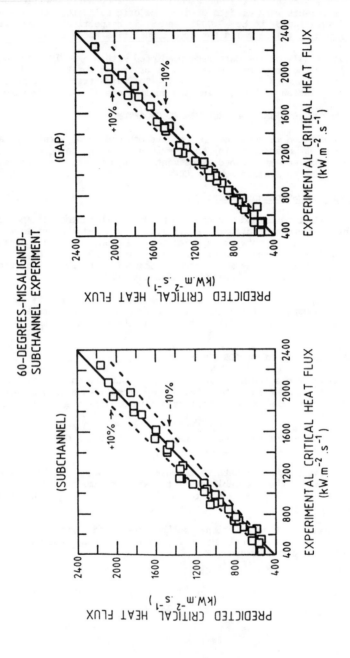

Fig. 9 – Comparison of Experimental and Predicted CHF.

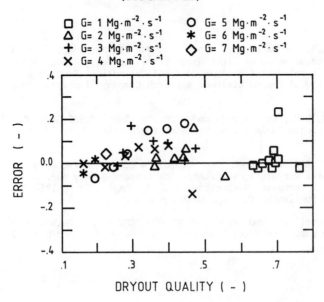

Fig. 10 – Effect of Quality and Mass Flux on the Prediction Error

(5) The data obtained during this study are suitable for use with subchannel codes. In particular, the CHF and intersubchannel mixing models of these codes can be verified using these data. Further work is planned in this area.

ACKNOWLEDGEMENTS

A large number of branch members cooperated in this project. A.H. Smith, K. Moore, W.C. Cameron, K.F. Rudzinski and S.T. Yin performed the experiments. D.E. Bullock, J.W. Martin and B.P. Shaw looked after the on-line data acquisition. W.N. Selander helped with the formulation of the Discrepancy Ratio. F. Mancini did the off-line computing. B.A. Blimkie prepared this report for issue. Their help is very much appreciated. Funding for this project by the CANDU Owners Group is gratefully acknowledged.

REFERENCES

1. Borodin, A.S. and MacDonald, I.P.L. (1984) "An Examination of some Separate Measurements of Light Water (H2O) and Heavy Water (D2O) Flow Boiling Critical Heat Flux in a Vertical Cylindrical Tube Geometry", Atomic Energy of Canada Limited, Research Company, Internal Report CRNL-2645.

2. Groeneveld, D.C., Cheng, S.C. and Doan, T. (1986) "1986 AECL-UO Critical Heat Flux Lookup Table", Heat Transfer Engineering, Vol. 7, Nos.1-2, p. 46-62.

3. Groeneveld, D.C. and Snoek, C.W. (1986) "A Comprehensive Examination of Heat Transfer Correlations Suitable for Reactor Safety Analysis", Multiphase Science and Technology, Vol. 2, Chapter 3, pp 181-274, Hemisphere Publishing Company, Washington D.C.

4. Mueller-Menzel, Th. and Zeggel, W. (1987) "CHF in the Parameter Range of Advanced Pressurized Water Reactor Cores", Nucl. Eng. and Des. 99, pp 265-273.

Simulation of the Two-Loop Test Apparatus (TLTA) Large Break Test 6423 Run 3 with the Goblin Computer Code

DEREK B. EBELING-KONING[1]
Westinghouse Electric Corporation
Pittsburgh, Pennsylvania 15230, USA

Abstract

Westinghouse, in conjunction with ASEA Brown Boveri Atom of Sweden, has developed a Loss of Coolant Accident (LOCA) evaluation model in compliance with Title 10 Part 50.46 and Appendix K of the Code of the Federal Regulations, for licensing BWR reload fuel. This paper briefly describes the GOBLIN thermal-hydraulic system analysis code and then presents an integral qualification simulation of the code against test 6423 run 3 of the Two-Loop Test Apparatus (TLTA) large break series 5A. This test simulated a design large break LOCA in a peak power fuel bundle of a BWR/6 with degraded emergency core cooling system performance.

The GOBLIN simulation of test 6423 run 3 shows excellent agreement with measured data for the important thermal-hydraulics parameters, including system pressure response, mass inventory distribution, regional flow rates and rod temperature response. In addition, a heat-up calculation is presented to demonstrate the bounding conservatism attributed to the use of heat transfer coefficients recommended by 10 CFR 50 Appendix K.

1. INTRODUCTION

For the last decade ASEA Brown Boveri Atom of Sweden, has been using a series of computer codes called GOBLIN, DRAGON, and CHACHA for performing Loss of Coolant Accident (LOCA) licensing analysis for Boiling Water Reactor (BWR) fuel in Sweden and Europe. In 1981, Westinghouse in conjunction with ABB ATOM, initiated development of these codes for LOCA licensing analysis in the United States. The subsequent BWR LOCA Evaluation Model presently is under review by the U. S. Nuclear Regulatory Commission.

The LOCA evaluation model consists of a systems response calculation using the GOBLIN code, a detailed hot fuel assembly calculation using the DRAGON code and a peak plane fuel rod response calculation using the CHACHA code.

Extensive qualification of the codes has been conducted by ABB ATOM and independently by Westinghouse. These qualification programs covered the breadth of separate effects and integral tests of interest to LOCA analysis, and summarized in Table I.

---

[1] Work presented in this paper was conducted between 1981 and 1987. Dr. Ebeling-Koning is currently working under subcontract to ABB ATOM.

TABLE I: GOBLIN/DRAGON/CHACHA EVALUATION MODEL QUALIFICATION

Test	Phenomena	Reference
FIX-II	Non-jet pump plant system blowdown response	[7]
TLTA/4	Jet pump reactor system blowdown response	[2]
TLTA/5A	Jet pump reactor system blowdown and ECC response	[1,3]
TLTA/5C	Jet pump reactor small break system response	[8]
TLTA Core Uncovery	Low flow core boiloff	[9]
Westinghouse G-2	Low pressure top down ECC heat transfer	[10]
FLECHT-SEASET	Reflood heat transfer	[11]
SVEA Spray Cooling	Spray cooling heat transfer	[12,13]
General Electric Level Swell	Level swell during depressurization	[14]
FRIGG	Rod bundle void distribution	[15]
EG&G 1/6 Scale Jet Pump	Jet pump performance	[16]
TVO II Reactor Pump Trip	Reactor neutron kinetics	--

As part of the evaluation model qualification program, GOBLIN simulations of several of the integral large break tests conducted in the Two-Loop Test Apparatus (TLTA) were performed by Westinghouse. One of these simulations was test 6423 run 3 conducted in the TLTA/5A series [1]. This test modelled a General Electric 8 x 8 rod assembly in a prototypical BWR/6 plant design at peak reactor power with degradated Emergency Core Cooling (ECC) system conditions. The focus of this paper is the GOBLIN code predictions of TLTA test 6423 run 3. However, before the simulation is presented, a brief description of the GOBLIN computer code and TLTA facility, shall be given.

2. CODE DESCRIPTION

The GOBLIN computer code utilizes one-dimensional control volumes and flow paths to model a BWR reactor system. The GOBLIN code solves fluid conservation equations for mass, energy, and momentum together with the equation of state for each control volume. The conservation equations in GOBLIN include all terms in the theoretical derivations for one-dimensional, drift-flux, thermal equilibrium formulation (with the exception of the kinetic and potential energy terms in the energy balance). The hydraulic models include constitutive correlations for the calculation of two-phase drift flux flow, two-phase water level tracking, emergency core cooling spray/fluid interaction, critical flow, pressure drops and countercurrent flow limitation (CCFL).

The fuel thermal model calculates the heat transferred from the fuel rods to the coolant. The radial form of the heat conduction equation in the fuel rod is solved by an implicit finite-difference technique using appropriate heat transfer coefficients as boundary conditions. Axial conduction is neglected. The heat transfer coefficients are calculated from the current coolant state as determined by the hydraulic model. Thermal radiation between rods, and between the rods and channel walls is also modelled.

The power generation models account for reactor fission power and decay heat. The reactor fission power is calculated by a point kinetics model which is coupled with the thermal-hydraulics calculation via void, fuel temperature and moderator temperature reactivity feedback. The spatial power distribution is specified to the GOBLIN code from appropriate core and fuel performance codes. The time-dependent power history also may be user specified for simulations of experimental facilities.

The thermal structural model calculates heat conduction through the vessel and internal structures. The model calculates conduction either between adjacent control volumes or between a control volume and a prescribed boundary environment. The structural model accounts for changing hydraulic fluid conditions including moving two-phase water levels.

The GOBLIN code incorporates models for all essential BWR systems and components. These include models for the jet pumps, main recirculation pumps, separators and dryers, the feedwater and steamline systems, reactor protection systems, reactor level measurement system, pressure relief systems, and the emergency core cooling spray and injection systems.

Some of the salient features of GOBLIN are the drift flux model, two-phase level tracking and noding flexibility.

The GOBLIN drift flux model transforms the drift flux correlation into terms of mass flux, and "folds" the correlation into the countercurrent flow correlation. The resultant formulation yields a continuous relation for relative phasic flow which smoothly complies with the countercurrent flow restriction.

The GOBLIN two-phase level tracking model can be designated for any series of vertical nodes. Once specified, the code relocates the nearest control volume boundary at the two-phase level and tracks the movement of the level. The movement of the two-phase level is determined by conserving the phasic mass flows above and below the two-phase boundary.

The GOBLIN code nodalization structure allows for ample flexibility in modelling, including: an unrestricted number of control volumes, flow paths, structural plates, and fuel rods; unlimited number of flow path connections for parallel channel modelling; and modelling of annular fuel rods.

3. FACILITY AND TEST DESCRIPTION

The original Two-Loop Test Apparatus (TLTA) facility was built as part of the BWR Blowdown Heat Transfer Program, sponsored by the U. S. Nuclear Regulatory Commission, Electric Power Research Institute, and General Electric Company. The program was completed in 1975 [2]. The TLTA/5A series of tests were part of the subsequent BWR Blowdown/Emergency Core Cooling Program [1,3], sponsored by the same parties. The objective of this test series was to obtain integral system thermal-hydraulic responses to a large break LOCA including the timing of events and phenomena following actuation of the emergency core cooling systems.

The TLTA/5A facility is a simulated BWR/6 design plant with a full scale General Electric 8 x 8 fuel assembly. A schematic of the facility is shown in Figure 1. The fuel assembly is electrically heated and enclosed in a scaled reactor vessel capable of simulating typical BWR operating pressures and temperatures. All salient features of a BWR reactor were incorporated including guide tube, jet pumps, upper plenum, ECC injection systems, recirculation loops, steam dome, downcomer, and steam separator. A detailed description of the facility is given in Reference [1].

The simulation presented here is of test 6423 run 3. This test modelled a peak power fuel assembly with degraded emergency core cooling performance. A summary of the test initial conditions are given in Table II. The bundle initial power was 6.46 MW, which corresponds to 1.4 times the average bundle power. The ECC flow in test 6423 run 3 was degraded to approximately 70 percent of that in the reference tests and the ECC water temperature was raised by 100 degrees to approximately 212 degrees F. The other key initial conditions were similar to those for the reference test.

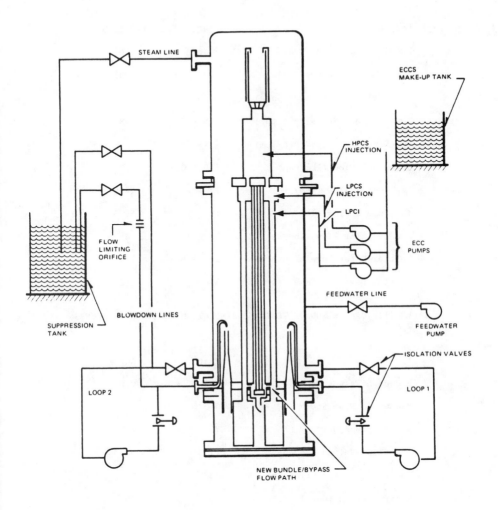

Fig. 1. - Two Loop Test Apparatus Configuration 5A (TLTA 5A) with Emergency Core Cooling Systems.

TABLE II: COMPARISON OF TEST AND SIMULATION INITIAL CONDITIONS

Parameter	Test	GOBLIN
Bundle Power (MW)	6.46	6.46
Steam Dome Pressure (psia)	1037 $\pm$ 5	1031
Lower Plenum Pressure (psia)	1065 $\pm$ 5	1057
Lower Plenum Enthalpy (Btu/lbm)	518 $\pm$ 5	518
Feedwater Enthalpy (Btu/lbm)	41 $\pm$ 2	41
Feedwater Flow (lbm/sec)	1.0 $\pm$ 0.3	1.2
Jet Pump 1 Flow (lbm/sec)	17 $\pm$ 2	18.5
Jet Pump 2 Flow (lbm/sec)	19 $\pm$ 2	19.0
Bundle Inlet Flow (lbm/sec)	33 $\pm$ 5	34
Initial Water Level (inch elev.)	123 $\pm$ 6	122
Initial Downcomer Mass (lbm)	310	312

TABLE III: COMPARISON OF TEST AND SIMULATION SEQUENCE OF EVENTS

EVENT	Time (Seconds)	
	Test	GOBLIN
Blowdown Values Open	0.0	0.0
Feedwater Flow Stops	0.5	1.0
Jet Pump Uncovery	7	7
Steamline Valves Closed	11.5	11.5
Lower Plenum Bulk Flashing	15	17
Recirculation Loop 1 Isolated	20	20
HPCS Injection Begins	27	21
LPCS Injection Begins	65	67
LPCI Flow Begins	72	72
End of Simulation	400	250

The test was initiated by opening two blowdown lines to simulate a double-ended guillotine break of the suction leg of the recirculation line. The vessel depressurized over approximately 100 seconds, during which the feedwater was terminated, the steamline valves were closed, recirculation loop 1 was isolated (to isolate the large, atypical water volume), and the safety injection system was activated. The test was run through a programmed decay power to 400 seconds, at which time the test was terminated. More details of the test transient are discussed in the section on the simulation results.

The test series demonstrated the ability of the ECC system to cool the bundle following a postulated LOCA. It also identified and demonstrated reflooding and recovery of the fuel bundle due to CCFL at the bundle inlet, before completed refilling of the lower plenum.

4. SIMULATION MODEL AND ASSUMPTIONS

The GOBLIN simulation model of the TLTA/5A facility consists of 56 hydraulic control volumes and 41 structural components (metal plates). Nine of the control volumes are in the core, each containing five rod groups for a total of 45 rod segments. The simulation includes models for the two external recirculation loops, two jet pumps, separator, orificed break lines, and ECC spray and injection. Vessel heat losses were modelled using a prescribed ambient environment boundary condition. The heater rods were modelled in a total of 11 radial nodes, comprised of mixed oxide filler, vapor gap, and Inconel cladding. Two-phase level tracking was incorporated in the downcomer annulus and upper plenum regions.

The GOBLIN simulation of TLTA case 6423 run 3 excluded several requirements of the Appendix K evaluation model in order to accurately evaluate the code. The Appendix K requirements eliminated from the system simulation are:

- Rewetting of the fuel rods was allowed

- A best estimate critical break flow model, based on experimental data for the TLTA test orifices [4] was used. Specifically the homogeneous equilibrium model was used, with a subcooled flow multiplier. This model replaced the Appendix K required Moody critical flow model.

- The actual test power history was used whereas Appendix K requires the ANS 1971 decay heat standard plus 20 percent conservatism.

The next section presents a comparison of the GOBLIN simulation with test data using the model described above. In addition a fuel heat-up calculation is presented to demonstrate the conservatism of the BWR LOCA Evaluation Model with respect to the variations in the individual rod temperature measurements.

425

5. SIMULATION RESULTS

Test 6423 run 3 was initiated from a power of 6.46 MW, however, the TLTA facility was equipped to maintain a cooling capacity of only 2.0 MW. Hence, the test was initialed from a transient condition. The GOBLIN simulation was initialized to a steady state condition at 6.46 MW and then the boundary conditions were adjusted to match the test initial conditions. The test and simulation initial conditions are summarized in Table II.

The basic phenomena of the large break LOCA and TLTA/5A series tests are described extensively in the literature [3], hence, only brief summaries are given in this paper. A summary of the transient events is given in Table III. The test simulation was initiated by opening the blowdown lines at time zero. The GOBLIN simulation system pressure response is shown in Figure 2. The simulation agrees very well with the measured pressure. The initial rapid depressurization is due to the subcooled liquid break flow. The pressure recovery from about 4 to 7 seconds is a consequence of the rapid steamline valve closure (see Figure 3). At 7 seconds the downcomer level uncovers the jet pump, allowing vapor to flow out the drive line side of the break and causing a return of the system depressurization. The jet pump uncovery is apparent in Figure 4, where the intact jet pump performance is severely degraded after 7 seconds. A more rapid depressurization occurs at about 10 seconds once the downcomer empties and vapor also flows out the recirculation line suction side of the break. The remainder of the depressurization follows the test data very closely with a slightly lower final pressure.

The bundle inlet flow for the initial phase of the transient is shown in Figure 5. The general agreement with the data is good. The initial drop in bundle flow is a little sharper in the test. This is a reflection of the initial flow reversal in the broken jet pump (Figure 6) which is attributed to the initial, rapid nonequilibrium break flow out the broken jet pump drive line. The GOBLIN equilibrium code cannot capture this small nonequilibrium effect. The start of lower plenum flashing also is visible in Figure 5 in the rise in bundle inlet flow at approximately 15 seconds.

The total vessel mass inventory is shown in Figure 7. The good agreement of the total mass inventory and system pressure responses confirm the calculation of the break flow throughout the transient. (Accurate direct measurements of the break flow in the test were not available, see Reference [3], Appendix H.)

A comparison of the mass inventory distribution throughout the transient is shown in Figures 8 and 9. The agreement in the trends and timing of events is quite good. The vessel mass inventory depletes as the break drains the downcomer, followed by the upper plenum, bundle, bypass and lower plenum. Fluid flashing expels the guide tube liquid inventory. Following actuation, the ECC systems start to replenish the upper plenum inventory, which subsequently drains through the bundle and bypass to the lower plenum and guide tubes. Once the vessel is depressurized inventory slowly recovers in the guide tubes, bypass, core, and lower plenum. The downcomer mass inventory agreement is excellent. The bypass and guide tube masses are also in good agreement when considering the offset in the

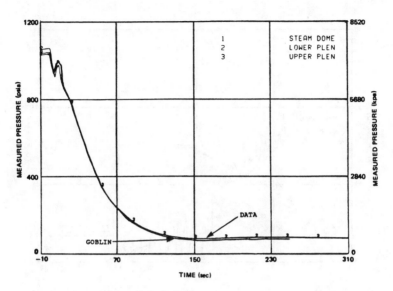

Fig. 2. - System Pressure Response.

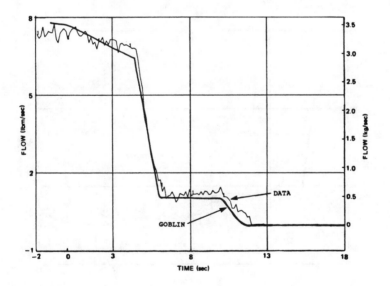

Fig. 3. - Steamline Mass Flow Rate Boundary Condition.

427

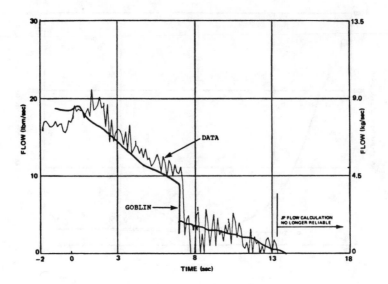

Fig. 4. - Intact Loop Jet Pump Mass Flow Rate.

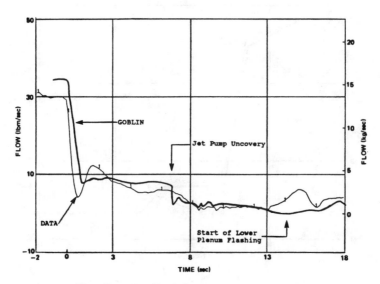

Fig. 5. - Bundle Inlet Mass Flow Rate.

428

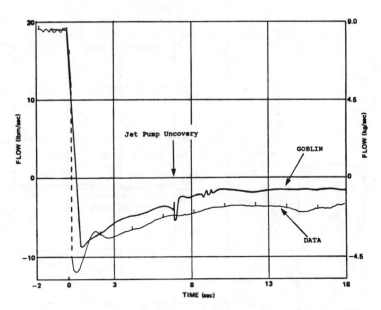

Fig. 6. - Broken Loop Jet Pump Mass Flow Rate.

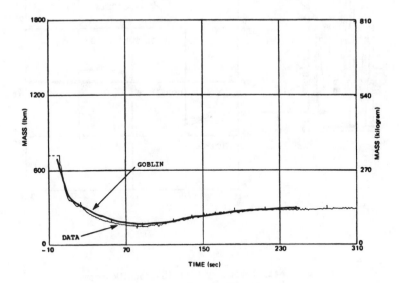

Fig. 7. - Total Vessel Fluid Mass Inventory.

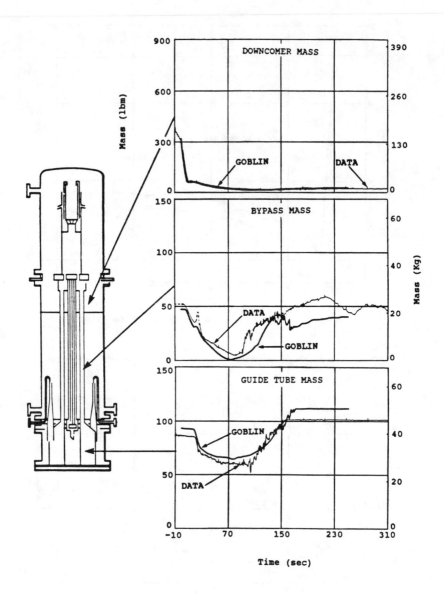

Fig. 8. - Mass Inventory Distribution (Downcomer
Bypass, and Guide Tubes).

430

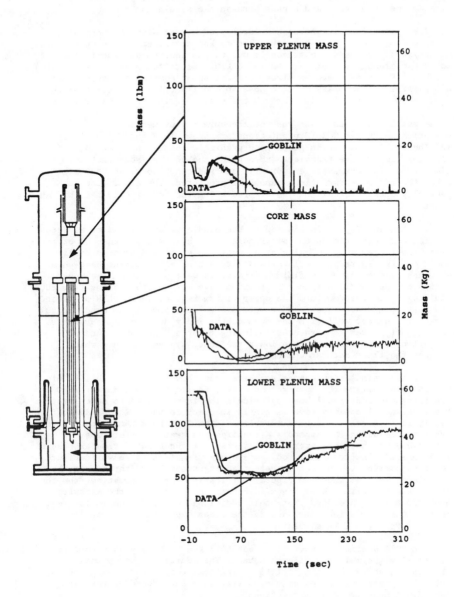

Fig. 9. - Mass Inventory Distribution (Upper Plenum,
Core, and Lower Plenum).

431

initial and final states. The larger initial guide tube mass and smaller bypass mass are due to a deviation in the definition of the boundary between the guide tube and bypass between the simulation and the test.

The upper plenum, bundle, and lower plenum mass distributions also have good agreement with the data. The marked deviation in the upper plenum/separator mass comparison is due to the continuous redefinition of the upper plenum control volumes in GOBLIN, due to two-phase water level tracking. For this reason a direct comparison of equivalent regions was not possible. In summary, GOBLIN does an excellent job of predicting the mass inventory distribution through the vessel during the LOCA transient.

Comparisons of the test rod thermocouple measurements at various elevations with the GOBLIN predictions are shown in Figures 10, 11, and 12. The GOBLIN simulation gives good prediction of the average rod cladding temperature transients throughout the bundle. Note that the simulation provides average hydraulic and rod conditions, so that all local thermocouple variations cannot be predicted. The simulation generally does an excellent job of predicting the rod dryout, heat-up, and rewet.

The initial dryout in the top of the bundle is calculated to be later in time and more pronounced (see Figure 12). This is a consequence of the later and longer drop in calculated bundle inlet flow discussed earlier and shown in Figure 5. The lower and middle bundle dryout and heat-up at 40 seconds, due to bundle fluid drainage, is well predicted. The upper region of the bundle does not heat-up as severely due to cooling from ECC fluid being introduced into the upper plenum region. The GOBLIN predicted rod heat up and subsequent rod rewet, due to recovery of the bundle mass inventory, agrees well with the mean of the individual rod measurements. The slightly lower temperatures following rewet at the lower bundle elevations are a result of the lower predicted system pressure which reduces the fluid saturation temperature.

The test simulation comparison presented above shows the ability of GOBLIN to calculate the average thermal-hydraulic response during a LOCA transient. An additional heat-up calculation is presented to demonstrate the substantial conservative margin is inherent in the BWR LOCA Evaluation Model. In this calculation, simulation of the TLTA peak temperature plane was repeated using the Appendix K required rod heat transfer. The prescribed rod heat transfer coefficients as a function of time are shown in Figure 13. The resultant rod temperature transient is shown in Figure 14. A comparison to the bundle peak cladding temperature throughout the transient, is also shown in Figure 15. Clearly, the conservatism from the Appendix K prescribed heat transfer bounds the scatter in the measured peak rod temperature data. Note that additional conservatism is inherent in the LOCA Evaluation Model due to the other Appendix K assumptions excluded from the TLTA simulation.

The GOBLIN simulation of TLTA test 6423 run 3 shows good prediction of all the key thermal-hydraulic phenomena. The additional hot plane calculation shows that the variations in local temperature measurements are well within the bounds of the conservatism introduced by the uses of Appendix K heat transfer assumptions.

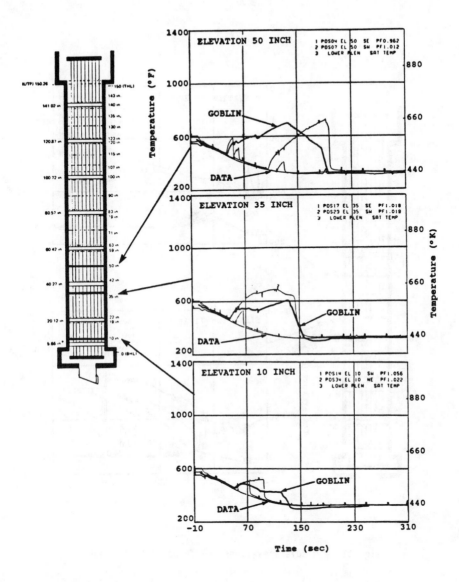

Fig. 10. - Lower Elevations Rod Temperature Response.

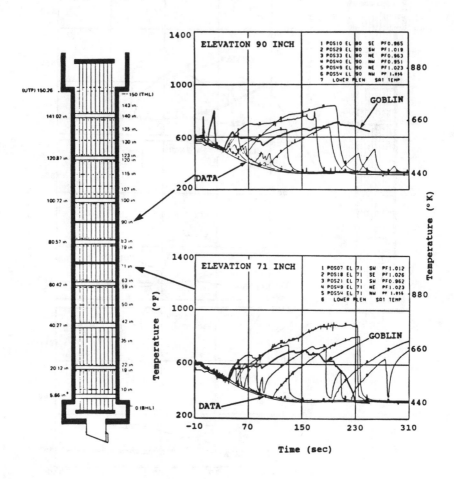

Fig. 11. - Middle Elevations Rod Temperature Response.

434

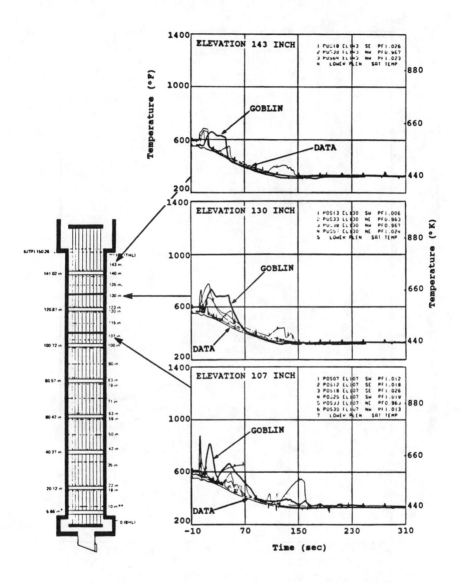

Fig. 12. - Top Elevations Rod Temperature Response.

435

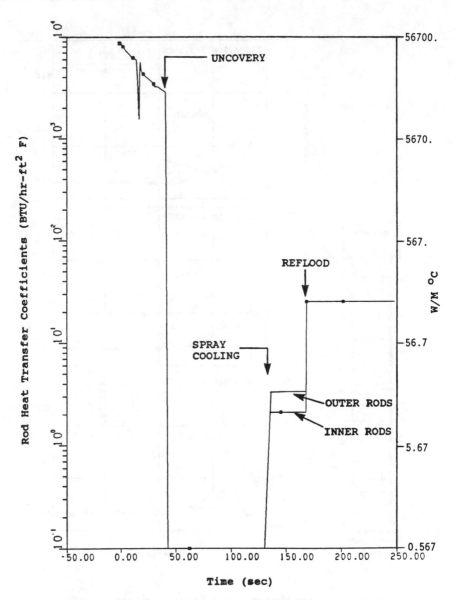

Fig. 13. - Conservative Hot Plane Calculation Prescribed
Rod Surface Heat Transfer Coefficients.

436

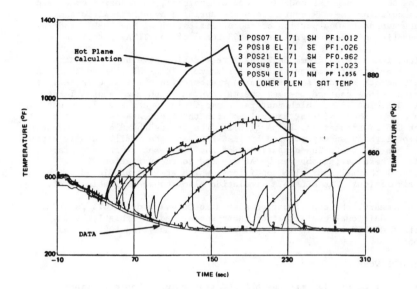

Fig. 14. - Conservative Hot Plane Calculation Cladding
Temperature Comparison.

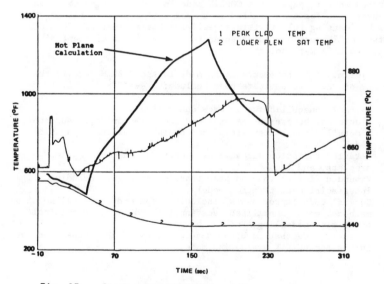

Fig. 15. - Conservative Hot Plane Calculation Bundle
Peak Cladding Temperature Comparison.

6. SUMMARY AND CONCLUSIONS

The GOBLIN computer code is a one-dimensional, equilibrium code used for calculating the BWR system response during a hypothetical loss of coolant accident. One portion of the qualification of the GOBLIN code was a simulation of the integral test 6423 run 3 of the TLTA/5A test series.

Individual tests from the TLTA/5A test series have been simulated using a variety of computer codes, for example TRAC-BWR [5] and RELAP5YA [6]. These other codes show similar agreement with the test data. All of the simulations capture the key thermal-hydraulic phenomena of a large break LOCA. TRAC-BWR, which features a two-fluid (6 conservation equation) formulation, shows similar deviations with the test data as were observed in the GOBLIN simulation. In particular, the drained bundle mass inventory is also slightly underpredicted by TRAC-BWR, and the refill rate is more rapid than the test. The comparisons of the rod temperatures are also similar. These deviations can be attributed to the variations in subchannel thermal-hydraulic conditions, which are not predicted by the code simulations. The RELAP5YA simulation also shows comparable results.

In summary, the qualification simulation presented in this paper along with the additional qualification tests summarized in Table I, have confirmed GOBLIN as a viable computer code for analyzing LOCA system responses.

7. FUTURE WORK

In order to utilize the new NRC ECC system rule change to improve the operational flexibility for nuclear power plants, ABB ATOM has initiated a best-estimate BWR LOCA methods development program. These methods are based on the GOBLIN, DRAGON, and CHACHA codes. As was discussed and demonstrated in this paper, the GOBLIN code included many of the advanced thermal-hydraulic features desired in a best-estimate code, although the Appendix K licensing rule restricts realizing these benefits. For this reason the GOBLIN code was selected as the basis for development of a best-estimate LOCA methodology.

The major model improvements being made to the GOBLIN code in the development of the best-estimate code include:

- Thermal nonequilibrium in each control volume;
- Liquid may be present as droplets and films on solid surfaces;
- A detailed rod, water cross, and canister quench propagation model;
- Detailed radiation exchange between rods, channels, water droplets, and liquid films;
- Expanded flow regime and heat transfer package, including interfacial heat transfer models;
- General code improvements, including advanced numerical solution methods, more generalized user inputs, and improved critical break flow modelling; and
- A single code that combines the functions of the three codes used in the Appendix K evaluation model.

This single consolidated code will enable more accurate calculations of the important LOCA phenomena, including subcooled countercurrent flow breakdown, core vapor superheat in the presences of droplets, rod rewet from quench front propagation, and feedback between the fuel rod heat-up, hot channel thermal-hydraulic, and system responses.

With these more accurate thermal-hydraulic models and the revised licensing requirements, it is expected that substantial LOCA margin can be realized. It is expected that 500 to 700 degrees F reduction in peak cladding temperature will be realized for most reactor applications. This margin will be available to relax many of the restrictive plant operation limits that were caused by the conservative Appendix K LOCA evaluation model and requirements.

Acknowledgements

The GOBLIN code was developed by ABB ATOM of Sweden. The author wishes to thank H. Wijkström of ABB ATOM, Sweden, and A. Cheung, M. Nissley, R. Jakub, J. Dederer, J. Besspiata and S. Saunders of Westinghouse for their assistance and consultation while performing this work.

REFERENCES

1. W. J. Letzing, "BWR Blowdown/Emergency Core Cooling Program Preliminary Facility Report for the BD/ECC1A Test Phase," GEAP-23592, General Electric Company, December 1977.

2. R. Muralidharan, "BWR Blowdown Heat Transfer Program Final Report," GEAP-21214, General Electric Company, February 1976.

3. L. S. Lee, et al., "BWR Large Break Simulation Tests - BWR Blowdown/Emergency Core Cooling Program, Volume 1 and 2," NUREG/CR-2229, U. S. Nuclear Regulatory Commission, July 1982.

4. A. F. Morrison, "Blowdown Flow in the BWR BDHT Test Apparatus," GEAP-21656, General Electric Company, October 1977.

5. Md. Alamgir, "BWR Refill-Reflood Program Task 4.8 - TRAC-BWR Model Qualification for BWR Safety Analysis Final Report," GEAP-22049, General Electric Company, July 1983.

6. R. K. Sundaram, et al., "Simulation of Large Break Tests in the Two-Loop test Apparatus (TLTA) with the RELAP5YA Computer Program," AIChE Symposium Series, Heat Transfer - Pittsburgh, No. 257, Volume 83, 1987.

7. L. Nilsson, P. Å. Gustavsson, "FIX-II - LOCA Blowdown and Pump Trip Heat Transfer Experiments," Studsvik Report NR-83/238, February 1983.

8. W. S. Hwang, "BWR Small Break Simulating Tests with and without Degraded ECC systems," GEAP-24963, March 1981.

9. "BWR Low-Flow Bundle Uncovery Test and Analysis," EPRI-NP-1781 (NUREG/CR-2231), June 1982.

10. "Heat Transfer Above the Two-Phase Mixture Level Under Core Uncovery Conditions in a 336-Rod Bundle," EPRI NP-1692, January 1981.

11. "PWR FLECHT SEASET 21-Rod Bundle Flow Blockage Task Data and Analysis Report," EPRI NP-2014 (NUREG/CR-2444), September 1982.

12. O. Nylund, "Spray Cooling Experiment on SVEA Water Cross BWR Fuel," Presented at the European Two-Phase Flow Group Meeting, Brussels, Belgium, May 30 - June 2, 1988.

13. R. Eklund and H. Wijkström, "SVEA Fuel Spray Cooling Experiments and Their Application in LOCA Analysis," Presented at the International Conference on Thermal Reactor Safety, ENS/ANS Topical Meeting, Avignon, France, 2-7 October 1988.

14. "Loss-of-Coolant Accident and Emergency Core Cooling Models for General Electric Boiling Water Reactors," NEDO-10329, April, 1971.

15. O. Nylund et al., "FRIGG Loop Project," FRIGG, R4-494/RL.

16. H.S. Crapo, "LOFT Test Support Branch Data Abstract Report: One Sixth Scale Model BWR Jet Pump Test," EGG-LOFT-5063, November 29, 1979.

Effect of Interfacial Mass and Momentum Transfer and Wall Heat Transfer Models on PWR ECC Bypass

N. K. POPOV*
University Kiril and Metodij
Department of Electrical Engineering
Skopje, 91000, Yugoslavia

U. S. ROHATGI
Brookhaven National Laboratory
Department of Nuclear Energy
Upton, New York, 11973, USA

Abstract

The effect of interfacial mass and momentum transfer and wall heat transfer on Lower Plenum refilling rate and Downcomer ECC bypass phenomena during the refill phase of a large break PWR loss of coolant accident (LBLOCA) is assessed. A two-dimensional transient two-phase two-fluid model of PWR Downcomer and Lower Plenum is developed. The results from this study indicate that the interfacial momentum transfer has the most influence on the Lower Plenum refilling rate, followed by interfacial mass transfer and wall heat transfer.

1. INTRODUCTION

A cold leg large break loss-of-coolant accident (LBLOCA) in PWR is a design basis accident which is analyzed for any PWR as part of licensing process. This hypothetical accident can be divided into three phenomeno-logically distinct phases [1], namely, blowdown, refill and reflood phase. During the blowdown phase the system depressurizes to around 40 bar, and core is completely voided. The fuel cladding continues to heat up due to decay heat generation in the fuel and poor heat transfer in the core. The refill phase begins with the initiation of Emergency Core Coolant (ECC) injection in the intact loops at the time when the system pressure is around 40 bar. The last phase of the accident, reflood phase, begins when the Lower Plenum is full and the ECC starts to fill the core.

The phenomena occurring during the refill phase of LBLOCA have strong influence on the eventual reflood peak clad temperature and on the likelihood of fuel damage. The steam produced in the Lower Plenum flows into the Downcomer and opposes the ECC (Accumulator and safety injection) flow. Initially all the ECC is bypassed through the broken cold leg. As the accident progresses, the steam flow to the Downcomer decreases due to the reduction in flashing and increase in condensation. The ECC is finally able

---

*Presently a post-doctoral fellow of the Natural Sciences and Engineering Research Council of Canada, with Atomic Energy of Canada Limited, at Whiteshell Nuclear Research Establishment, Pinawa, Manitoba, ROE 1L0, Canada.

to penetrate the Lower Plenum. The schematic view of the reactor interior in case of ECC bypass phenomena is shown in Figure 1. The flow in the Downcomer during the refill phase is multidimensional and goes through many flow regimes. There is strong coupling between the thermal and hydraulic processes.

In this paper an effort has been made to assess the effect of three most important processes, i.e. correlations, namely, interfacial friction, interfacial mass transfer by phase change, and wall heat transfer on Lower Plenum refilling rate.

2.   OUTLINE OF MATHEMATICAL MODEL

Applying a standard mathematical-numerical technique [2], a transient two-dimensional two-fluid mathematical model has been developed and used for analysis of ECC bypass/refill phenomenon [3].

Modeling approximations

Several important approximations have been introduced into the model, to develop an efficient numerical procedure [3]. One of the most important approximations was two-dimensional representation of the PWR Downcomer and Lower Plenum. Figure 2 shows the transformation from three-dimensional real experimental facility into two-dimensional model. The radial variation of all parameters has been neglected, thus the flow was modelled only in the vertical and azimuthal directions. The width of the two-dimensional representation was assumed to be equal to the Downcomer gap width. Thus, as seen from Figure 2, the length of the Lower Plenum was greatly enlarged to preserve the Lower Plenum volume. However, this spatial distortion of the Lower Plenum was considered to have negligible influence on the results, as the bypass phenomenon takes place in the Downcomer.

The Downcomer flow was assumed azimuthally symmetrical with respect to the broken cold leg entrance. This assumption allows the modeling of one half of the Downcomer annulus, as indicated on Figure 2. That greatly reduced computer time.

The heat conduction in the walls, surrounding the Downcomer and Lower Plenum, was modeled only in radial direction. The walls were considered "thermally thick" in the radial direction, i.e. they were modeled as infinite thermal reservoir with respect to the total time of interest. Thus, the temperature deep in the wall interior remained constant through the transient and equal with the initial temperature [3]. To simplify the numerical solution of heat conduction in the walls, an exponential radial temperature variation was assumed in the wall. It was further assumed that the wall heat flux during the transient never reached the critical heat flux (CHF), and that there was no heat generation on the wall.

In this model, the steam coming in reverse direction from the core was assumed to be always at saturation, and the heat transfer between steam and walls was neglected. Hence, the steam heat conservation equation was omitted in the model. All fluid thermal and transport properties were considered

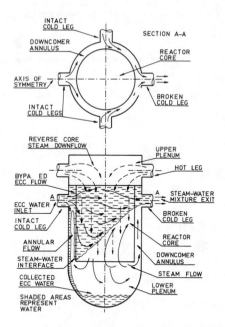

Fig. 1. – Schematic View of Reactor Interior
During ECC Bypass Phenomenon

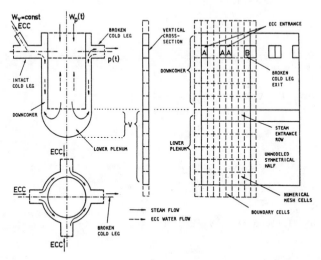

Fig. 2. – Transformation of PWR Downcomer and Lower Plenum
into a Two-dimensional Numerical Model

variable in time, but constant in space. The water properties were estimated at the pressure and temperature at the ECC inlet in the Downcomer, while the steam properties were estimated in terms of pressure at the break.

Conservation Equations

The liquid and vapor mass conservation equations have the following form [4]:

$$\frac{\partial}{\partial t}[(1-\alpha)\rho_f] + \rho_f \nabla[(1-\alpha)\vec{V}_f] = -\Gamma \tag{1}$$

$$\frac{\partial}{\partial t}[\alpha\rho_g] + \rho_g \nabla[\alpha\vec{V}_g] = \Gamma \tag{2}$$

The liquid and vapor momentum conservation equations [4] are as follows

$$\frac{\partial}{\partial t}[(1-\alpha)\rho_f\vec{V}_f] + \rho_f\nabla[(1-\alpha)\vec{V}_f\vec{V}_f] = -(1-\alpha)\nabla p + \mu_f[\nabla(1-\alpha)\nabla]\vec{V}_f$$
$$+ (1-\alpha)\rho_f g - \Gamma\vec{V}_{fi} - C_{wf}|\vec{V}_f|\vec{V}_f - C_{if}|\vec{V}_f - \vec{V}_g|(\vec{V}_f - \vec{V}_g) \tag{3}$$

$$\frac{\partial}{\partial t}[\alpha\rho_g\vec{V}_g] + \rho_g\nabla[\alpha\vec{V}_g\vec{V}_g] = -\alpha\nabla p + \mu_g[\nabla\alpha\nabla]\vec{V}_g$$
$$+ \alpha\rho_g g + \Gamma\vec{V}_{gi} - C_{wg}|\vec{V}_g|\vec{V}_g - C_{ig}|\vec{V}_g - \vec{V}_f|(\vec{V}_g - \vec{V}_f) \tag{4}$$

The simplified liquid energy conservation equation [4] is as follows,

$$\frac{\partial}{\partial t}[(1-\alpha)\rho_f c_p,^{T}_f] + \rho_f c_{pf}\nabla[(1-\alpha)\vec{V}_f T_f] = -\Gamma \cdot c_{pf}T_{if}$$
$$+ k_f{}^{T}(1-\alpha)\nabla T_f] + q_{wf}/L_{wf} + q_{if}/L_{if} \tag{5}$$

The vapor generation rate [4] consistent with saturated vapor assumption is defined as follows,

$$\Gamma = -\frac{q_{if}}{h_{lat}L_{if}} = -\frac{h_{if}(T_s - T_f)}{h_{lat}L_{if}} \tag{6}$$

The liquid energy equation (5) is written in terms of liquid superheat T_{fs} is as given here,

$$\frac{\partial}{\partial t}[(1-\alpha)T_{fs}] + \nabla[(1-\alpha)\vec{V}_f T_{fs}] = \frac{k_f}{\rho_f c_{pf}}\nabla[(1-\alpha)\nabla T_{fs}] \tag{7}$$
$$+ \frac{q_{wf}}{\rho_f c_{pf}L_{wf}} - \frac{\Gamma \cdot i_{lat}}{\rho_f c_{pf}}$$

The wall interior heat conduction conservation equation has the usual form [5]:

$$\rho_w c_w \frac{\partial T_w}{\partial t} = k_w \frac{\partial^2 T_w}{\partial r^2} \tag{8}$$

Exponential temperature variation was assumed in the walls, such as,

444

$$T_w(r) = T_{w\infty} + (T_{wf}-T_{w\infty}) \, e^{-r/R} \tag{9}$$

where $R(x,y,t)$ is wall cooling depth. Substituting equation (9) into equation (8), along with corresponding minor approximations [3], the wall cooling depth differential equation is obtained as follows,

$$\frac{\partial R^2}{\partial t} = \frac{k_w}{\rho_w c_w} \tag{10}$$

Heat flux is continuous at the wall-fluid interface. The heat flux from the liquid side is defined as,

$$q_{fw} = h_w(T_{wf}-T_f) \tag{11}$$

while the heat flux from the wall side, using equation (9), is defined as,

$$q_{wf} = k_w \left.\frac{\partial T_w}{\partial r}\right|_{r=0} = \frac{k_w}{R} (T_{w\infty}-T_{wf}) \tag{12}$$

The wall surface temperature is obtained by equating heat fluxes from equations (11) and (12) as,

$$T_{wf} = T_f + \frac{k_w(T_{w\infty}-T_f)}{R \cdot h_w + k_w} \tag{13}$$

Finally, substituting equation (13) into equation (11), the wall heat flux term (which is needed in the liquid heat conservation equation) is obtained as:

$$q_{wf} = \frac{k_w}{R + \dfrac{k_w}{h_w}} (T_{w\infty} - T_f) \tag{14}$$

Numerical procedure

Equations (1),(2),(3),(4),(7) and (10) constitute a set of coupled partial differential equations. They are solved numerically, by using a standard numerical semi-implicit finite-difference technique [2], with the Newton-Raphson iterative procedure [3]. As shown in Figure 2, the PWR Downcomer and Lower Plenum were unwrapped in two-dimensional geometry and subdivided in numerical mesh cells of equal axial and azimuthal step length. Corresponding boundary conditions were applied at the ECC inlet, steam inlet, and broken cold leg. A detailed explanation of the numerical procedure is given in [3]. Based on this numerical procedure, a computer program, named REFILL, was created, and used for this analysis.

3. CONSTITUTIVE CORRELATIONS

A number of constitutive correlations for mass, momentum and heat transfer are needed in the model, and they have been explained in detail elsewhere. In this paper, only those correlations are given, whose effect on the Lower Plenum filling rate has been assessed, such as correlations for interfacial momentum transfer by interfacial friction, interfacial mass transfer by phase change, and wall heat transfer.

In the present study, a simple two-phase flow regime map was used. It consisted of three flow regimes: bubbly, intermittent and annular [3]. Bubbly flow regime was assumed for void fraction $\alpha < 0.3$, while annular flow regime for void fractions $\alpha > 0.75$. In the range for $0.3 \leq \alpha \leq 0.75$, a transient intermittent regime was assumed, in which all flow parameters, i.e. correlations, were linearly interpolated between the corresponding boundary values belonging to bubbly and annular flow regimes.

Interfacial Friction

For bubbly flow regime, the interfacial friction parameter in the equations (3) and (4) as used in TRAC-PF1 [6], and explained in [7] is defined as:

$$(C_i)_b = \frac{3(f_i)_b \alpha \rho_f}{4 \cdot D_b} \tag{15}$$

For the interfacial friction coefficient, $(f_i)_b$, the correlation obtained from the standard formula for a sphere ([8], p. 366), used and modified in the TRAC-PF1 code [6], and explained in [7], is also used here as follows,

$$
\begin{aligned}
(f_i)_b &= 240 & \text{for} & & Re_b &< 0.1 \\
(f_i)_b &= 24/Re_b & \text{for} & & 0.1 \leq Re_b &< 2.0 \\
(f_i)_b &= 18.7/Re_b^{0.68} & \text{for} & & 2.0 \leq Re_b &\leq 248.13 \\
(f_i)_b &= 0.44 & \text{for} & & 248.13 \leq Re_b &
\end{aligned}
\right\} \tag{16}
$$

where the the Reynolds number for bubbles is defined as $Re_b = \rho_f V_r D_b / \mu_f$, and the bubble diameter is defined as $D_b = \sigma We_b / (V_r^2 \rho_f)$.

For annular flow regime, the interfacial friction parameter in the equations (3) and (4) as used in TRAC-PF1 [6], and explained in [7] is defined as,

$$(C_i)_a = \frac{2(f_i)_a \rho_g}{d_{hy}} \tag{17}$$

The interfacial friction coefficient, $(f_i)_a$, is defined by two correlations. The Wallis correlation [9] has the following form:

$$(f_i)_a = 0.005 \ (1 + 300 \cdot \delta/d_{hy}) \tag{18}$$

The Popov and Rohatgi correlation [10] has been developed for annular counter-current flow. It takes into account the effect of interfacial mass transfer by droplet entrainment from the film, and has the following form:

$$(f_i)_a = \left(0.005 + 1.5 \cdot \frac{\delta}{d_{hy}}\right) \cdot \max\left\{1, \left[\frac{|V_r|}{(V_r)_{crit}}\right]^n\right\} \tag{19}$$

where n=2.5, and the critical relative velocity for droplet entrainment initiation [10] is defined as,

$$(V_r)_{crit} = \frac{2.6}{\mu_f} \cdot \sqrt{\frac{\rho_f}{\rho_g}} \cdot N_\mu^{0.8} \cdot Re_f^{-0.2} \qquad (20)$$

with dimensionless viscosity number of the following form:

$$N_\mu = \mu_f \left[\rho_f \sigma \sqrt{\sigma/g(\rho_f - \rho_g)} \right]^{-0.5} \qquad (21)$$

Interfacial Mass Transfer by Phase Change

In this model, the interfacial mass transfer by phase change has been modeled using the modified Nigmatulin correlation [11,3], where the interfacial mass transfer is proportional to the magnitude of phase mixing, and has the following form:

$$\Gamma = \Gamma_c \alpha (1-\alpha) \frac{T_s - T_f}{T_s} \qquad (22)$$

where Γ_c is semi-empirical constant, and its optimal value was estimated in [3] as $\Gamma_c = -4400$ kg/m^{3}s.

Correlation (22) is particularly suitable for condensation dominated processes, which occur during refill phase of LBLOCA in the PWR vessel. The liquid phase is subcooled, and there is net interfacial condensation in the system. However, in this paper, correlation (22) was also used for evaporation, at those locations where liquid temperature was slightly higher than the saturation temperature.

In an alternate approach, the interfacial mass transfer by phase change was defined using the relation (6). In that case, the liquid interfacial heat transfer coefficient is needed. Several different correlations were assessed for liquid interfacial heat transfer coefficient, such as the correlation used in the TRAC-PF1 code [6], Kim and Bankoff [12], Segev et.al. [13], and Wilke correlation [14]. None of these correlations has produced results in acceptable agreement with experimental data. They seem to overpredict the interfacial mass transfer. This was probably due to the assumption of thermal non-equilibrium only at the liquid side, while otherwise a more general form of interfacial heat jump condition (6) would apply. Therefore, these correlations are not presented, neither further discussed in this paper.

Wall Heat Transfer

In accordance with the approximations in this model, only two wall heat transfer regimes were modeled, namely the forced convection to single phase liquid, and nucleate boiling.

Forced convection to single phase liquid was assumed [3] when wall temperature was less than saturation temperature. In this regime, for laminar flow ($Re_f \leq 2000$), the Rohsenow and Choi correlation [15] was used,

$$h_{wSPL} = 4 \frac{k_f}{d_{hy}} \tag{23}$$

In the case of turbulent flow regime ($Re_f > 2000$), the Dittus and Boelter correlation [15] was used,

$$h_{wSPL} = 0.023 \frac{k_f}{d_{hy}} Re_f^{0.8} Pr_f^{0.4} \tag{24}$$

where the liquid Prandtl and Reynolds numbers are defined as $Pr_f = \mu_f c_{pf}/k_f$, and $Re_f = \rho_f V_f d_{hy}/\mu_f$.

When local wall surface temperature was equal or higher than the saturation temperature, i.e. $T_{wf} \geq T_s$, then the nucleate boiling regime was assumed, and the Chen correlation [15] for wall heat transfer coefficient was applied:

$$h_{wTP} = h_{wSPL} \cdot F + \min \left[1, \frac{T_{wf} - T_s}{T_{wf} - T_f} \right] \cdot h_c \tag{25}$$

where

$$h_c = 0.00122 \frac{k_f^{0.79} c_{pf}^{0.45} \rho_f^{0.49}}{\sigma^{0.5} \mu_f^{0.29} i_{lat}^{0.24} \rho_g^{0.24}} (T_{wf} - T_s)^{0.24} \cdot (p_w - p)^{0.75} \cdot S \tag{26}$$

and p_w is saturation pressure at wall surface temperature. The Lockhart-Martinelli parameter is defined as follows:

$$x_{TT}^{-1} = \left[\frac{x_f}{1 - x_f} \right]^{0.9} \cdot \left[\frac{\rho_f}{\rho_g} \right]^{0.5} \cdot \left[\frac{\mu_g}{\mu_f} \right]^{0.1} \tag{27}$$

while parameters F and S have the following form:

$$\left. \begin{array}{lll} F = 1.0 & \text{for} & 1/x_{TT} \leq 0.1 \\ F = 2.35 \left(x_{TT}^{-1} + 0.213 \right)^{0.736} & \text{for} & 1/x_{TT} > 0.1 \end{array} \right\} \tag{28}$$

$$\left. \begin{array}{lll} S = \left(1 + 0.12 \, Re_{TP}^{1.14} \right)^{-1} & \text{for} & Re_{TP} < 3.25 \\ S = \left(1 + 0.42 \, Re_{TP}^{0.78} \right) & \text{for} & 3.25 \leq Re_{TP} \leq 70 \end{array} \right\} \tag{29}$$

$$Re_{TP} = 10^{-4} |V_f| \rho_f \alpha_f d_{hy} F^{1.25}/\mu_f \tag{30}$$

4. DISCUSSION OF RESULTS

REFILL code [3] was used to perform a large number of calculations, simulating PWR Lower Plenum refilling, i.e. Downcomer bypass phenomenon. The model was verified by comparing the results with selected CREARE-1/15 scale experimental data [16]. The most important parameters for these tests are given in the Table I. Figure 3 shows a comparison of the predicted Lower Plenum liquid inventory as a function of time, with CREARE data. Note that the calculations as well as the data denoted as H85 belong to the adiabatic

TABLE I: EXPERIMENTAL PARAMETERS FOR CREARE ECC REFILLING TESTS

Facility scale	1/15
Downcomer gap width	0.0127 m
Downcomer length	0.457 m
Downcomer circumference	0.88 m
ECC injection flow rate (constant)	3.75 kg/s
ECC dimensionless flux	0.116
Reverse core steam domensionless flux: - variable with time, decreasing from 0.3 to 0. - slightly different between the tests H1, H85, and H157	
System pressure: - variable with time, decreasing from 2 10^{-5} to 0.84 10^{-5} Pa - slightly different between the tests H1, H85 and H157	

	TEST H1	TEST H85	TEST H157
ECC inlet temperature	300.5 °K	391. °K	300.5 °K
Wall initial temperature	391. °K	391. °K	450. °K

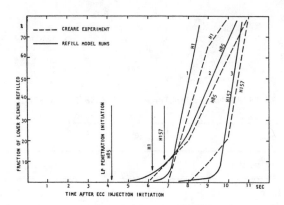

Fig. 3. – Comparison of the REFILL Predictions
with CREARE Experimental Data

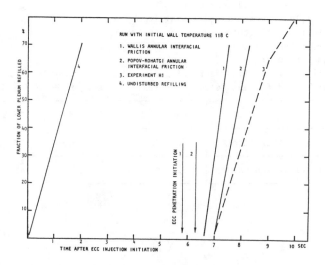

Fig. 4. – Effect of Interfacial Friction Correlation
on Refilling Predictions in case of
Initially Saturated Walls

case, H1 to the case with initially saturated walls, and H157 to the case with initially superheated walls.

As evident from Figure 3 very good agreement was obtained between the data and calculations. For each calculation two-dimensional plots have been obtained [3], showing void fraction, liquid and steam velocities in both directions, liquid temperature, wall surface temperature and pressure, but they are not presented in this paper.

Effect of Interfacial Friction Correlation

Figure 4 shows predicted Lower Plenum liquid inventory in case of initially saturated walls. Curve 4 on this figure, and on the other figures as well, shows the maximum Lower Plenum ECC accumulation rate in case no interfacial and wall friction was present in the model. As evident from Figure 4, refilling delay between 2 to 3 seconds was obtained when Wallis, or Popov and Rohatgi correlations for annular interfacial friction were alternatively used in the model. However, with both correlations the refilling rate was almost the same. Much better agreement with data was obtained with the Popov and Rohatgi correlation, which predicted higher annular interfacial friction than the Wallis correlation.

The same comparison for the case with initially superheated walls is shown on Figure 5. In this case, the model was more sensitive to annular interfacial friction correlation. With Wallis correlation the refilling delay was greatly underpredicted, indicating much earlier refilling initiation and completion. With Popov and Rohatgi correlation for annular interfacial friction (higher friction), given with the relation (19), much better agreement with data was obtained [17]. However, less accurate prediction of refilling rate was obtained at the initiation of Lower Plenum refilling when compared to data in this case, which is probably due to some deficiencies in the wall heat transfer model at higher heat fluxes.

Figures 6 and 7 show predicted Lower Plenum liquid inventory in case of adiabatic calculations. These were done to assess sensitivity of the model to bubbly and annular interfacial friction, without being affected by wall heat transfer. By TRAC model for bubbly interfacial friction on Figure 6, the correlation (16) is assumed. On the Figures 6 and 7, calculations were performed with unchanged interfacial friction correlation until 4.2 seconds, and then the interfacial friction coefficient was perturbed there after, using a restart option in the program REFILL. As evident from Figures 6 and 7, the predicted refill delay was more sensitive to a decrease in bubbly interfacial friction, than to an increase by the same amount. In contrary, the predicted refill delay was more sensitive to an increase in annular interfacial friction than to a decrease. This behavior was a result of distribution of bubbly and annular flow regimes in the Downcomer annulus. It was observed [3] that bubbly flow dominated in the upper portion of Downcomer, around the cold leg entrances, while annular flow prevailed in the lower portion of Downcomer, along the ECC penetration path. Figure 6 clearly shows that unless sufficient annular interfacial friction is provided, the model tends to underpredict the refilling delay even in the case when the bubbly interfacial friction is highly increased (for an order of magnitude).

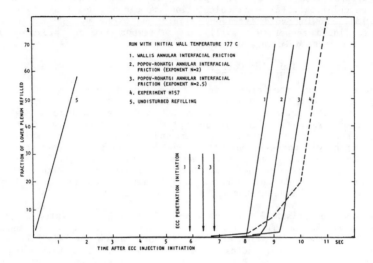

Fig. 5. - Effect of Interfacial Friction Correlation
on Refilling Predictions in case of
Initially Superheated Walls

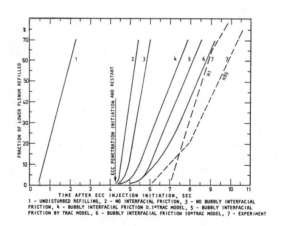

1 - UNDISTURBED REFILLING, 2 - NO INTERFACIAL FRICTION, 3 - NO BUBBLY INTERFACIAL
FRICTION, 4 - BUBBLY INTERFACIAL FRICTION 0.1*TRAC MODEL, 5 - BUBBLY INTERFACIAL
FRICTION BY TRAC MODEL, 6 - BUBBLY INTERFACIAL FRICTION 10*TRAC MODEL, 7 - EXPERIMENT

Fig. 6. - Model Sensitivity to Bubbly Interfacial
Friction Correlation in case of
Adiabatic Calculations

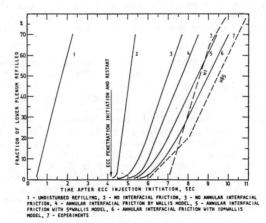

1 - UNDISTURBED REFILLING, 2 - NO INTERFACIAL FRICTION, 3 - NO ANNULAR INTERFACIAL FRICTION, 4 - ANNULAR INTERFACIAL FRICTION BY WALLIS MODEL, 5 - ANNULAR INTERFACIAL FRICTION WITH 5*WALLIS MODEL, 6 - ANNULAR INTERFACIAL FRICTION WITH 10*WALLIS MODEL, 7 - EXPERIMENTS

Fig. 7. – Model Sensitivity on Annular Interfacial
Friction Correlation in case of
Adiabatic Calculations

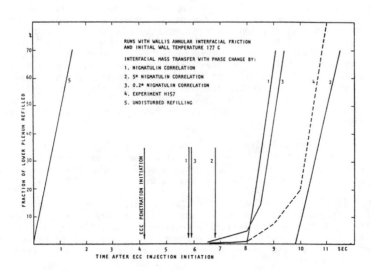

Fig. 8. – Effect of Interfacial Mass Transfer Correlation
on Refilling Predictions in case of
Initially Superheated Walls

453

Effect of Interfacial Mass Transfer by Phase Change

The effect of the interfacial mass transfer was strongly coupled with the effect of wall heat transfer, and therefore, with the ECC temperature, which varied with time and space in the Downcomer. It was also found [3], that this effect is dependent on the direction of interfacial mass transfer (condensation or evaporation) and on the two-phase flow regime distribution in the Downcomer.

Figure 8 shows the effect of interfacial mass transfer in case of initially superheated walls. In this case, the ECC was significantly heated up, specially in the lower portion of Downcomer, where it was even slightly superheated. Therefore, in case of refilling calculations with increased interfacial mass transfer coefficient (curve 2 on Figure 8), significant increase of evaporation rate in the lower Downcomer was predicted, and generation of more steam which supplemented the reverse core steam flow. All that resulted in increase of steam velocity in the lower portion of the Downcomer and in increase of interfacial friction. As shown on Figure 8 (curve 2) this caused very significant refilling delay. It can be concluded that the effect of increased evaporation rate in the lower portion of Downcomer surpassed the effect of increased condensation rate taking place simultaneously in the upper portion of Downcomer. Furthermore, a decrease of interfacial mass transfer coefficient resulted in loosening of the coupling between heat transfer and momentum transfer, and thus curve 3 on Figure 8 is similar to the adiabatic curve (curve 2 on Figure 3). Further reduction of interfacial mass transfer resulted in complete resemblance of the adiabatic case.

In case of initially saturated walls, the ECC water remained subcooled longer and over a larger area in the Downcomer, and therefore, the total condensation rate was much greater than in the case with initially super-heated walls. Figure 9 shows ECC refilling predictions for this case. It is evident from this figure that the effects of condensation and evaporation processes at different locations (elevations) were almost balanced, as the model showed very little sensitivity to interfacial mass transfer correlation. Nevertheless, when the interfacial mass transfer coefficient was increased (curve 3), slightly earlier refilling was predicted, indicating that larger area was in condensation mode of interfacial mass transfer, and therefore, its effect on reducing the net steam mass flux was significant. Calculations were also performed in case of adiabatic walls, when due to the increased importance of condensation, further decrease in ECC refilling delay was predicted.

According to the refilling predictions shown on Figure 4 (curve 2) and Figure 5 (curve 3), the modified Nigmatulin model for interfacial mass transfer, given by the relation (22), when used with the Popov and Rohatgi model for annular interfacial friction, agreed well with CREARE data.

Figure 10 shows the effect of interfacial velocity assumption on Lower Plenum refilling predictions in case of initially superheated walls. Curves 1 and 2 on Figure 9 (in case of initially superheated walls) also indicate the effect of interfacial velocity. It is evident that the effect of interfacial mass transfer term in the momentum equation, or the effect of interfacial velocity is negligible.

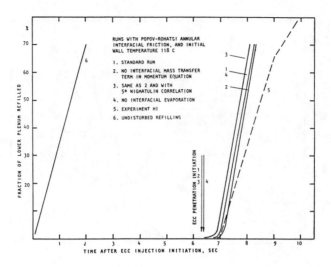

Fig. 9. - Effect of Interfacial Mass Transfer Correlation
on Refilling Predictions in case of
Initially Saturated Walls

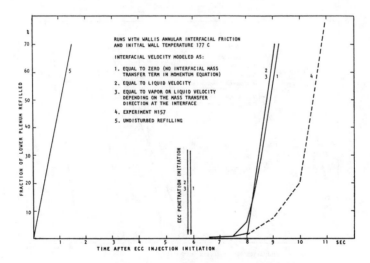

Fig. 10. - Effect of Interfacial Velocity Correlation
on Refilling Predictions in case of
Initially Superheated Walls

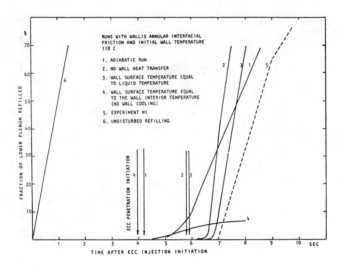

Fig. 11. – Effect of Wall Heat Transfer Correlation
on Refilling Predictions in case of
Initially Saturated Walls

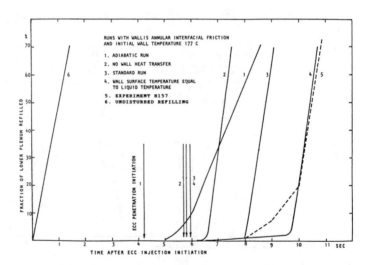

Fig. 12. – Effect of Wall Heat Transfer Correlation
on Refilling Predictions inb case of
Initially Superheated Walls

Effect of Wall Heat Transfer Correlation

As there was assumed no vapor generation at the wall-fluid interface in this model, the effect of wall heat transfer was manifested through liquid temperature on the interfacial mass transfer. Figure 11 shows the sensitivity of refilling predictions to wall heat transfer coefficient in case of initially saturated walls. Note that due to pressure drop in the system during the transient, they became slightly superheated later on. Also, note that although the walls were initially saturated, the ECC was highly subcooled at the inlet, resulting in significant wall heat transfer. It is evident from Figure 11 that the Lower Plenum refilling rate was not very sensitive to wall heat transfer coefficient. There was no appreciable difference in refilling delay between the extreme cases of infinite heat resistance at wall-fluid interface (no heat transfer), as shown with the curve 2 on Figure 11, and the case with no heat resistance at the wall-fluid interface, as shown with the curve 3 on Figure 11. It can be concluded that with any wall heat transfer model, negligible difference in Lower Plenum refilling prediction is obtained for this case.

Figure 12 shows the sensitivity of refilling predictions on wall heat transfer coefficient for initially superheated walls. Here, significant difference was observed between the cases of no wall heat transfer and wall heat transfer fully determined by heat conduction in the wall (curves 2 and 4 respectively on Figure 12). Obviously, the magnitude of wall heat transfer coefficient played important role in this case. The refilling prediction obtained with the Chen's correlation for nucleate boiling is shown by curve 3 on Figure 12. However, when the wall heat transfer coefficient was decreased (as for example by applying Dittus-Boelter correlation in the nucleate boiling region), insignificant reduction of refilling delay was predicted.

5. CONCLUSIONS

This study indicated that the Lower Plenum refilling rate was very sensitive to interfacial momentum transfer, specially for the case of initially superheated walls. Furthermore, the results were more sensitive to interfacial momentum transfer in the annular flow regime than in the bubbly flow regime. Based on the comparisons with CREARE tests, it can be concluded that for the model used in this study, the correlations for interfacial friction in bubbly flow regime, as used in the TRAC-PF1 code [6], and in annular flow regime, as defined by Popov and Rohatgi [10], constitute a suitable set for predicting interfacial momentum transfer.

The effect of the correlation for interfacial mass transfer by phase change was found to be quite significant, specially in the case of initially superheated walls. The model sensitivity on interfacial mass transfer strongly depended on spatial distribution of local interfacial condensation and evaporation rates. In case of initially superheated walls an increase in interfacial mass transfer coefficient resulted in high overprediction of refilling delay, while for initially saturated walls or adiabatic walls an increase in interfacial mass transfer coefficient resulted in moderate decrease in predicted refilling delay.

The wall heat transfer model indirectly affected the Lower Plenum refilling delay by increasing liquid temperature and thus, affecting the interfacial mass transfer. This effect was more pronounced for initially high superheated walls than for initially saturated walls.

6. NOMENCLATURE

C_i	interfacial friction parameter, (N s^2/m^5)
C_w	wall friction parameter, (N s^2/m^5)
c	specific heat capacity, (J/kg K)
D_b	bubble diameter, (m)
D_d	droplet diameter, (m)
d	diameter, (m)
f	friction coefficient
g	gravity acceleration, (m/s^2)
h	heat transfer coefficient, (J/m^{2}s K)
i	enthalpy, (kJ/kg)
k	heat conduction coefficient, (J/m s K)
L^{-1}	heat transfer area concentration, (m^{-1})
N_μ	dimensionless viscosity number
Pr	Prandtl number
p	pressure, (Pa)
q	heat flux, (J/m^{2}s)
R	wall cooling depth, (m)
Re	Reynolds number
r	radial coordinate, radius, (m)
T	temperature, (K)
$T_{w\infty}$	wall deep interior temperature, (K)
t	time, (s)
V	velocity, (m/s)
We	Weber number
x	phasic quality
x_{TT}	Lockhart-Martinelli parameter

Greek notations

α	volumetric fraction
Γ	interfacial mass transfer, steam generation rate, (kg/m^{3}s)
Γ_c	interfacial mass transfer coefficient, (kg/m^{3}s)
δ	film thickness in annular flow, (m)
μ	dynamic viscosity, (N s/m^2)
σ	surface tension, (N/m)
ρ	density, (kg/m^3)

Subscripts

crit	critical
f	liquid
g	vapor
hy	hydraulic
i	interfacial
lat	latent

r   relative
s   saturated
SPL   single phase liquid
TP   two-phase
w   wall

Acknowledgements

The research work presented in this paper has been financially supported by the USA-Yugoslav Joint Fund for Scientific Cooperation, Grant No.
JFP-559, in the period September 31, 1985 through August 31, 1988.

The authors wish to express their gratitude to the cooperating insti-
tutions in the project this paper resulted from, namely the Macedonian
Academy of Sciences and Arts from Yugoslavia, and the Brookhaven National
Laboratory from USA.

REFERENCES

1.   Fabic S., "Accident Analysis", in 'Handbook of Multiphase Systems',
     edited by G. Hatsroni, McGraw-Hill Book Company, New York, 1988.

2.   Harlow F.H., Amsden A.A., "Numerical Calculation of Multi-Phase Flow",
     Journal of Computational Physics, Vol. 17, pp. 19-52, 1975.

3.   Popov N.K., "Applicability Assessment of Some Constitutive Correlations
     and Sensitivity of ECC Bypass Flow Model in case of Large Break LOCA in
     PWRs", PhD thesis, University of Zagreb, Mechanical Engineering Depart-
     ment, Zagreb, Yugoslavia, 1988.

4.   Ishii M., "Thermo-Fluid Dynamics Theory of Two-Phase Flow", Eyrolles,
     Paris, France, 1975.

5.   Bird R.B., Stewart W.E., Lightfoot E.N., "Transport Phenomena", John
     Wiley & Sons Inc., New York, 1960.

6.   Liles D.R., Mahaffy J.N., "TRAC-PF1: An Advanced Best-Estimate Computer
     Program for Pressurized Water Reactor Thermal-Hydraulics Analysis", Los
     Alamos Scientific Laboratory, NUREG/CR-3567, LA-994-MS, 1986.

7.   Rohatgi U.S., Jo J.H., Slovik G.C., "A Comparative Analysis of
     Constitutive Relations in TRAC-PF1 and RELAP/MOD1, Brookhaven National
     Laboratory report, NUREG/CR-4292, BNL-NUREG-51898, 1985.

8.   Govier G.W., Aziz A., "The Flow of Complex Mixtures in Pipes", Van
     Nostrand-Rhienhold Co., New York, 1972.

9.   Wallis G.B., "One-Dimensional Two-Phase Flow", McGraw-Hill Book Com-
     pany, New York, 1969.

10. Popov N.K., Rohatgi U.S., "Effect of Interfacial Shear and Entrainment Models on Flooding Predictions", Journal of AIChE, Vol. 32, No. 6, pp. 1027-1035, 1986.

11. Rivard W.C., Torrey M.D., "Numerical Calculations of Flashing from Long Pipes Using a Two-fluid Model", Los Alamos Scientific Laboratory Report, LA-NUREG-6330-MS, 1976.

12. Kim H.J., Bankoff S.G., "Local Heat Transfer Coefficient for Condensation in Stratified Countercurrent Steam-Water Flows, ASME paper 82-WA/HT-24, 1982.

13. Segev A., Flannigan L.J., Kurth R.E., Collier R.P., "Experimental Study of Countercurrent Steam Condensation", Journal of Heat Transfer, Vol. 103, p. 307, 1981.

14. Rohsenow W.M., Hartnett J.P., Ganic E.N., "Handbook of Heat Transfer Foundamentals", 'Boiling' by Rohsenow W.M., pp. 12-76, McGraw-Hill Book Company, New York, 1985.

15. Delhaye J.M., Giot M., Riethmuller M.L., "Thermohydraulics of Two-Phase Systems for Industrial Design and Nuclear Engineering", McGraw-Hill Book Company, New York, 1981.

16. Crowley C.J., Block J.A., Cary C.N., "Downcomer Effects in a 1/15-Scale PWR Geometry - Experimental Data Report", NUREG-0281, Creare Inc. 1977.

17. Popov N.K., Rohatgi U.S., "Assessment of Some Interfacial Shear Correlations in ECC Bypass Flow Adiabatic Model for PWR Downcomer", 4th Miami International Symposium on Multi-Phase Transport & Particulate Phenomena, Miami, Florida, 1986.

Effect of Various Thermal-Hydraulic Parameters on PWR ECC Reflooding

N. K. POPOV*
University Kiril and Metodij
Department of Electrical Engineering
Skopje, 91000, Yugoslavia

D. HADZI-MISEV
The City Central Heating Company
Skopje, 91000, Yugoslavia

Abstract

A transient numerical model has been developed to study the effect of thermal-hydraulic parameters on PWR fuel rod ECC quenching. A two-dimensional model for heat conduction within the fuel rod, was coupled with a semi-empirical one-dimensional model of heat and mass transfer in the fluid region. Predicted cladding temperatures during quenching were analyzed. It was found that the effect of stored energy in the fuel prior to quenching initiation was more important than the decay heat generation rate. The effect of ECC flow rate and subcooling was also very significant. The effect of cladding gap varied with elevation and was less important.

1. INTRODUCTION

A cold leg large break loss-of-coolant accident (LBLOCA) in PWR is a design basis accident which is analyzed for every PWR as part of licensing process. This hypothetical accident can be divided into three phenomenologically distinct phases [1], namely, blowdown, refill and reflood phase. During the blowdown phase the system depressurizes to around 40 bar, and core is completely voided. The fuel clad continues to heat up due to the decay heat generation in the fuel and poor heat transfer in the core. The refill phase begins with initiation of Emergency Core Coolant (ECC) injection in the intact loops at the time when the system pressure is around 40 bar. The last phase of the accident, reflood phase, begins when the Lower Plenum is full and the ECC starts to fill the core. However, as the fuel clad is overheated, the liquid can not contact it. A thin vapor blanket is formed between the cladding and the liquid, i.e. inverted annular two-phase flow regime is established. The drag between the vapor and the rising liquid causes liquid break up into droplets, which are carried upward. The entrained liquid droplets provide significant precooling of the fuel rods downstream, which combined with axial heat conduction in the cladding, reduces cladding temperature downstream and enables quenching. Usually, quenching is first observed at the fuel rod bottom. It propagates upward with a variable speed which depends on the wall heat flux, pressure, ECC mass flow rate, ECC inlet subcooling, decay heat generation rate in the fuel, amount of stored energy in the fuel prior of reflooding (fuel initial temperature), etc., [2], [3].

*Presently a post-doctoral fellow of the Natural Sciences and Engineering Research Council of Canada, with Atomic Energy of Canada Limited, at Whiteshell Nuclear Research Establishment, Pinawa, Manitoba, ROE 1L0, Canada.

During the quench front propagation upward, there are three distinct zones along the fuel rod, as shown on Figure 1, zone of already rewetted cladding, quenching zone, and zone of still dry overheated cladding. Typically, in the rewetted zone the heat transfer is through forced convection to single phase liquid or subcooled boiling. In the zone of quench front, along a very small length, various heat transfer regimes may occur, such as fully developed boiling, transition (unstable) boiling or film boiling in inverted annular flow. In the dry zone, typically, the regimes of dispersed droplets boiling or heat convection to steam are encountered.

There have been two major types of quenching experiments reported in the literature: with electrically heated fuel rods, like the FLECHT experiments [4], and with hollow pipes, like the Kabanov experiments [5],[6] (heat generation in the pipe wall). The quenching experiments with simulated fuel rods provide more accurate data and are more suitable for analyzing the PWR core reflooding but are also more expensive in comparison to quenching experiments with hollow pipes, which are mostly used to gain physical understanding about heat transfer phenomena during quenching.

This paper shows that the stored energy in the fuel prior of reflooding initiation, has significant effect on the clad-to-coolant heat flux in all three zones during quenching. It has been shown [7] that the effect of stored energy can not be taken into account with any heat generation distribution or transfer rate in the hollow pipe wall. Therefore, significant discrepancies may occur when applying data from quenching hollow pipes for PWR reflooding predictions. In this paper, the effect of the magnitude of stored energy in the fuel and the influence of thermal-hydraulic parameters on quenching predictions was assessed.

2.   MATHEMATICAL MODELING

Modeling Approximations

Several approximations have been introduced in this model to simplify numerical solution. The most important was the assumption of quenching a single fuel rod in a single fuel channel. Thus, any influence of surrounding fuel rods occurring in a real reactor, has been neglected.

In the fluid region, besides the assumption of azimuthal symmetry, any flow parameter variation in the radial direction was also neglected. Therefore, the heat transfer was modeled only in axial (vertical) direction. Some flow parameters, such as vapor void fraction and phasic qualities, were estimated by semi-empirical correlations, and the phasic mass balance equation was omitted from the model. The pressure drop along the fuel channel was considered to be very small [6], and was neglected. Therefore the momentum balance equation was also omitted from the model. In the energy balance equation, one of the phases was always assumed to be saturated.

Conservation Equations

As depicted on Figure 1, the fuel rod consisted of three material zones, namely fuel, gap (filled with helium), and cladding, surrounded by two-phase ECC flow.

The heat conduction in the fuel rod was modeled by applying the standard heat conduction partial differential equation in radial and axial directions [8] as follows:

$$\rho_k c_k \left(\frac{\partial T_k}{\partial t} \right) = \nabla(k_k \nabla T_k) + Q_k(r,z,t) \tag{1}$$

where k is any material zone, such as fuel or cladding.

The heat conduction equation was not solved in the cladding gap. Instead, the effect of heat resistance (heat conduction) in the gap was taken into account by introducing corresponding heat transfer coefficient at the fuel-cladding boundary.

In radial direction, at the fuel-cladding interface, the following boundary condition was applied:

$$- k_f \left(\frac{\partial T_f}{\partial r} \right)_{r=r_f} = h_g [T_f(r_f,z,t) - T_c(z,t)] \tag{2}$$

while at the cladding-fluid interface,

$$- k_c \left(\frac{\partial T_c}{\partial r} \right)_{r=r_c} = h_w [T_c(z,t) - T_{ECC}(z,t)] \tag{3}$$

Note that in the relationships (2) and (3) only one radial numerical node in the cladding was assumed.

In axial direction, at the lower and upper end of the fuel rod, following heat transfer boundary conditions were applied:

$$- k_c \left(\frac{\partial T_c}{\partial z} \right)_{z=0} = h_w [T_c(0,t) - T_{ECC}(0,t)] \tag{4}$$

$$- k_c \left(\frac{\partial T_c}{\partial z} \right)_{z=H} = h_w [T_c(H,t) - T_{ECC}(H,t)] \tag{5}$$

In the fluid region the following heat transfer equation was solved [7]:

$$h_w [T_c(z,t) - T_{ECC}(z,t)] \cdot dA = \tag{6}$$
$$= \left[x \cdot di_v + dx \cdot (i''-i') + (1-x) \cdot di_l \right]$$

where: $di_v = (\partial i_v / \partial z)dz$ is the enthalpy increment of the vapor phase in an elementary fluid volume, $di_l = (\partial i_l / \partial z)dz$ is the enthalpy increment of the liquid phase in an elementary fluid volume, dA is elementary contact area between the cladding and the coolant, dx is the vapor quality increment, and i" and i' are the enthalpy of saturated vapor and the enthalpy of saturated liquid respectively.

Numerical Procedure

A standard explicit finite-difference numerical technique has been used to solve the heat conduction equation (1) in the fuel rod and the heat balance equation (6) in the coolant. The numerical procedure is explained in detail in [7].

It is important to stress that two different axial numerical steps were used along the fuel rod. As shown on Figure 1, in the rewetted zone and dry zone, larger axial numerical step was used, due to the lower and smoother wall heat fluxes. In the zone of quench front, as the wall heat flux was quite high and sharply changing, axial numerical step which is ten times smaller than in other regions, was applied.

Initial axial and radial power and temperature profiles were obtained from experimental tests being simulated. A computer program REWET was created based on the equations described in sections 2 and 3.

3. CONSTITUTIVE RELATIONS

Phasic Quality

Figure 2 shows phasic qualities as a function of axial distance during quenching, with several zones in general [7]. Note that on Figure 2, as well as in the later discussions, vapor quality is defined as "true" quality [9], i.e. $x_v=G_v/(G_v+G_1)$. Phasic quality expressed in terms of enthalpies [9] is used for the balance quality, i.e. $x_b=(i_{TP}-i')/(i''-i')$, and for the liquid quality, i.e. $x_1=(i_1-i')/(i''-i')$.

The zone $z < z_A=z_{SB}$ on Figure 2, denotes a zone of single phase sub-cooled liquid, where z_{SB} is the point of incipience of subcooled boiling defined by the Bergles and Rohsenow criterion (16). There is no vapor generation in this zone, and therefore $x=x_v=0$. Consequently, $di_v=0$ in the equation (6), which is solved for i_1.

The zone $z_A \leq z < z_B=z_{NVG}$, denotes the subcooled boiling zone, with bulk liquid still subcooled, i.e. $T_1 < T_s$, where z_{NVG} is the point of net vapor generation initiation defined by the Saha and Zuber criterion (21). There is no net vapor generation in this zone, i.e. $x=x_v=0$, $di_v=0$, and i_1 is be calculated from equation (6).

Downstream of the point of net vapor generation, i.e. in the zone $z_{NVG}=z_B \leq z < z_C=z_{CR}$, $T_1<T_s$, and $T_v=T_s$, where z_{CR} is the point of critical heat flux defined by the Zuber criterion (19), or by the Ivey and Morris criterion (20). In this zone liquid enthalpy is obtained as $i_1=i'-x_1(i''-i')$, where the following relation is used for x_1 [10]:

$$x_1 = \frac{2 \cdot x_{NVG}}{1 + \exp[2(1-x_b/x_{NVG})]} \tag{7}$$

where x_{NVG} is the balance vapor quality at the initiation of net vapor generation. In this zone $di_v=0$, and equation (6) is solved for $x(z,t)$.

In the zone $z_{CR}=z_C \leq z < z_D$, both phases were considered saturated, i.e. $T_v=T_1=T_s$, where z_D is the point of $\alpha=0.45$ [7]. Therefore $di_1=di_v=0$, and equation (6) is solved for $x(z,t)$. Note that this zone may not exist, depending on the flow parameters.

In the zone $z_D \leq z < z_E$, liquid is assumed saturated and vapor super-heated, i.e. $T_1=T_s$, $T_v>T_s$, thus $x_1=0$, $di_1=0$, where z_E is the point of $x=1$.

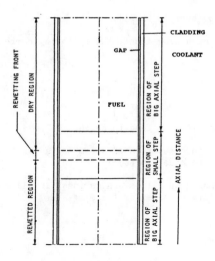

Fig. 1. - Schematic View of a Fuel Rod During
Quenching with Two Axial Numerical Steps

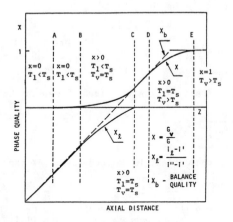

Fig. 2. - Definition of Vapor and Liquid Qualities by
the Principle of Non-equilibrium at One Side

465

Vapor quality $x(z,t)$ in the vicinity of the quench front is calculated using the Barzzoni and Martini correlation [11] in the following form:

$$x = 1 - (1-x_{CR}) \cdot \exp[-(x_b-x_{CR})] \tag{8}$$

while away from the quenching front, the following correlation was used [7]:

$$x = x_{CR} + (x_b-x_{CR}) \cdot a \tag{9}$$

where

$$a = 0.402 + 0.0674 \cdot \ln\left[G_{ECC}\left(\frac{d_{hy}}{\rho_v}\right)^{0.5} \cdot \left(1-x_{CR}\right)^5\right] \tag{10}$$

Equation (6) was used in this zone to calculate i_v.

In the zone of pure superheated steam for $z \geq z_E$, clearly $x_1=0$, $di_1=0$, and $x(z,t)=1$. Therefore equation (6) is solved for i_v.

Finally, as phasic qualities are defined in all zones, the mass fluxes of vapor and liquid are obtained as follows:

$$G_v = x \, G_{ECC} \tag{11}$$

$$G_1 = (1-x) \, G_{ECC} \tag{12}$$

Vapor void Fraction

Two correlations were used to estimate vapor void fraction. For void fraction less than 0.99, the Cunningham and Yeh correlation was used in the following form [3]:

$$\alpha = 0.925 \left(\frac{\rho_v}{\rho_1}\right)^{0.239} \cdot \left(\frac{u_v}{u_{bcR}}\right)^a \cdot \left[\frac{u_v}{u_v+u_1}\right]^{0.6} \tag{13}$$

where:
$$a = 0.67 \quad \text{if} \quad u_v/u_{bcR} < 1$$
$$a = 0.47 \quad \text{if} \quad u_v/u_{bcR} \geq 1$$

$$u_{bCR} = \frac{2}{3}(g \, R_{bcR})^{0.5}, \qquad R_{bcR} = 5.267\left[\frac{\sigma}{g(\rho_1-\rho_v)}\right]^{0.5}$$

$$u_v = x \cdot G_{ECC}/\rho_v \qquad u_1 = (1-x) \cdot G_{ECC}/\rho_1$$

For vapor void fraction greater than 0.99, the Hein correlation [12] was used:

$$\alpha = \frac{1}{1+\frac{(1-x)}{x}\frac{\rho_v}{\rho_1}S} \tag{14}$$

where the phase slip coefficient is defined as $S=1+4/[\exp(7.5 \, x_b)]$.

Gap Heat Transfer Coefficient

In the gap between the fuel and cladding, the heat transfer coefficient is determined by using the Westinghouse correlation [13]:

$$h_g = \frac{k_{He}}{b_g/2 + 4.4 \cdot 10^{-6}} \tag{15}$$

where k_{He} is the heat conduction coefficient in the helium for a certain pressure, and b_g is gap width.

Wall-Fluid Heat Transfer

Figure 3 shows the wall heat transfer logic used in this model to select different wall heat transfer regimes. The heat transfer and transition criteria are described in this subsection in order of appearance on Figure 3.

For the transition criterion between the convective heat transfer to single phase liquid and the subcooled boiling regime, the Bergles and Rohsenow correlation [3] was used in the following form:

$$q_{SB} = 1083 \cdot p^{1.156} \left[1.8(T_w - T_s) \right]^{2.16/p^{0.0234}} \tag{16}$$

The heat transfer coefficient for single phase liquid in turbulent flow was estimated using Dittus and Boelter correlation [3]:

$$h_w = 0.023 \frac{k_1}{d_{hy}} Re_1^{0.8} Pr_1^{0.4} \qquad \text{for} \qquad Re_1 > 2000 \tag{17}$$

while for laminar flow it was estimated using the Rohsenow and Choi relation [3]:

$$h_w = 4 \frac{k_1}{d_{hy}} \qquad \text{for} \qquad Re_1 \leq 2000 \tag{18}$$

For a transition criterion between pre- and post-critical heat flux regimes the Zuber correlation [3] was used in the following form:

$$q_{CR} = 0.13 \, \rho_v(i''-i') \left[\frac{\sigma g(\rho_1 - \rho_v)}{\rho_v} \right]^{0.25} \left[\frac{\rho_1}{\rho_1 - \rho_v} \right]^{0.5} \tag{19}$$

or when taking into account the effect of coolant subcooling, the Ivey and Morris relation [3] in the following form:

$$q_{CRS} = q_{CR} \left[1 + 0.1 \left(\frac{\rho_1}{\rho_v} \right)^{0.75} \frac{k_1(T_s - T_1)}{i'' - i'} \right] \tag{20}$$

The Saha and Zuber correlation [3] was used to distinguish between subcooled boiling and fully developed subcooled boiling, i.e. to define the point of net vapor generation:

$$\Delta T_{sub} = \frac{q_{FDB} \, d_{hy}}{455 \cdot k_1} \tag{21}$$

where ΔT_{sub} is coolant subcooling at the elevation where net vapor generation is initiated, and q_{FDB} is the heat flux at the point of fully developed boiling. However, in this model, for both, subcooled boiling, and fully developed boiling, the same correlation was applied, namely the Thom correlation [3]:

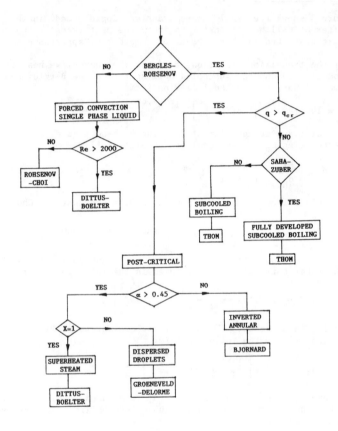

Fig. 3. - Wall Heat Transfer Regimes in the
Model REWET for ECC Quenching

$$q_w = 1.97 \ 10^{-3} \ (T_w-T_s)^2 \ \exp(0.23p) \tag{22}$$

The transition between the inverted annular and dispersed droplets boiling was defined by the void fraction of $\alpha=0.45$ [7].

For transition boiling and inverted annular boiling regimes, the same correlation was used, namely the Bjornard correlation [14]:

$$h_w = f \cdot h_{THOM} + (1-f)(h_{MB} + \alpha \cdot h_{sv}) \tag{23}$$

where h_{THOM} is the heat transfer coefficient at the point of critical heat flux estimated by the Thom correlation (22), and the modified Bromley correlation [3] is defined as:

$$h_{MB} = 0.62 \left[\frac{k_v \rho_v (\rho_1-\rho_v)(i''-i')g}{L_c \mu_v (T_w-T_s)}\right]^{0.25} \tag{24}$$

$$L_c = \frac{1}{2\pi} \left[\frac{g(\rho_1-\rho_v)}{\sigma}\right]^{0.5} \tag{25}$$

and h_{sv} was estimated by Dittus and Boelter correlation for superheated steam [3]:

$$h_{sv} = 0.023 \ \frac{k_v}{d_{hy}} \ Re_v^{0.8} \ Pr_v^{0.4} \tag{26}$$

The factor f in the relation (23) was defined as [7]:

$$f = \frac{T_L - T_w}{T_L-(T_w)_{CR}} \tag{27}$$

where T_L is the rewetting temperature defined in [14] at the point of minimum transition boiling heat flux and $(T_w)_{CR}$ was critical cladding temperature.

The vapor quality of $x=1$ was used as a transition criterion between dispersed droplets and superheated vapor regimes. For dispersed droplets heat transfer regime, the heat transfer coefficient was estimated by the following correlation [7]:

$$h_w = h_{GD} + h_R \tag{28}$$

where the Groeneveld and Delorme correlation for dispersed droplets regime was defined as [3]:

$$h_{GD} = 0.008348 \left(\frac{k_v}{d_{hy}}\right) \left\{Re_v\left[x + \frac{\rho_v}{\rho_1}(1-x)\right]\right\}^{0.8774} Pr_v^{0.6112} \tag{29}$$

while the heat radiation coefficient was defined as:

$$h_R = \varepsilon_R \ 1.7 \ 10^{-9} \ \frac{T_w^4-T_s^4}{T_w-T_s} \tag{30}$$

469

In the superheated steam regime, the heat transfer coefficient was estimated by the Dittus and Boelter correlation, as given by equation (26).

4. DISCUSSION OF RESULTS

Typical cladding temperature predictions during quenching of fuel rods obtained with this model are compared on Figure 4 with COBRA/TRAC [15] results, as well as with FLECHT data [4]. Note that these results, and others presented in this paper as well, are obtained with high ECC inlet mass flux. Relatively good agreement between FLECHT data and this model is evident from Figure 4. A comparison of typical cladding temperature predictions in case of quenching hollow pipes, with Kabanov data [5], [6], is shown on Figure 5. Excellent agreement is evident at this figure between the model predictions and data.

In Table I, the most important geometrical and thermal-hydraulic parameters for quenching calculations are given, which served as a default set of parameters for all calculations presented on the following figures. However, calculations for comparisons with FLECHT and Kabanov data were done with corresponding set of parameters, different from those given in Table I.

Effect of Stored Energy in the Fuel and Decay Heat Generation Rate

Figure 6 shows an interesting comparison of predicted quenching speeds for rewetting fuel elements and hollow pipes. Note that the total decay heat generation rate per fuel element was kept the same for both cases, meaning that the heat generation density in case of hollow pipes was much higher than in the case of fuel rods, to compensate the difference in volumes. Calculations for three different initial temperatures are shown. As the initial temperature was increased, lower quenching speed was predicted. Note that the quenching speed was much higher for hollow pipes than for fuel rods, regardless the decay heat generation rate being the same. Also, the predicted quenching speed decreased downstream, due to the effect of ECC heating up in the rewetted region, i.e. due to decreasing ECC subcooling downstream. Similar comparison is shown on Figure 7, but in terms of quench front location along the fuel element. Note that quenching was completed much earlier in case of hollow pipes than in the case of fuel rods.

Figure 8 shows a comparison of cladding temperature histories at two elevations during quenching hollow pipes and fuel elements. A distinct difference is evident, as cladding temperatures were much lower in the case of hollow pipes than in the case of fuel rods. Apparently, the energy stored in the fuel played a very important role. It was found in [7], that the heat flux upstream of the quench front (in the rewetted zone) was higher in case of fuel elements than in the case of hollow pipes, as most of the energy stored in the fuel was transferred to the coolant after the quenching. There was obviously an effect of "thick walls" with quenching fuel elements, i.e. delayed heat transfer to the coolant.

Figure 9 shows the effect of the initial fuel temperature, i.e. the level of initially stored energy in the fuel. When the initial temperature was increased, the quenching delay was also increased for about the same amount of time at all elevations.

470

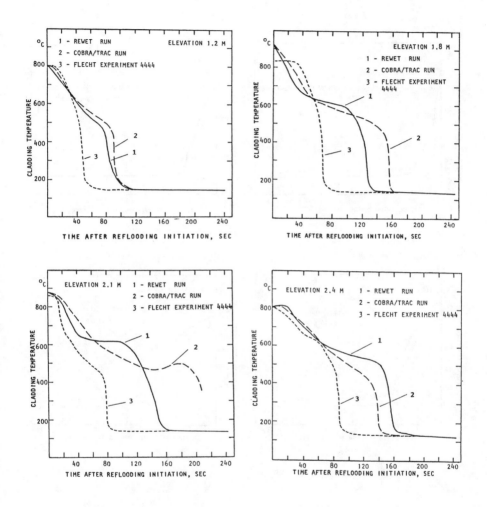

Fig. 4. – Comparison of ECC Quenching Predictions Obtained with
the Model REWET, FLECHT Data and COBRA/TRAC Runs

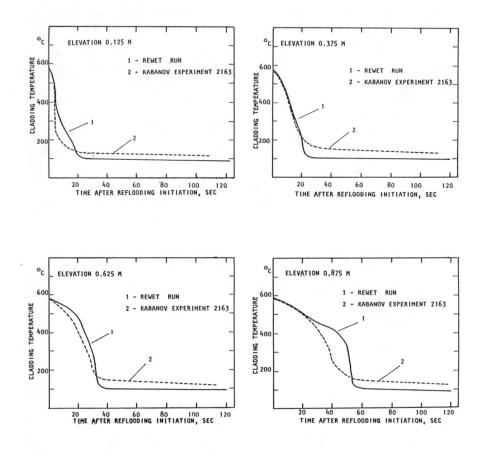

Fig. 5. - Comparison of ECC Quenching Predictions Obtained with
the Model REWET and Kabanov Hollow Pipe Quenching Data

TABLE I: MOST IMPORTANT NOMINAL QUENCHING PARAMETERS

Power generation rate	21 MW/m^3
Initial fuel temperature	700 °K
ECC inlet mass flux	150 kg/m^{2}s
ECC inlet subcooling	50 °K
Pressure	0.1 MPa
Fuel element length	3.2 m
Cladding outer diameter	9.15 mm
Cladding gap width	0.1 mm
Hydraulic diameter	8.5 mm
Constant axial power profile (except for FLECHT)	
Constant initial temperature axial profile	
Parabolic initial radial temperature profile	

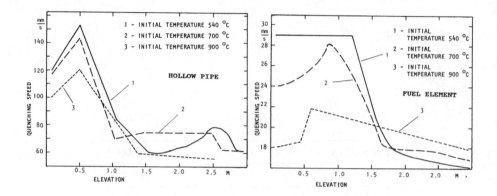

Fig. 6. – Comparison of Quenching Speed Predicted with the
Model REWET in case of Hollow Pipes and Fuel Rods

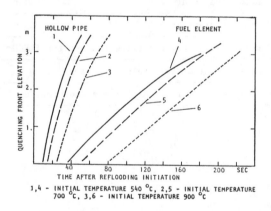

Fig. 7. – Comparison of Quenched Elevations Predicted with
the Model REWET for Hollow Pipes and Fuel Rods

474

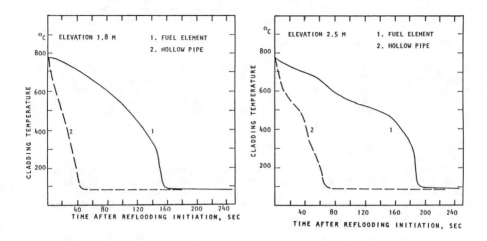

Fig. 8. – Comparison of Cladding Temperature Predictions
During Quenching Hollow Pipes and Fuel Rods

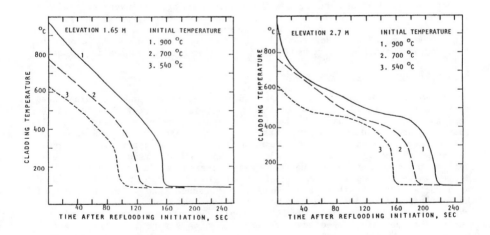

Fig. 9. – Effect of Initial Temperature on Quenching Speed

475

Figure 10 shows the effect of decay heat generation rate on cladding temperature predictions during quenching. It is evident that the heat generation rate also played a significant role. With doubled heat generation rate (curve 4), moderate cladding temperature increase at the middle elevation and considerable increase of quenching delay was predicted, as the ECC mass flow rate was not sufficient to remove the generated heat in the fuel. When the heat generation rate was decreased, the cladding temperature and quenching delay were decreased, but in this case the model was less sensitive than in the case of increased decay heat generation rate.

Comparing the cladding temperature with no heat generation (curve 3 on Figure 10), with the case of quenching a hollow pipe (curve 2 on Figure 8), it is evident that in the case of fuel element with no heat generation quenching was still predicted later than in the case of hollow pipes. That is a clear evidence of the strong effect of the initially stored energy in the fuel.

Effect of ECC Thermal-Hydraulic Parameters

Figure 11 shows cladding temperature predictions during quenching, with different ECC inlet mass flow rates. The ECC mass flux directly influenced the ECC cooling capability. Obviously, when the ECC mass flux was decreased the quenching delay was increased. The effect of the ECC mass flux increased with elevation.

As shown on Figure 12, the effect of ECC initial subcooling was extremely significant, as it directly influenced the ECC cooling capability. When the ECC initial subcooling was reduced, the quenching delay was highly increased. The effect of ECC initial subcooling on quenching speed increased at high elevations.

The effect of ECC pressure is closely connected with the effect of ECC subcooling, as saturation temperature is changed with pressure. The cladding temperature predictions during quenching shown on Figure 13 were carried out with constant ECC inlet temperature; therefore ECC subcooling was changed. It is evident from this figure that when the pressure was increased, the quenching delay was reduced. However, this effect was almost constant along the fuel rod, quite different compared with the effect of ECC subcooling alone.

Effect of Fuel-Cladding Gap Width

The effect of the gap width on cladding temperature and quenching speed is shown on Figure 14. Due to relatively low conduction heat transfer in the gap, any significant change in gap width presumably results in significantly changed quenching delay. However, as seen from Figure 14, the net effect of changed gap width was not appreciable, as the effect of gap width on quenching speed varied with elevation. When the gap width was reduced, at low elevations the predicted quenching speed was decreased, while at high elevations it was increased. Apparently, the effect of stored energy in the fuel played an important role in this case. Namely, with reduced gap width, the rate of heat transfer from the fuel to the coolant upstream and in the quench zone was increased. Due to that, the ECC subcooling downstream was decreased, reducing the quenching speed downstream. As seen on Figure 14, the net effect was not significant.

476

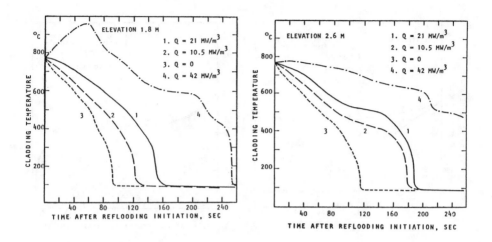

Fig. 10. - Effect of Decay Heat Generation Rate on Quenching Speed

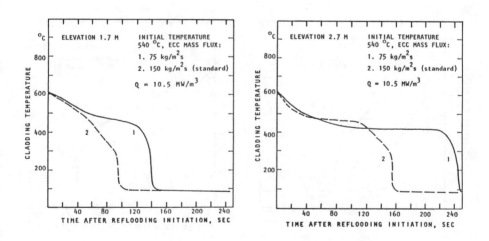

Fig. 11. - Effect of ECC Mass Flux on Quenching Speed

477

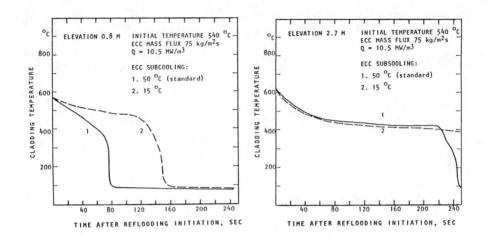

Fig. 12. - Effect of ECC Subcooling on Quenching Speed

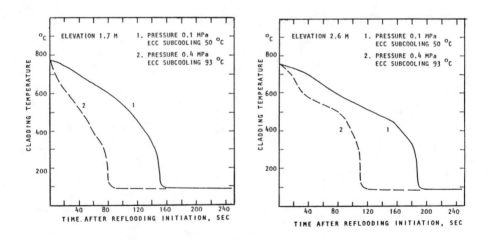

Fig. 13. - Effect of Pressure in the Channel on Quenching Speed

478

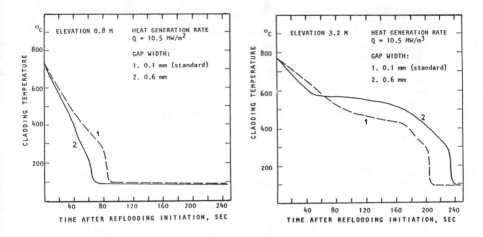

Fig. 14. - Effect of Cladding Gap Width on Quenching Speed

5. CONCLUSIONS

Mainly due to the effect of stored energy in the fuel, much less quenching speed was predicted in case of quenching fuel elements in comparison with quenching hollow pipes. Even with high ECC mass fluxes, the effect of stored heat in the fuel could not be eliminated. This effect was basically caused by increased wall heat flux upstream of the quenching point (in the already rewetted zone) in case of fuel rods, which reduced ECC inlet subcooling downstream. It was also found that with a few different profiles and rates of decay heat generation rates in the cladding (in case of hollow pipes), it was not possible to compensate the effect of initially stored energy in the fuel. With increased fuel initial temperature, the predicted quenching delay was also increased.

The effect of decay heat generation rate in the fuel was found to be significant, but much less important than the effect of the stored energy in the fuel. On the other hand, the effects of ECC subcooling and mass transfer were found to be very important. Significant increase of quenching delay was predicted when ECC subcooling or mass flux was decreased. The effect of ECC pressure in the fuel channel was also appreciable, as pressure is indirectly coupled with ECC subcooling through saturation temperature.

The effect of fuel–cladding gap was moderate and dependent on elevation. Increased gap width resulted in a decrease of quenching delay at low elevations, and in an increase of quenching delay at high elevations. This was caused by the change in heat flux distribution along the fuel rod, i.e. by a rearrangement of the amount of heat transferred to the ECC water at different elevations.

479

6. NOMENCLATURE

A	cross section area of the channel, (m^2)
c	specific heat capacity, (J/kg K)
d	diameter, (m)
f	coefficient
G	mass flux, (kg/m^2s)
g	gravity acceleration, (m/s^2)
h	heat transfer coefficient, (J/m^2s K)
i	enthalpy, (kJ/kg)
i'	enthalpy of saturated water, (kJ/kg)
i"	enthalpy of saturated steam, (kJ/kg)
k	heat conduction coefficient, (J/m s K)
Pr	Prandtl number
p	pressure, (Pa)
Q	decay heat generation rate, (MW/m^3)
q	heat flux, (J/m^2s)
Re	Reynolds number
r	radial coordinate, radius, (m)
T	temperature, (K)
t	time, (s)
u	velocity, (m/s)
W	mass flow rate, (kg/s)
x	phasic quality
z	axial coordinate, (m)

Greek notations

α	volumetric fraction
μ	dynamic viscosity, (N s/m^2)
ε	heat radiation constant
σ	surface tension, (N/m)
ρ	density, (kg/m^3)
ΔT_{sub}	ECC inlet subcooling, (K)

Subscripts

c	cladding
CR	critical
ECC	emergency core coolant
FDB	fully developed boiling
f	fuel
g	gap
He	helium
hy	hydraulic
in	inlet
k	any material zone
lam	laminar
l	liquid
NVG	onset of net vapor generation
s	saturated
SB	subcooled boiling
SPL	single phase liquid
SV	superheated vapor

```
SUB      subcooled
TP       two-phase
tur      turbulent
v        steam
w        wall
```

Acknowledgements

        The research work presented in this paper has been financially sup-
ported by the USA-Yugoslav Joint Fund for Scientific Cooperation, Grant No.
JFP-559, in the period September 31, 1985 through August 31, 1988.

        The authors wish to express their gratitude to the cooperating insti-
tutions in the project this paper was part of, namely the Macedonian Academy
of Sciences and Arts from Yugoslavia, and the Brookhaven National Laboratory
from USA. Valuable suggestions and assistance in preparing the paper provided
by Upendra S. Rohatgi from Brookhaven National Laboratory are specially
appreciated.

REFERENCES

1.   Fabic S., "Accident Analysis", in the 'Handbook of Multiphase Systems',
      edited by G. Hetsroni, McGraw-Hill Book Company, New York, 1986.

2.   Mayinger F., "Reflood and Rewet Heat Transfer", Summer School 'Nuclear
      Reactor Safety Heat Transfer', paper 12, Dubrovnik, Yugoslavia, 1980.

3.   Delhaye J.M., Giot M., Riethmuller M.L., "Thermohydraulics of Two-Phase
      Systems for Industrial Design and Nuclear Engineering", McGraw-Hill
      Book Company, New York, 1981.

4.   Cadek F.F., et.al., "PWR FLECHT (Full Length Emergency Cooling Heat
      Transfer) Final Report", WCAP-7665, April 1971.

5.   Zemyanuhin V.V., Kabanov L.P. et.al., "Quenching speed in a 7-pin Fuel
      Element with Zirconium Alloy Cladding and Parameters Typical for PWR
      Accidental Cooling", Institute I.V. Kurchatov report, Moscow, 1985.

6.   Kabanov L.P., "Experimental Research of Thermalhydraulics Processes
      During Accidental Cooling of PWRs", Institute of MEI report, Moscow,
      1980.

7.   Hadzi-Misev D., "Analysis of the Influence of Stored Heat in the Fuel
      on Quenching Speed During a PWR Core Reflooding Following LOCA", PhD
      thesis, University Kiril & Metodij, Mechanical Engineering, Skopje,
      Yugoslavia, 1988.

8.   Bird R.B., Stewart W.E., Lightfoot E.N., "Transport Phenomena", John
      Wiley & Sons Inc., New York, 1960.

9.   Hewitt G.F., "Liquid-Gas Systems", in the 'Handbook of Multiphase
      Systems', edited by G. Hetsroni, McGraw-Hill Book Company, New York,
      1986.

10. Zuber N., Shtaub F., Bayrod G., "Steam Mass Quality in Subcooled and Saturated Boiling", 'Achievements in Heat and Mass Transfer', edited in 'Mir', Moscow, 1970.

11. Kabanov L.P., "Heat Transfer Processes During a PWR Core Reflooding Following LOCA", Contributions to Nuclear Science and Technology, Series: Nuclear Reactor Physics, Vol. 7, p. 36, Moscow, 1983.

12. Hein D., "Modell Vorstellungen zum Weiderbentren Durch Fluten", PhD thesis, University of Hanover, FR Germany, 1980.

13. Tong L.S., Weisman J., "Thermal Analysis of Pressurized Water Reactors", American Nuclear Society, LaGrande Park, Illinois, 1979.

14. Bjornard T.A., "Blowdown Heat Transfer in a Pressurized Water Reactor", Phd thesis, Department of Mechanical Engineering, Massachusetts Institute of technology, 1977.

15. Turgood M.J., et.al., "COBRA-TRAC – A Thermal-Hydraulic Code for Transient Analysis of Nuclear Reactor Vessels and Primary Coolant Systems", NUREG/CR-3046, 1983.

Source Term and Two-Phase Flow Phenomena in Connection with Hazardous Vapor Clouds

H. K. FAUSKE and M. EPSTEIN
Fauske & Associates, Inc.
16W070 West 83rd Street
Burr Ridge, Illinois 60521, USA

Abstract

Source term and two-phase flow phenomena involving high momentum jet releases are presented in terms of simple models. These include methods for assessing vapor versus two-phase jet releases, the discharge rate, expanded jet behavior, aerosol formation versus liquid rainout, and turbulent mixing of the jet with the atmosphere. The illustrated strong sensitivity of downwind concentration profile to minor source term variations suggests that the accuracy of the simple models are adequate for most hazard assessment purposes.

1. INTRODUCTION

The rapidly growing emphasis on risk and hazard evaluation in the chemical process industry involving emergency releases of flashing liquid jets and vapor cloud formation (see Figure 1) calls for simple physical models to allow consequence assessment to be carried out in a timely and cost effective manner. Simple models are particularly desirable to address the following questions:

- Will the concentration downwind of the plant site exceed critical toxicity levels or explosive limits?

- When will the cloud disperse to a safe level?

Answers to these questions will largely determine the needs for further considerations related to location, prevention, mitigation and emergency planning.

In this paper we summarize models with a level of detail which seems appropriate for the purpose of carrying out risk evaluations in connection with high momentum jet releases. This approach seems justified considering the strong sensitivity of the downwind concentration profile to minor variations in the source term. The recommended methodology includes considerations related to:

- Release Type: Consideration of vapor disengagement to distinguish between vapor and two-phase jet releases.

- Jet Expansion: Consideration of equilibrium jet expansion parameters, including jet velocity, jet density and jet radius.

- Aerosol Formation: Consideration of jet breakup to distinguish between aerosol formation and liquid rainout based upon initial release conditions.

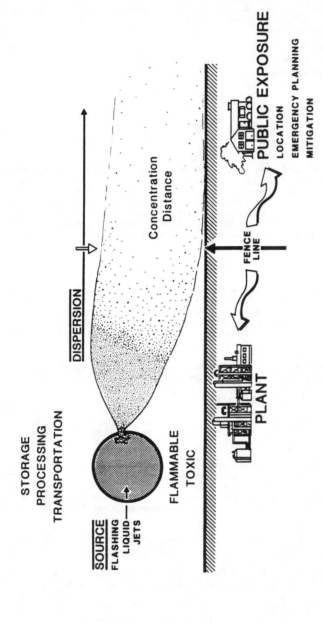

Fig. 1. – Major Hazard and Risk to Public – Vapor Clouds from Flashing Liquid Releases.

• Jet Dispersion: Consideration of turbulent mixing of the jet with the atmosphere leading to downwind concentration profile.

The above methodology is used to illustrate the strong sensitivity of downwind concentration profile to source term variations, including release type, variation in discharge rate from non-equilibrium, equilibrium, sub-cooling and frictional effects, and aerosol formation versus liquid rainout.

2. RELEASE TYPE

Accidental releases of material stored under their own vapor pressures can occur as vapor or flashing two-phase flows. For the same size break the flashing two-phase flow rate is much larger than that of the all-vapor flow, and following expansion to ambient conditions, the low quality mixture has the potential to become aerosolized followed by complete vaporization. The simple disengagement rule presented in Figure 2 is suggested for assessing release type.

For foamy-like liquids (many chemical systems exhibit this behavior) where the liquid phase remains continuous to essentially 100% void fraction, the discharge should be assumed to be two-phase at all times, i.e., the release type is independent of break location [1].

In the case of non-foamy systems, the transition from a liquid to vapor continuous regime generally occurs around 50% void fraction. In the vapor continuous regime, vapor disengagement becomes very effective leading to a vapor-like release for break diameters much less than the vessel diameter (see Figure 2). Vapor-like release can also occur with the liquid continuous regime, if the vessel superficial vapor velocity based on maximum vapor flow through the break, $(D_B^2/D_V^2)(0.6)(R\,T_o/M)^{1/2}$, is less than the characteristic two-phase drift velocity for the so-called bubbly regime, U_∞ $[(L - H)/L]/(H/L)$ where $U_\infty = 1.2 \left[\sigma g \left(\rho_f - \rho_g\right)\right]^{1/4}/\rho_f^{1/2}$. Here we note that it is difficult to credit significant vapor disengagement beyond the bubbly regime. Relatively small quantities of impurities tend to favor this regime as compared to the churn turbulent flow regime which is typical for clean water-like conditions.

3. JET EXPANSION

Following jet release and expansion we are interested in determining the equilibrium jet velocity, jet density and jet radius prior to sig-nificant interactions with the atmosphere. As we will see later these parameters are of principal interest in determining the potential for aerosol formation (requiring a knowledge of the jet velocity and jet density) and downwind jet dispersal (requiring a knowledge of the jet density and jet radius).

The fully expanded jet properties at the end of depressurization zone (see Figure 3), can be determined by applying the principles of conservation of momentum and mass fluxes [1-3]. The well known relationship for the jet velocity, u_j, in terms of the break (or choke) plane quantities u_E, $G = u_E$ ρ_E, and P_E, is illustrated in Figure 3. A summary of possible exit condi-tions are summarized in Table I, including various release types (non-flashing, vapor and flashing flows) as well as different flashing flow

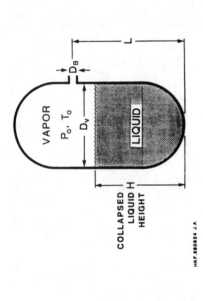

HKF.880824 J.A

FLOW TYPE	FOAMY	NON-FOAMY
FLASHING LIQUID	$H > 0$	$\left[\dfrac{D_B^2}{D_V^2}\right](0.6)\left(\dfrac{RT_0}{M}\right)^{1/2} > v_\infty\left[\dfrac{L - H}{L}\right]\left[\dfrac{H}{L}\right]$
VAPOR FLOW		$H < L/2$

Fig. 2. – Simple Vapor Disengagement Rule for Determining Release Type with High Pressure Storage.

486

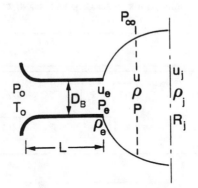

$$U_J = U_E + \frac{P_E - P_\infty}{U_E \, \rho_E} = U_E + \frac{P_E - P_\infty}{G}$$

$$\rho_J = \frac{\rho_G}{X} = \frac{\rho_G \, H_{FG}}{\Delta T \, C}$$

$$R_J = R_B \left(\frac{G}{\rho_J \, U_J} \right)^{1/2}$$

Fig. 3. - Summary of Jet Expansion Model.

TABLE I: SUMMARY OF DISCHARGE RATES AND EXIT CONDITIONS

Flow Condition	G	u_e	P_e
Non-Flashing Flow	$\sqrt{2\left[P_o - P_\infty\right]\rho_f}$	G/ρ_f	P_∞
Vapor Flow	$0.6\sqrt{P_o\,\rho_{go}}$	$\sqrt{(RT)/M}$	$0.6\,P_o$
FLASHING FLOWS			
Non-Equilibrium, L ~ 0	$\sqrt{2\left[P_o - P_\infty\right]\rho_f}$	G/ρ_f	$P_v\!\left(T_o\right)$
Subcooled, L ~ 0.1 m	$\sqrt{2\left[P_o - P_v\!\left(T_o\right)\right]\rho_f}$	G/ρ_f	$P_v\!\left(T_o\right)$
Saturated-No Friction, L ~ 0.1 m	$h_{fg}\,\rho_g\left(\dfrac{1}{Tc}\right)^{1/2}$	G/ρ_f	$P_v\!\left(T_o\right)$
Saturated-Friction*	$F\,h_{fg}\,\rho_g\left(\dfrac{1}{Tc}\right)^{1/2}$	$\dfrac{G}{\left(\rho_g/x\right)FP_v\!\left(T_o\right)}$	$FP_v\!\left(T_o\right)$

*F is <u>flow reduction factor</u> and <u>critical pressure ratio</u> (homogeneous flow).

L/D	F
0	1
50	0.8
100	0.7
200	0.6
400	0.5

conditions including non-equilibrium, subcooled, saturated equilibrium and frictional effects [1]. The expression for the jet density, ρ_j, shown in Figure 3 is obtained by assuming an isenthalpic expansion. Finally, by conserving mass, an expression for the jet radius, R_j, is obtained, which is also shown in Figure 3.

4. AEROSOL FORMATION

During the depressurization, the liquid can disintegrate into fine droplets. A simple model is summarized in Figure 4 which allows an estimate of whether these droplets are likely to remain airborne or will rain out. The basic premise of the model is that the final drop size is determined by hydrodynamic considerations. This droplet size will then remain airborne if the rate at which the jet expands by entrainment exceeds its terminal settling velocity, viz

$$u_j \frac{dR}{dx} > u_t \tag{1}$$

Ricou and Spalding [4] and others have verified through measurements that for high-momentum gaseous jets the entrainment velocity is well-represented by the expression

$$u_{en} = 0.08 \left(\frac{\rho_j}{\rho_\infty}\right)^{1/2} u_j \tag{2}$$

Comparison of jet decay calculations using this expression with experiments employing two-phase jets have shown good agreement [1,5,6]. One can show that to first order the rate of growth of the plume is related to u_{en} by the following expression:

$$u_j \frac{dR}{dx} = \frac{\rho_\infty}{\rho_j} u_{en} \tag{3}$$

The terminal sedimentation velocity of the droplets within the jet is, from Stokes law,

$$u_t = \frac{g\ D^2\ \rho_f}{18\ \mu_g} \tag{4}$$

Using the well-known Weber number criterion to predict the stable drop size at the completion of jet depressurization, gives

$$D \approx \frac{10\ \sigma}{\rho_\infty\ u_j^2} \tag{5}$$

Eliminating D between Equations (4) and (5) and substituting the result, together with Equations (2) and (3), into criterion (1) gives the following expression for the velocity of the fully depressurized jet above which rainout is precluded:

$$u_j \approx 2.3 \left(\frac{\rho_j}{\rho_\infty}\right)^{1/10} \left(\frac{\rho_f\ g\ \sigma^2}{\rho_\infty^2\ \mu_g}\right)^{1/5} \tag{6}$$

489

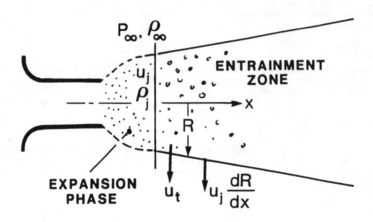

HKF.880824 B.A

No Liquid Rainout :

$$u_j \frac{dR}{dx} > u_t$$

Fig. 4. - High Momentum Jet Liquid Rainout Model (Aerosol Criteria).

For most materials of interest the group $\left[\rho_f \, g/\rho_\infty^2 \, \mu_g\right]^{1/5}$ is approximately equal to 52 in mks units. So that the critical jet velocity required for the droplets to remain airborne within the jet is

$$u_j \approx 120 \; \sigma^{2/5} \; \left(\rho_j/\rho_\infty\right)^{1/10} \tag{7}$$

where σ is in units of N m^{-1} and u_j is in units of m s^{-1}.

Equation (7) is the key to determining the vessel (stagnation) conditions for which rainout does not occur. For example, consider a cold liquid release (jet temperature < liquid saturation temperature at atmospheric pressure). In this case the aerodynamic interaction at the liquid-air interface leads to the breakup of the liquid jet into droplets. The velocity of the jet just prior to mechanical atomization is given by the Bernoulli equation (see Table I):

$$u_j \approx \left[2\left(P_o - P_\infty\right)/\rho_f\right]^{1/2} \tag{8}$$

Eliminating u_j between Equations (7) and (8) and solving the result for the stagnation (vessel) pressure gives

$$P_o = P_\infty + 0.72 \times 10^4 \; \rho_f \; \sigma^{4/5} \; \left(\rho_j/\rho_\infty\right)^{1/10} \tag{9}$$

We shall illustrate this result by a few figures, referring to water. Putting in Equation (9) $\rho_j = \rho_f = 10^3$ kg m^{-3}, $\sigma = 0.07$ N m^{-1}, $\rho_\infty = 1$ kg m^{-3}, we get

$$P_o = 1.8 \times 10^6 \; \text{Pa} = 18 \; \text{atm}$$

This means that very large vessel pressures are required to produce a high degree of dispersion of the cold liquid jet and prevent rainout.

The level of vessel pressure required to prevent the rainout of a saturated liquid release can be calculated in a similar way as for the cold liquid release. Starting with the expression for G for saturated, friction-less, flashing flow, as given in Table I, and the momentum balance equation for jet depressurization (see Figure 3), it can be shown that

$$u_j \approx \Delta T(c/T)^{1/2} \tag{10}$$

where ΔT is the difference between the temperature of the pressurized liquid in the vessel and its saturation temperature at atmospheric pressure (liquid "superheat" relative to its boiling point at one atmosphere), and c is the specific heat of the liquid. Substituting Equation (10) in Equation (7) and solving for ΔT, we get

$$\Delta T \approx 78 \; \sigma^{4/11} \; (T/c)^{5/11} \left[\frac{\rho_g \, h_{fg}}{\rho_\infty \, c}\right]^{1/11} \tag{11}$$

where we used the second equation of Figure 3 to eliminate ρ_j in favor of ΔT. For most hazardous liquids of interest the group $\rho_g \, h_{fg}/\left(\rho_\infty \, c\right)$ is a number near 10^3 in mks units. Thus Equation (10) simplifies to the dimensional relation

491

$$\Delta T \approx 140 \; \sigma^{4/11} \; (T/c)^{5/11} \tag{12}$$

For the purpose of illustration we shall again consider the case of water. With $T = 373K$ and $c = 4 \times 10^3$ J kg^{-1} K^{-1}, Equation (12) gives

$$\Delta T \approx 15°C$$

which corresponds to a vessel overpressure of only 1.6 atm, compared with the previously calculated value of 18 atm for the dispersal of a cold jet. This calculated ΔT for water is given in Table II, together with predicted ΔT's for some hazardous materials of practical interest.

For subcooled liquids the calculation of incipient rainout conditions is slightly more involved. The velocity u_i is determined from the momentum equation in Figure 3 and the appropriate expression for G in Table I for subcooled flashing flows. If this value of u_i exceeds the value calculated with Equation (7), then complete dispersal (i.e. no rainout) of the sub-cooled flashing liquid jet is expected. Table III demonstrates how successful this method is when compared with the reported observations on aerosol formation and rainout for subcooled flashing jets.

Using the methodology presented in this section a superheat versus excess vessel pressure $[P_o - P_v(T_o)]$ map for flashing water jets has been prepared and is shown in Figure 5, indicating the region in which rainout occurs. Interestingly enough, the rainout region is seen to expand as the excess vessel pressure is increased.

5. JET DISPERSION AND DILUTION

In the previous sections our attention was focused mainly on the region of jet depressurization. The expansion of gas or initially liquid jets in the depressurization zone is so rapid that their interaction with the atmosphere is negligible and very little mixing between jet and ambient occurs. The region of jet dispersion or, equivalently, entrainment, begins when the jet is depressurized to atmospheric pressure. Typically this occurs within one or two break diameters (see Figure 4). The jet behavior in the entrainment region is strongly influenced by mixing with the atmosphere. The jet visibly "grows" by entrainment of ambient air.

Let $\dot{m}_E$ denote the total rate of air entrainment by the jet through its boundary from the source to the downstream position x, then

$$\dot{m}_E = \int_0^x 2\pi R \; \rho_\infty \; u_{en} \; dx \tag{13}$$

Using Equation (2) to eliminate u_{en} from the above integral gives

$$\dot{m}_E = 2\pi \; (0.08) \int_0^x \left(\rho_\infty \; \rho \right)^{1/2} u \; R \; dx \tag{14}$$

TABLE II: PREDICTED FLASHING SATURATED LIQUID
JET ΔT FOR AEROSOL FORMATION

Fluid	ΔT, °C	x
H_2O	~ 15	~ 0.03
HF	~ 7	~ 0.06
CH_3NH_2	~ 11	~ 0.025
NH_3	~ 13	~ 0.02
Cl_2	~ 23	~ 0.08
F-11	~ 20	~ 0.1

TABLE III: COMPARISON BETWEEN PREDICTION AND
EXPERIMENTAL OBSERVATIONS RELATIVE
TO AEROSOL FORMATION AND RAINOUT

Fluid	Required Jet Velocity [Eq. (7)]	Experimental Jet Velocity (Momentum Equation, Fig. 3)	Observation
NH_3 Tortoise 3	~ 35	~ 86	Complete Aerosol Formation [7]
NH_3 (Sweden)	~ 36	~ 155	Complete Aerosol Formation [8]
HF1 (Goldfish)	~ 23.4	~ 27	Complete Aerosol Formation [9]
HF2 (Goldfish)	~ 23.4	~ 27.6	Complete Aerosol Formation [9]
CH_3NH_2 (No. 4)	~ 34	~ 33	Complete Aerosol Formation [10]
CH_3NH_2 (No. 8)	~ 38	~ 21	Partial Rainout [10]

493

where ρ, u and R are the density, velocity and radius of the jet at any location between the break at x = 0 and x. From the conservation of momentum we know that

$$\rho \, u^2 \, R^2 = \rho_j \, u_j^2 \, R_j^2 = const \tag{15}$$

Thus the integrand in Equation (14) is constant and $\dot{m}_E$ reduces to the algebraic form

$$\dot{m}_E = 2\pi \, (0.08) \left(\rho_\infty \, \rho_j\right)^{1/2} u_j \, R_j \, x \tag{16}$$

Now, let the symbol Y denote the mass fraction of the hazardous jet material at position x. A total mass balance for the jet material is

$$\pi \, R_j^2 \, \rho_j \, u_j = Y\left[\pi \, R_j^2 \, \rho_j \, u_j + \dot{m}_E\right] \tag{17}$$

which simply states that the source rate of flow equals the total rate of flow of jet material through position x. Solving Equation (17) for Y gives the sought result

$$Y = \frac{1}{1 + 0.16 \left(\dfrac{\rho_\infty}{\rho_j}\right)^{1/2} \dfrac{x}{R_j}} \tag{18}$$

This simple expression is in excellent agreement with the far field concentration predictions obtained with a somewhat detailed computer model that accounts for the effect of "laminar" wind velocity, jet trajectory, aerosol evaporation, and the condensation of the entrained water vapor [11]. Moreover, as can be seen from Figure 6, the model can reproduce field observation data from a high momentum release of liquid ammonia [7] to a degree of accuracy that is more than adequate for most hazard assessment purposes. The only input data used for this calculation comparison are the known initial conditions of P_o, T_o and the break diameter, D_b. The model tracks the NH_3 concentration versus distance trend of the data in a manner that is superior to available, but significantly more complex jet dispersion models [12-14]. These complex models apparently fail in this regard because they do not account for the momentum of the source.

An examination of Equation (18) indicates that the dilution rate of the jet material is sensitive to the jet properties ρ_j and R_j at the end of the depressurization zone. This, in turn, implies that jet behavior is a strong function of the geometry of the break and the release type, vapor or two-phase. The concentration versus downwind distance curves for a typical liquid chemical are plotted in Figure 7 as a function of the release geometry and type. This figure was constructed using Equation (18) and the simple methodology displayed in Figure 3 and Table I for calculating the fully expanded jet properties at the end of the depressurization zone. In Figure 7 one sees the extreme sensitivity of jet dilution to release geometry and type. Variations in the figure include:

• all vapor flow,

• equilibrium flashing flow with complete rainout,

494

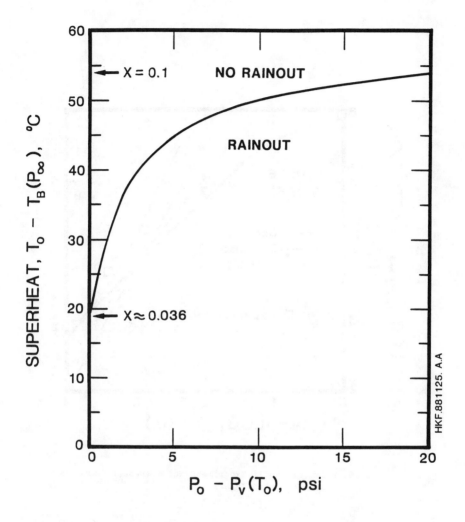

Fig. 5. – Relationship Between Superheat Resulting in No Rainout and the Subcooling (i.e., Excess Pressure Relative to the Vapor Pressure) (Example: H_2O).

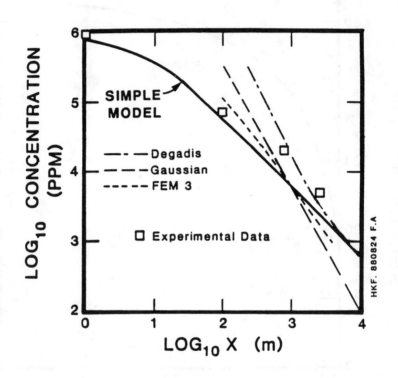

Fig. 6. – Ammonia Spill Experiment No. 4 (Desert Tortoise)
and Predicted Values.

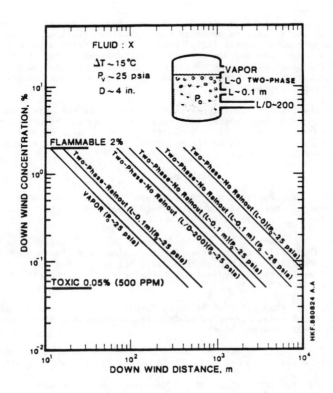

Fig. 7. – Illustration of Source Term Sensitivity
and Hazard Evaluation.

497

- equilibrium flashing flow with no rainout, length-to-diameter ratio of 200,

- equilibrium flashing flow and no rainout with negligible friction,

- slightly subcooled stagnation condition, vapor pressure exceeded by 1 psi, and

- non-equilibrium flow, i.e., no time for flashing.

6. CONCLUDING REMARKS

A large number of dispersion models involving various assumptions are available to estimate downwind concentration profiles following emergency releases of dangerous chemicals [15]. However, the connection between these models and the actual source term is not well defined. For many cases involving significant risks, the source can be related to high momentum jet releases. Furthermore, in the majority of these cases, the source term involves flashing two-phase flows.

Based on the limited sensitivity study presented in the previous section, we conclude that major attention should be placed on definition of the source term and, more specifically, mitigation of the source term. Further refinement of the simple dispersion model for the purposes of risk assessment does not appear warranted.

REFERENCES

1. Fauske H. K. and Epstein, M., "Source Term Considerations in Connection with Chemical Accidents and Vapor Cloud Modeling", J. Loss Prevention Process Ind., 1, pp. 75-83 (1988).

2. Epstein, M., Henry, R. E., Midvidy, W., and Pauls, R., "One-Dimensional Modeling of Two-Phase Jet Expansion and Impingement", in Thermal-Hydraulics of Nuclear Reactors, Volume II, 2nd Int'l. Mtg. on Nuclear Reactor Thermal Hydraulics, Santa Barbara, California (January, 1983).

3. Moody, F. J., "Prediction of Blowdown Thrust and Jet Forces", Presented at ASME-AIChE Heat Transfer Conference, Minneapolis, Minnesota, ASME Paper 69-HT-31 (August, 1969).

4. Ricou, F. P. and Spalding, D. B., "Measurements of Entrainment by Axisymmetric Turbulent Jets", J. Fluid Mech., 11, pp. 21-32 (1961).

5. Cheung, F. B. and Epstein, M., "Two-Phase Gas Bubble-Liquid Boundary Layer Flow Along Vertical and Inclined Surfaces", Nucl. Eng. and Design, 99, pp. 93-100 (1987).

6. Hussain, N. A. and Siegel, R., "Liquid Jet Pumped by Rising Gas Bubbles", J. Fluids Engrg. 98 (1976) 49-56.

7. Goldwire, H. C., Jr., "Large-Scale Ammonia Spill Tests", Chemical Engineering Progress, (April, 1986), pp. 35-41.

8. Nyren, K. and Winger, S., "Two-Phase Discharge of Liquefied Gases Through Pipes, Field Experiments with Ammonia and Theoretical Model", Symposium Series of Inst. Chem. Engng. on Loss Prevention, Harrogate, England, (1983).

9. Blewitt, D. N., Yohn, J. F., Koopman, R. P., and Brown, T. K., "Conduct of Anhydrous Hydrofluoric Acid Spill Experiments", in International Conference on Vapor Cloud Modeling, November 2-4, 1987, Cambridge, Massachusetts (Woodward, J., ed.), pp. 1-38.

10. Lantzy, R. J. and Myers, R. D., "Atmospheric Release Tests of Monomethylamine", Rohm & Haas Company, (June 15, 1988).

11. Epstein, M., Fauske, H. K., and Hauser, G. M., "A Dispersion Model of Flashing Two-Phase High Momentum Releases", to be presented at 5th Miami International Symposium on Multiphase Transport & Particulate Phenomena (December 12-14, 1988).

12. Spicer, T. O. and Havens, J., "Modeling HF and NH_3 Spill Test Data Using DEGADIS", presented at 1988 Summer Nat'l. Mtg. of AIChE, Paper No. 87b.

13. Pasquill, F., and Smith, F. B., Atmospheric Diffusion, 3rd ed., Halstead Press, New York, (1983).

14. Chan, S. T., "FEM3 - A Finite Element Model for the Simulation of Heavy Gas Dispersion and Incompressible Flow" User's Manual", Lawrence Livermore National Laboratory, UCRL-53397, (February, 1983); see also Ref. [7].

15. Hanna, S. R. and Drivas, P. J., "Guidelines for Use of Vapor Cloud Dispersion Models", Center for Chemical Process Safety of the American Institute of Chemical Engineers, (1987).

NOMENCLATURE

c   Heat capacity of liquid jet material.

D   Droplet diameter.

D_B Diameter of break (Figures 2, 3).

D_v Diameter of vessel (Figure 2).

F   Flow reduction factor (Table 1).

g   Gravitational acceleration.

G   Mass flux release rate.

H   Collapsed liquid height (Figure 2).

L   Height of vent above vessel bottom (Figure 2) or flow length to break (Figure 3).

M   Molecular weight.

$\dot{m}_E$ Mass rate of air entrainment by jet.

P   Pressure.

P_v Saturation vapor pressure.

R   Radius of circular plume or ideal gas constant (Figure 2).

R_B Radius of break.

T   Temperature.

ΔT Liquid superheat relative to boiling point at one atmosphere.

u   Mean velocity inside plume.

u_∞ Terminal bubble rise velocity of vapor phase during liquid swell.

U_T Terminal sedimentation velocity of droplets.

u_{en} Entrainment velocity.

We   Weber number; Equation (2).

x   Distance measured from the source along the ground or vapor quality.

Greek Letters

μ Viscosity.

ρ Density.

σ Surface tension.

Subscripts

e   Break or choke plane quantities.

f   Liquid phase.

fg   Liquid-gas phase transition.

g   Gas phase.

j   At end of depressurization zone.

o   Stagnation or vessel condition.

∞ Pertains to the atmosphere.

ENERGY/THERMAL
APPLICATIONS

Heat Pump Supported by Heat Pipe Solar Collector

SUMER SAHIN, ALI YUCEL UYAREL, and HUSEYIN USTA
Gazi Universitesi
Teknik Egitim Fakultesi
Besevler-Ankara, Turkey

ABSTRACT

In an experimental set-up, the solar heat pump supported by a heat pipe was investigated. Heat pipe evaporators were connected to a flat plate collector. The heat pipe condenser with a diameter of 150 mm was constructed as an isolated pipe with glass-wool. The heat pump evaporator coil was placed into the heat pipe condenser. As working fluids, ethanol was used in the heat pipe and R-12 was used in the heat pump cycle. Collector was tilted 45° and oriented to South. A single glased collector plate with an area of 1 m^2 was used.

During the experiments, the performance coefficient of the heat pump could reach a value of approximately 4 by a mean solar radiation flux of 500 W/m^2. The evaporator surface area and the thermal loop of the system were smaller than the heat pump used in commercial applications. Furthermore, the system eliminates the freezing problem, because the evaporator takes the heat from the condensing ethanol, and not from the water tank, as it is the case in commercial systems. Therefore, the system can operate with a higher efficiency and safety even in lower temperature conditions, and during the night.

1. INTRODUCTION

Heat pump is an efficient device for low temperature heating purposes. It is videly used for space heating in winter, and cooling for summer. The efficiency of a heat pump can be increased substantially by increasing the evaporator temperature, and decreasing the temperature difference between condenser and evaporator. In the present study, this is done in a very elegant and simple way as follows:

A closed solar collector system is used to increase the evaporator temperature of the heat pump. The solar energy is transfered from the solar collector to the evaporator with the help of a heat pipe system. This allows one to elevate the evaporator temperature, practically to the same level of the solar collector temperature, because of the very superior heat transfer capabilty of the heat

503

pipe through condensation.

In a more conventional option, one could transfer the solar heat via classical circulation with a working fluid to a storage tank in order to heat the evaporator of the heat pump. However, this classical way would have some handicaps:

    a. There is a significant temperature drop from the solar collector to the storage tank,

    b. During start-up, the response time is quite high, i.e., it needs some time before the storage tank reaches the operation temperature,

    c. The working fluid can have freezing problems if the balance of the heat transfer from solar collector to the heat pump evaporator goes beyond narrow operation limits of the system. Thus, anti-freezing is necessary,

    d. Natural circulation is very slow. Hence, an additional pump is required.

All these handicaps are eliminated by using a heat pipe system between the solar collector and heat pump evaporator. As the heat pipe is transfering the heat energy with a high capacity via the latent heats of condensation and evaporation the heat transfer surfaces will be very small. The system becomes compact and elegant. Heat pipe continuous to take heat energy from environment at a lower vapor pressure even during the night.

2. DETERMINATION OF THE DIMENSIONS

The aim of the experiments is to obtain a higher performance coefficient β of the heat pump than the commercial ones. In this study a value of 4 is desired.

During the experiments, a heat pump compressor with a power of $\dot{W}_C$=245 watts is used, which was immediately available at the Department. All other dimensions are adjusted to these data, as described in this section in detail.

The heat power of condenser $\dot{Q}_K$ becomes

$$\dot{Q}_K = \beta\, \dot{W}_C = 980 \text{ watts} \tag{1}$$

The required heat flux from the solar collector $\dot{Q}_E$ is

$$\dot{Q}_E = \dot{Q}_K - \dot{W}_C = 735 \text{ watts, (see Figures 1 and 2)} \tag{2}$$

The solar flux in Ankara in November (during the experiments) is about I_s=500 W/m^2 on a horizontal surface. The flux on the collector becomes

$$\dot{Q}_E = \eta_c \cdot I_s \cdot \varphi \cdot F_c \tag{3}$$

where ;

    η_c : collector efficiency

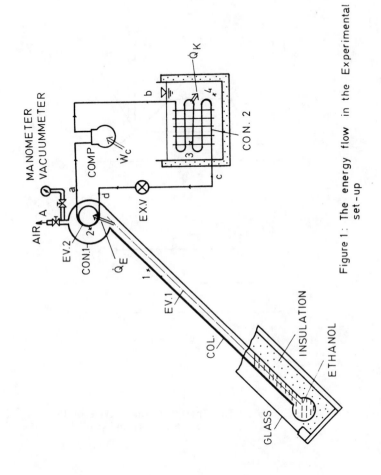

Figure 1: The energy flow in the Experimental set-up

505

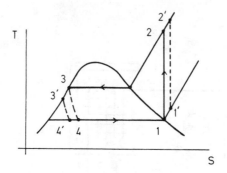

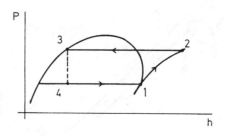

Figure 2: The Thermodynamic cycle of the
heat pump in (T,s) and (p,h) diagrams.

φ : collector tilt factor

F_c : collector surface area

For Ankara in December, one can assume [1,2]

η_c=0.80 (for collector to an ambient temperature difference of 20°C)

φ =1.8 (by an inclination of 45 ^{o})

so that one can obtain

F_c=1.0 m^2

The collector efficiency η_c is higher than that of conventional systems, as expected by a heat pipe system.

2.1. The dimensions of condenser

The dimensions of the condenser of the heat pump is evaluated as follows, see Figure 3:

$$\dot{Q}_K= \eta_F . A. \ \alpha_o . (T_K - T_{fo}) \tag{4}$$

$$\eta_F = \frac{Tanh(ml)}{ml} \qquad : \text{fin efficiency} \tag{5}$$

$$m= \sqrt{\frac{\alpha_o . P}{\lambda . F_f}} \tag{6}$$

where,

α_o : convection heat transfer coefficient of the cooling fluid

P : cross-section perimter of the fin

λ : conduction heat transfer coefficient of the fin material

F_f : cross-section area of the fin

Ref. [3] suggests the following relation for the heat transfer in condenser ;

$$\dot{Q}_K = K.A_o.\Delta T_o \tag{7}$$

where,

$$\frac{1}{K} = \frac{1}{\eta . \alpha_o} + \frac{A_o}{\alpha_i . A_i} \tag{7a}$$

$$\eta =1 - \frac{A_f}{A_o} . (1 - \eta_F) \tag{7b}$$

where,

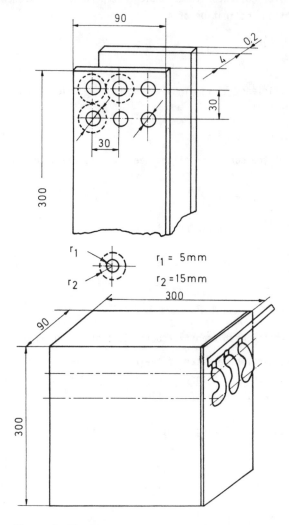

Figure 3 : The detailed sketch of the heat
Pump condenser

η : fin efficiency factor

A_f: fin area

A_o: total outside area of the condenser

ΔT_o:temperature of the base of the fin-temperature of the fluid difference

Table I shows the most important data in the design of the condenser.

It turns out that, the Equations (4) and (7) are leading to similar surface areas for the condenser.

2.2. The dimensions of evaporator

The evaporator is made of copper pipes with an outer diameter of 10 mm. The fluids inside and outside of the pipes are R-12 and ethanol, respectively.

The heat transfer coefficient for the condensing ethanol is calculated, as [4]

$$\alpha_o = 0.555 \left[\frac{k_f^3 \cdot \rho \cdot (\rho - \rho_v) \cdot g \cdot r}{d \cdot \mu \cdot \Delta T_o} \right]^{0.25} \tag{8}$$

The evaporation rate of R-12 is calculated, as

$$\dot{m}_r = \dot{Q}_E / r \tag{9}$$

and the evaporativ heat transfer coefficient inside the pipe is also calculated, as [4]

$$\alpha_i = \frac{c^2 \cdot \dot{m}_r \cdot \Delta T_i}{d^2} \tag{10}$$

where,

k_f^3 : liquid condensing ethanol conductive heat transfer coefficient

ρ, ρ_v: densities of liquid and vapor of ethanol

r : enthalpy of evaporation of ethanol

d : diameter of the pipe

μ : dynamic viscosity of liquid ethanol

ΔT_o : fluid to surface temperature difference (outside)

ΔT_i : fluid to inside surface temperature difference

C : a constant, according to Ref[3] its value is C=1.52

Table II shows the relevant data for the evaporator.

Table I The Dimensions of the Heat Pump Condenser
($\dot{Q}_K$=980 watts)

	Air Cooling		Water Cooling
	Natural Convection	Forced Convection u_a=10 m/s Re=1600 $St.Pr^{2/3}$=0.012 × [4]	
α_{o_2} (W/m²C)	8	40	380
ml (-)	0.218	0.426	1.98
η_F	0.96	0.92	0.65
η	0.98	0.95	0.72
ΔT_o (°C)	25	8	1.1
A_o (m²)	3.92	3.58	3.79

Table II The Dimensions of the Evaporator
($\dot{Q}_E$=735 watts, d_E=10 mm)

Specifications of condensing ethanol (outside of the pipe)						Specifications of R-12 (inside of the pipe)					CALCULATIONS		
λ_f	ρ_f	ρ_v	r	μ	ΔT_o	T_i	P_i	r	$\dot{m}/A_c$	ΔT_i	α_o	α_i	l_E
(W/m C)	(kg/m³)	(kg/m³)	(kJ/kg)	(mkg/s)	(°C)	(°C)	(bar)	(kJ/kg)	(kg/m²s)	(°C)	(W/m²C)	(W/m²C)	(m)
0.286	1264	0.789	940	1.207	2.5	20	5.2	140	66.87	3.8	820	580	12.0

3. EXPERIMENTS

After evaluation of the main dimensions, an experimental set-up is constructed, as shown in Figures 4 and 5. The heat pump and heat pipe are filled with the working fluids, R-12 and ethanol, respectively, as mentioned before.

The heat pipes are made of 6 pipes with an outer diameter of 20 mm and lenght of 1 meter, and welded to a condenser pipe having a diameter of 50 mm. The latter contains the heat pump evaporator with the main data in Table I.

During the experiments, the temperature measurements are performed with the help of Cu-Cons thermocouples and are read electronicallly over 12 channels.

During the experiments, the condenser was cooled with air, and later with water. The air cooling is realised in an air chamber having the internal dimensions of 0.86x1.10x1.00 m. The chamber is made of 1 mm steel plates and isolated with 50 mm glass-wool so that the total heat transfer coefficient is kept by about 0.75 $W/m^2 C$.

The water cooling is realised in a tank with a volume of 51 liters having a 30 mm glass-wool isolation.

In the experiments, the heat transfer from the condenser is evaluated, as follows

$$\dot{Q}_K = (m_f \cdot c_{pf} + m_t \cdot c_{pt}) \cdot \frac{\Delta T}{\Delta \tau} \qquad (11)$$

where,

m_f : mass of the cooling fluid

c_{pf}: specific heat at constan pressure of the fluid

m_t : mass of the tank

c_{pt}: specific heat of the tank (or chamber)

$\Delta \tau$: time of the experiment

Table III and IV show the experimental results with air cooling and water cooling of condenser, respectively.

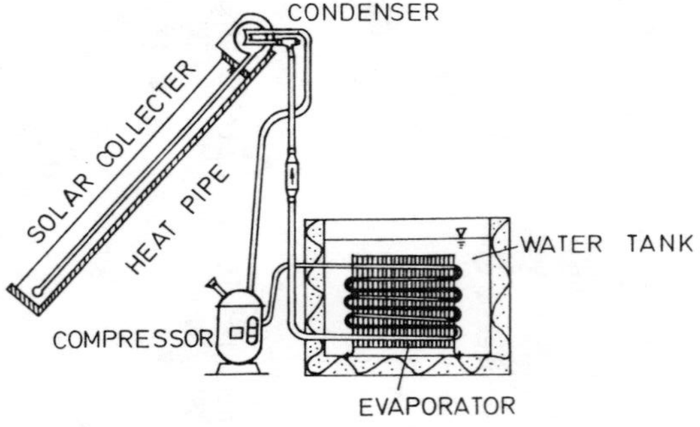

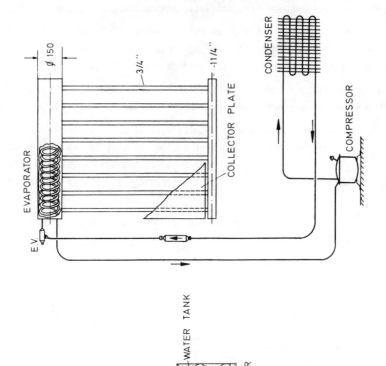

Figure 5: The main components of the experimental set-up

Figure 4: The experimental set-up

512

Table III The Experimental Results With Air Cooled Condenser

Local Time	T_1	T_2	T_3	T_4	T_5	Solar Radiation	$\dot{Q}_K/\dot{W}_C$ β
10:45	43.0	35.0	16.0	10.6	12.5	323	-
10:55	41.8	35.5	56.2	23.0	12.7	326	3.64
11:05	41.0	33.0	59.8	30.6	13.2	327 W/m^2	3.46
11:15	40.5	31.0	60.6	33.0	13.6	330	3.30
11:25	39.8	29.8	60.9	33.6	13.8	332	3.26
11:35	41.1	31.6	62.4	36.2	14.0	335	2.54
11:45	42.8	32.7	63.4	38.1	14.2	337	2.46

Table IV The Experimental Results With Water Cooled Condenser

Local Time	T_1	T_2	T_3	T_4	T_5	T_a	T_b	T_c	T_d	I_s W/m^2	$\dot{Q}_K/\dot{W}_C$ β
15:40	73.6	61.4	26.0	25.5	36.9	62.5	73.8	27.8	3.1	488	1.97
15:50	69.6	55.3	29.1	26.8	38.5	57.4	87.1	29.0	3.1		1.97
16:00	67.6	49.4	31.1	28.4	37.4	52.4	83.0	29.2	2.9		2.42
16:10	58.9	41.9	33.6	29.9	35.0	45.1	86.6	31.9	2.8	509	2.27
16:20	54.0	41.5	36.2	32.3	36.0	43.8	87.2	32.8	2.6		3.68
16:30	50.3	35.6	38.0	34.8	34.5	38.1	89.2	33.8	2.5		3.78
16:40	48.6	32.5	39.4	37.5	34.0	36.2	91.0	34.0	2.2		4.09

Temperature measurement points:

1: collector surface   a: compressor inlet
2: evaporator surface   b: compressor outlet
3: condenser surface   c: condenser outlet
4: cooling fluid   d: evaporator inlet
5: ambient

4. CONCLUSIONS

During the experiments, it has been possible to obtain relatively higher performance coefficients, such as, 4 to 5, as compared to the conventional heat pumps for which this coefficient (COP) is about 2.5 to 3. Water cooling of condenser allows one to obtain higher COP's than the air cooling.

All the expectations, cited in Section 1 could be realised in the course of experiments. The superiority of the heat pipe supported, solar heated heat pumps, as compared to the conventional ones is clearly demonstrated.

In the new system, the temperature difference between the evaporator and the condenser is smaller than that of the classical system, and also the evaporator temperature is higher. These are essentially the main reasons, why the COP is higher than in a classical system.

REFERENCES

[1] Kreider, J.F. and F. Kreith, Solar Heating and Cooling, Mc. Graw Hill
      Book Comp., New York, 1982

[2] Duffie, A.J. and W.A. Beckman, Solar Engineering of Thermal Processes
      John Wiley & Sons, Inc., New York, 1980

[3] Dunn, P.D. and D.A. Reay, Heat Pumps, 2 nd Edt., Pergamon Press,
      New York, 1978

[4] Holman, J.P., Heat Transfer, Mc. Graw Hill Book Comp., 5 th Edt.,
      New York, 1981

Effectiveness of Heat Pipes
to Electronic Component Cooling

H. KÜLÜNK
Kocaeli Engineering Faculty
İzmit, Turkey

Abstract

   In this work, the importance of electronic component cooling
in particular that of diodes and transistors are studied and in
conclusion heat pipe together with its pecularities is proposed
as an effective and contemporary technique to overcome the cooling problem and improve efficiency of electronic circuits.

1.   INTRODUCTION

   It is a fact that power dissipation in semi-conductor components such as diodes and transistors, as well as other electric
machines, used in a circuit is strongly dependent on working temperature, [1] , [2]. In spite of this reality, very little attention
(sometimes even with one sentence) is paid to electronic component
cooling problem in the books used in electrical and electronical
engineering departments of many developed countries, including
U.S.A., [3]. This drawback is also valid for the universities in
Turkey. However, recently, numerous workers studied the effectiveness of heat pipes to electronic component cooling [4] and electric machines, [5]. According to Author's knowledge, very few or
possibly non paper is published on this problem in Turkey by now.
This situation is the basic motivation behind the preparation of
the present article.

   The relation between Nyquist noise and temperature, the effect of temperature on the power dissipation of semi-conductor
components (diodes, LEDs and transistors), general character-

515

istics of heat pipes, a case study to cool BUZ 23 transistor by means of metal fin and heat pipe and finally an analog circuit of thermal resitances of heat pipe are studied in the following sections of this paper.

2. SENSITIVITY OF ELECTRONIC COMPONENTS TO TEMPERATURE

Nyquist Noise and Temperature

Due to thermally activated free motions of electrons in a resistance coupled with a signal generator a noise potential difference occures between the ends of resistance, [3]:

$$<v^2> = 4 \cdot k \cdot T \cdot R \qquad (1)$$

Since v is proportional with the square root of T and noise is not a desired phenomena, cooling operation is needed to decrease working temperature of the component. Same problem arises in the circuit which consists of a resistance, a capacitance and a signal generator since total energy stored in the capacitance is k.T/2 by equpartition theorem of thermodynamics. Especially multy-grid vacuum tubes are effected severely by Nyquist noise.

Effects of Temperature on Transistors

Temperature sensitivity of transistors can be reviewed by means of current-gain equation, [6]:

$$I_{CEO} = K \cdot T^3 \cdot e^{-(E/k \cdot T)} \cdot (\beta +1) \qquad (2)$$

According to equation (2) as the temperature T increases collector emitter cut-off current I_{CEO} also increases which is not tolerable. Hence cooling of the transistor is needed and it will increase efficiency of it.

On the other hand, power dissipation of a given transistor plotted as a function of temperature is a useful tool to study the thermal sensitivity of the component. Such characteristics are often offered by the manufacturing companies. As an example, the thermal behaviour of a high power transistor BUZ 23 is given on Fig.1., [7]. In the temperature region of 0-25°C this particular transistor operates with maximum and constant efficiency whereas

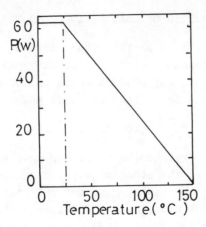

Fig.1.- Power Dissipation As a Function of Temperature For
Transistor BUZ 23 (diameter 19.5 mm).

after 25°C useful power of transistor tends rapidly to zero as
the working temperature reaches 150°C. Hence keeping working temperature in the region of 0-25°C requires effective cooling.Light
Emitting Diodes (LED) also predicts similiar trends and needs
cooling. And this operation implies that transistor is function-
ing properly with higher efficiency compared with no cooling case.

Effects of Temperature on Diodes

One of the basic pecularity of electronic components made by
semi-conductor materials is their sensitivity to light and heat
energies. This is the natural result of the electrons configurations and Fermi energy levels in the given material. In a typical
diode which consists of P and N types of materials the total a-
mount of current passing through the junction is given by rec-
tifier equation , [3] :

$$I = I_0 \cdot (e^{q \cdot v / k \cdot T} -1) \qquad (3)$$

Therefore, as the working temperature T increases then net current
tends to zero which implies nullfunction for the diode and not desired. Hence cooling operation is needed for the diode in circuit.

3. GENERAL PROPERTIES OF HEAT PIPES

It is a pipe filled partly with a working fluid (such as water, amoniac, alchohols etc) and the air inside is evacuated at a certain level (about 2000 Pascal). One of the best way to select working fluid is to look at Merit number, Me, [8]:

$$Me = \sigma \cdot L \cdot \rho / \mu \qquad (4)$$

Fluid with higher Merit number is prefered to be used in heat pipes in addition to the other requirements such as compatibility with wick and wall materials, good thermal stability, wettability of wick, high latent heat, high thermal conductivity, high surface tension. Inside wall of heat pipes are often covered with a wick (Copper or Nickel of 100-400 mesh) and it has three sections: e-vaporator, adiabatic, condenser. Working fluid is vaporised in evaporator by input heat energy and moved towards the condenser section via capillary surface tension and gravitational forces where it condences and gives out its latent heat and returns back to evaporator section through the space between wick and pipe wall to complete its closed cycle.Heat transferred to condenser by means of working fluid can be taken out through water jacket ambient air or alike system. In electronic component cooling op-eration input heat energy for evaporator section can be supplied by transistor chips. Such systems are shown schematically on Fig. 2, and Fig.3.

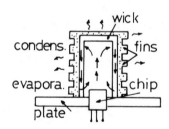

Fig.2.- Cooling of power transistor by heat pipe

One more unusual property of heat pipe is that its pressure and temperature inside is nearly uniform in axial direction.That's why heat pipe is very sensitive to temperature changes and can op-

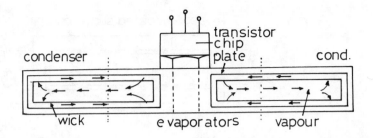

Fig.3.- Cooling of transistors by plane heat pipes

erate with even $1-2^\circ C$ temperature difference between ends. Due to
this fact, heat pipes can transfer heat energy as much as 1000-
10000 times that of metal conductors of the same dimensions and
working under similiar conditions. If a suitable and pure working
fluid is selected for the heat pipe system then only heat but not
mass transfer take place in the tube. From this point of view,
heat pipes are ideal systems for geothermal power stations,[9].
Due to their stable structure heat pipes can operate for years
with very little meintenance.

Maximum thermal power carried axially from evaporator section
to condenser of a heat pipe can be determined by using equilibrium
equation of pressure drop and Darcy's Law,[8]:

$$Q= \dot{m}.L =\left[\left(\frac{\rho.K.A}{\mu.h}\right).\left(\frac{2.\sigma.\cos\theta}{r} - \rho.g.h.\sin\phi\right)\right].L \qquad (5)$$

On the other hand, in an experimental work carried by the
Author, at slope angles in the interval of about $40-45^\circ$, thermal
power transfer nearly attains a constant value regardless of the
evaporator and condenser temperatures. When evaporator temperature
is doubled (from $30^\circ C$ to $60^\circ C$) thermal power is increased by a
factor of about 2.7,[10].

4.   THERMAL RESISTANCE ANALOG CIRCUIT OF HEAT PIPE

In order to study heat transfer phenomena of a heat pipe
thermal resistance analog circuit of it is drawn as shown on Fig.
4. Temperatures and corresponding thermal resistances are all defined on Fig.4. Here, input thermal power which is supplied by
transistor chips or electric motor at T_1 entered partly into e-

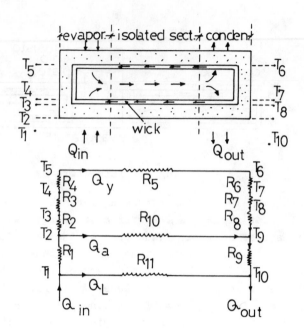

Fig.4.- Thermal resistances analog circuit of a heat
pipe system

vaporator section by means of convection. Other portions of input thermal power Q_y passes through several thermal resistances
and finally reaches ambient air of temperature T_{10}. Equivalent
thermal resistance of heat pipe is:

$$R_{eq} = (T_1 - T_{10})/Q_i \qquad (6)$$

and can also be found from series and parallel connections of
thermal resistances as:

$$R_{eq} = R_{11} \cdot \left[R_1 + R_9 + R_{28} \cdot R_{10}/(R_{28} + R_{10})\right] / \left[R_{11} + R_1 + R_9 + R_{28} \cdot R_{10}/(R_{28} + R_{10})\right]$$

$$R_{28} = R_2 + R_3 + R_4 + R_5 + R_6 + R_7 + R_8 \qquad (7)$$

In principle, all of thermal resistances can be determined
using conduction, convection and Clapeyron equations and further
some of these are negligible compared with the others to give
simple and practical solutions.

At this stage, in order to materilize the cooling of transistor by means of a heat pipe one example is studied here. Using Fig.1., one can easily determine heat flux of BUZ 23 transistor as 62.5 w/ 3.10^{-4} m^2 $=2.1$ 10^5 w/m^2 where transistor cross-sectional area is calculated from it's diameter. Heat flux for a typical heat pipe u-tilizing water as a working fluid is about 6.7 10^6 w/m^2, [11]. By comparison of these two heat fluxex one can conclude that 6.7 10^6/ 2.1 $10^5 =$ 32 times BUZ 23 transistors can be cooled by means of this heat pipe system in principle. On the other hand, thermal resistance of BUZ 23, from Fig.1., is $(150-25)^{\circ}C/62.5$ w $= 2$ $^{\circ}C$/w.

5. CONCLUSIONS

Heat pipes with relatively simple design and construction can easily be used to cool electronic components in particular diodes transistors and electric machines. Therefore, comprehensive re-search efforts should be made on this extraordinary heat transfer device.

6. NOMENCLATURE

A	: wick area (m^2)	;	E	: 1.79 10^{-19}(coul) ;
g	: 9.8 (m/s^2)	;	h	: pipe length (m) ;
I	: current (Ampere)	;	I_o	: saturation current(A) ;
I_{CEO}	: cut-off current(A);		k	: 1.38 10^{-23}($\dot{J}$/molecul$^{\circ}$K);
K	: permeability of	;	L	: Latent heat ($\dot{J}$/kg) ;
	wick (m^2) and a constant		$\dot{m}$	: maximum fluid flow rate in wick (kg/s) ;
Q_{in}	: input thermal power (w)	;	Q_{out}	: output thermal power (w) ;
Q_a	: thermal power bet-ween temperatures T_2 and T_9 (w)		Q_y	: thermal power between temperatures T_5,T_6 (w);
			r	: pore radius of wick(m);
Q_L	: lost thermal power between T_1,T_{10} (w);		R	: resistance ( ) ;
			R_{eq}	: equivalent thermal resistance ($^{\circ}$C/w) ;
$R_1,..R_{11}$	: thermal resistances of heat pipe ($^{\circ}$C/w)		T	: Temperature ($^{\circ}$C or $^{\circ}$K) ;
$T_1,..T_{10}$	: Temperatures at certain places($^{\circ}$C);		$\langle v^2 \rangle$	: mean square noise vol-tage per unit cycle of bandwidth (v^2/sec) ;

β : current gain coefficient ; θ : contact angle of heat pipe
ϕ : slope angle of heat pipe   ($^{\circ}$)   ;
    with horizontal ($^{\circ}$)   ; σ : surface tension of working
ρ : density of fluid in heat   fluid in pipe (n/m)   ;
    pipe (kg/m^3)   ; μ : viscosity of fluid in heat
                                       pipe (n s/m^2)   ;

REFERENCES

1. Dutcher,C.H.,Burke,M.R.,"Heat Pipes, A Cool Way to Cool Circuits", Electronics, pp.93-100, 16 February 1970.

2. Abdel Aziz,M.M.,"Thyristor Cooling by Heat Pipe", Proceedings of Melecon 83, Mediterranean Electrotechnical Conference 1983, Athens.

3. Brophy,J.J.,"Basic Electronics for Scientists" Mc Graw-Hill Book Company, 1966, New York.

4. Eldridge,J.M.,"Heat Pipe Cooling Etched Silicon Structure", IBM Technical Disclosure Bulletin, V.25,No.8,p.4118,Jan.1983.

5. Kukharskii,M.P.,Ivannıkov,V.A.,"Effectiveness and Applications of Centrifigal Heat Pipes In Electric Machines", Soviet Electr.Eng.(U.S.A), V.53,No.9,p.94, 1982.

6. Pastacı,H.,"Electronic", Yıldız University Press, 1985, Istanbul.

7. Components Technical Description and Characteristics for Students, SIEMENS, Munich, 1986.

8. Dunn,P.D.,Reay,D.A.,"Heat Pipes", Pergamon Press,1982,London.

9. Akyurt,M.,Basmacı,Y.,"Heat Pipe Application for Geothermal Wells", Journal of Thermal Sciences and Technology,V.6,No.1, June 1983, Ankara.

10. Kulunk,H.,"Experiences With Electrically Activated Heat Pipe", Workshop On Materials Science and Physics of non-Conventional Energy Sources, ICTP,11-29 Sep.1989, Trieste, Italy.

11. Holman,J.P.,"Heat Transfer",Mc Graw-Hill Kogakusha Ltd., 1976, Tokyo, Japan.

Heat Transfer Phenomena in Corrugated Water Trickle Collectors

A. GÜRSOY and E. TAŞDEMIROĞLU
Mechanical Engineering Department
Middle East Technical University
Ankara, Turkey

Abstract

In a corrugated sheet collector, the corrugated solar energy absorbing surface transfers the absorbed energy to a fluid by direct contact as the tubes are the integral part of the absorber plate. From the heat transfer point of view, the advantage of corrugated sheet collectors is evident over the conventional flat-plate collectors having the tubes contacted by any mean on the absorber plate. The corrugated water trickle collector is a self-draining one and, therefore does not require either an expensive antifreeze solution or a complex control system. It is an open, non-pressurized system which minimizes expensive and sophisticated plumbing. Although there are some problems such as overheating of absorber plate at high temperatures, and construction details, all the necessary principles can be easily studied. The required components are widely available and relatively inexpensive.

Due to these reasons, a corrugated water trickle collector seems to be an attractive alternative for domestic hot water applications in Turkey. This collector is therefore studied in detail. The related equations to express the heat transfer phenomena are developed. The thermal performance of water trickle collectors are theoretically compared with that of the conventional flat-plate collectors. As the results verify the suitability of water-trickle collectors, a prototype design is developed.

1. INTRODUCTION

Corrugated sheet solar water heaters are of two types:

i) Conventional corrugated sheet-flat sheet collectors
ii) Corrugated water-trickle collectors.

i) Conventional corrugated sheet-flat sheet collectors

In this type, the absorber is soldered by two galvanized iron or aliminum sheet forming a closed channel. The upper one has corrugative form and coated with flat black point and the lower one is a flat plate. The absorber is housed in a box and insulated with glass wool. Single or double glazed transparent cover may be used. Forced or natural circulation may be employed to circulate the heating medium.

Solar radiation fallen on waterheater is partly reflected and partly absorbed by the black corrugated absorbing surface and causing the surface temperature to rise. The surface transfers the absorbed energy to the fluid and a part of the energy is lost to the atmosphere due to the radiation and convection.

ii) The corrugated water-trickle collector

It employs an inclined corrugated aliminum plate with flat black coating. The plate is heated by solar radiation passing through a transparent cover. The circulated fluid is distributed by a header pipe at the top of the collector. The water is heated by the plate as it flows down the corrugations. The absorber is housed in a box and insulated with glass wool.

One characteristics of this type which is not found in the first type results from the presence of the open water surface on the collector plate. Water will evaporate from this surface and condense on the relatively cool inner surface of the glazing as a dropwise condensate film. As a result energy is transferred in the form of the mass transfer in addition to the radiation and convection between the absorber and the cover [1] .

2. PROTOTYPE DESIGN

The dimensions of the collector box and the absorber plate is given in Fig.1 and Fig.2. The collector box is made of galvanized iron profile. It is resistant to the environmental effects therefore no needed to paint.

The box is covered with a single layer of 3mm thick commercial flat window glass. The absorber plate is 1mm thick trapezoidal plate made of NT53 aliminum allay. Fig.2 gives the dimensions of the absorber plate. The back side of the collector is insulated with 6cm of glass wool to reduce the back losses.

Operational System

A manifold at the top of the plate consisting of a tube with small holes drilled in it, distributes water to each corrugation. The water is heated by the plate as it flows down the corrugations. A through at the bottom of the plate directs water from the corrugations to a single outlet.

3. ENERGY BALANCE

A theoretical model which described the short term performance of the water-trickle collector has been developed by May [2] . The energy transfer between the plate and condensate film occurs as radiation, convection and mass transfer. Radiative and convective transfers are assumed to occur vertically from plate to the cover and to condensate film. Mass transfer is assumed to occur from the plate valley to the cover.

The plate receives an energy input equal to I $(\tau\alpha)$, where the insolation rate, I, is an input variable. The reflectivity of the water film is also included as a variable.

The basic equation for the radiative transfer is

$$q_{r_{p-g}} = \frac{(T_p - T_g)}{(\frac{1}{\varepsilon_p}) + (\frac{1}{\varepsilon_w}) - 1}$$

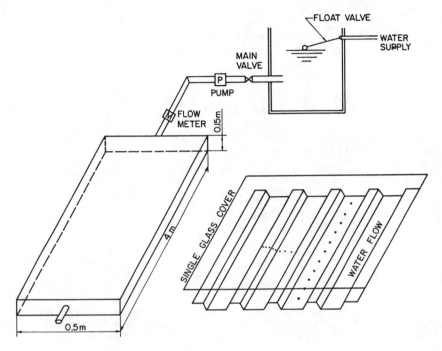

Fig. 1. Schematic Diagram of the Experimental Apparatus.

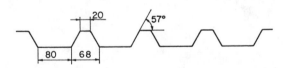

Fig. 2. Crosssectional View and Dimensions of the Absorber Plate.

where T_p and T_g are in degrees Kelvin.

The convective transfer can be expressed as

$$q_{c_{p-g}} = Nu \frac{k_{air}}{x} (T_p - T_g)$$

The Nusselt number, Nu, can be expressed as

$$Nu = 1 + 1.44 \left[1 - \frac{1708}{Ra \ Cos \ \theta} \right] (1 - \frac{(Sin \ 1.80)^{1.6} 1708}{Ra \ Cos\theta})$$

$$+ \left[(\frac{RaCos\theta}{5830})^{1/3} - 1 \right]^*, \ \theta \leq 60^o$$

where the term within $[\ \ ]^*$ is set equal to zero if its value is negative. The Rayleigh number, Ra, is defined as

$$Ra = \frac{x^3 g \beta (T_p - T_g)}{D_a \nu}$$

The mass transfer can be expressed by

$$q_{m_{p-g}} = \left[\frac{D_v P}{xRT} \ ln \ (\frac{P - P_g}{P - P_\beta}) \right] \lambda$$

The outer cover and the plate also experiences loss to the ambient through convection, radiation and conduction. The convective transfer from the glazing to the ambient is wind dependent and can be expressed as

$$q_{c_{g-a}} = h_w (T_g - T_a)$$

from Ref. 3 $h_w = 5.7 + 3.8V$ with V in m/s and h_w in W/m^2C^o. The radiative transfer can be described by

$$q_{r_{g-s}} = \varepsilon_g \sigma (T_g^4 - T_{sky}^4)$$

where

$$T_{sky} = 0.0552\ T_a^{1.5}$$

4. PROPOSED EXPERIMENT

Experimental system is given in Fig.1. In order to control the fluid flow precisely, a valve is installed in front of the pump. A float valve is also used for maintaining the water height in the storage tank. The flow of solar collector is controlled by a flow-meter manually. The solar radiation is measured with a pyronometer.

The inlet and outlet of water, the temperature of the absorber plate, glass cover and ambient are measured with T-type thermocouples. Some of the thermocouples are located in the absorber plate as shown in Fig.1. They are arranged in vertical and horizontal directions. Other thermocouples are included in the inlet and outlet pipes to measure the inlet and outlet collector fluid temperatures, and one thermocouple is centered on the outside surface of the glazing to measure its temperature.

Efficiency of Solar Collector

For outside test condition, it is very difficult to maintain constant design conditions (radiation, flow, exit and entrance temperature, etc.) Therefore the data are integrated and integral average values are used in the analysis.

The collector efficiency is defined as follows

$$\eta_c = \frac{\int_{t_{st}}^{t_{et}} mC_p\ \Delta T\ dt}{\int_{t_{st}}^{t_{et}} I\ dt}$$

where t_{st} and t_{et} are the start and end times of the measurement respectively. T is the water temperature difference between exit and entrance of the solar collector and I is the total insolation on the solar collector surface.

5. COMMENTS

In the tube and sheet collector, solar energy is absorbed mainly by the sheet, if the tube and sheet are not welded together, the thermal resistance between tube and sheet would be increased and performance would sharply decline as proved by Whillier [4] . In the corrugated sheet collector, the corrugated black solar energy absorbing surface transfers the absorbed energy to a fluid by direct contact. Therefore, from the view point of heat transfer, the advantage is evident.

Another advantage of the trickle collector over other types of liquid systems is that it is self-draining and therefore does not require either an expensive antifreeze solution or a complex control system. It is open, non-pressurized system which minimizes expensive and sophisticated plumbing. The required components are widely available and relatively inexpensive.

527

Besides the advantages of trickle collectors mentioned above, it has also some disadvantages. Since the water trickle collector has an open water surface on the collector plate, water will evaporate from this surface and condense on the relatively cool inner surface of the glazing as a dropwise condensate film. This phenomenon has two effects. Collector losses are increased relative to those of enclosed-fluid solar collectors because latent heat is transferred to the glazing through the evaporation condensation process. Also the optical losses are higher since the condensate film which forms on the inner glazing surface reduces the transmittance by about 10 to 20 per cent [2,5,6] . In spite of these losses, water trickle collectors have a good operating collection efficiency when operated at low temperatures i.e. below $50^{\circ}C$ [7] , and when the fluid temperature differences between inlet and outlet and the fluid inlet temperature are not too high since they may cause higher condensation and evaporation losses between the absorber and the glazing of the collector [8] .

NOMENCLATURE:

C_p = specific heat of water, J/kg . $^{\circ}C$

D_a = thermal diffusivity of air, m^2/s

D_v = diffusivity of water vapor in air, m^2/s

g = gravitational constant, m/s^2

h_w = outer glazing convective coefficient, $W/m^2 \, ^{\circ}C$

I = insolation rate, W/m^2

k_{air} = thermal conductivity of air, $W/m \, ^{\circ}C$

m = mass flow rate , kg/s

Nu = Nusselt number

P = atmospheric pressure, N/m^2

P_g = partial pressure of water vapor at T_g, N/m^2

P_p = partial pressure of water vapor at T_p, N/m^2

$q_{c_{p-g}}$ = convective transfer; plate-to-glazing, W/m^2

$q_{r_{p-g}}$ = radiative transfer; plate-to-glazing, W/m^2

$q_{m_{p-g}}$ = mass transfer; plate-to-glazing, W/m^2

$q_{c_{g-a}}$ = convective transfer; glazing-to-ambient, W/m^2

$q_{r_{g-s}}$ = radiative transfer; glazing-to-sky, W/m^2

R = gas constant, 8.314 J/g mole.K

R_a = Rayleigh number

T = average of T_p and T_g , $^{\circ}$C

T_a = ambient temperature, $^{\circ}$C

T_g = glazing temperature, $^{\circ}$C

T_i = inlet fluid temperature, $^{\circ}$C

T_p = average plate temperature, $^{\circ}$C

T_{sky} = sky temperature, K

V = wind speed, m/s

X = average distance from plate to glazing, m

α = absorptance

β = coefficient of thermal expansion of air, 1/k

ε = emittance of glass

ε_p = emittance of plate

ε_w = emittance of water

ε_c = collection efficiency

θ = tilt angle of collector from horizontal

λ = latent heat of evaporation of water, W.s/mol

ν = kinematic viscosity

Γ = Stefan Boltzman constant, 5.669×10^{-8} W/m^2

τ = glazing transmittance

$(\tau\alpha)_e$ = effective transmittance–absorptance product

REFERENCES

1. J.T. Beard et al. "Design and operational influences on thermal performance of Solaris solar collector", Transactions of the A.S.M.E. Journal of Engineering for Power 100, 497–502 (1978).
2. W.B. May, "Theoretical Model of the Solaris Open Water Flow Solar Collector", M.S. Thesis, University of Virginia, School of Engineering and Applied Science, Charlottesville, Virginia (May 1977).
3. J.A. Duffie and W.A. Beckman, Solar Engineering Thermal Process, Wiley, New York (1974).
4. A. Whillier and G.Saluja, "Effect of materials and construction details on the thermal performance of solar water heaters", Solar Energy 9, (1), 21–26 (1965).

5. J.T. Beard et al. "Analysis of Thermal Performance of Solaris Water-Trickle Solar Collector", Contributed by the Solar Energy Division of the A.S.M.E. for presentation at the Winter Annual Meeting. New York Dec. 5, 1976.

6. K.G.T. Hollands et al." Free convection heat transfer across inclined air layer", A.S.M.E. J.Heat Transfer 98, 189-193 (1976).

7. R.L.San Martin and G.J. Field, "Experimental performance of three solar collectors", Solar Energy 17, 345-349 (1975).

8. H.L. Jong et al. "An experimental and Theoretical Study on the Corrugated Water-Trickle Collector", Solar Energy 38, 113-123 (1987)

Thermal Characteristics of Packed Bed Storage System

M. S. ABDEL SALAM and S. L. ALY
Faculty of Engineering
Cairo University
Guiza, Egypt

A. I. EL-SHARKAWY and Z. ABDEL REHIM
National Research Center
Dokki, Guiza, Egypt

The thermal behavior of a packed bed storage system charged with hot air is modeled using two partial differential equations representing the energy conservation in the air and solid phases constituting the bed. These two equations are coupled through the heat exchange process between the two phases.

A fully implicit numerical scheme based on forward, upwind and central differencing for the time, first and second space derivatives, respectively, is used to solve the modeling equations. Marching technique is used for the air equation while a tri-diagonal matrix solver is employed to solve the solid equation. The solution yields the thermal structure of the bed, namely the air and solid temperature distribution inside the bed at any particular time, and the variation of total energy stored in bed with time. The effect of bed length, solid diameter and void fraction on the thermal characteristics of the packed bed is studied. Further, the performance of the bed under variable inlet air temperature and mass flow rate is investigated.

INTRODUCTION

Energy storage is vital in systems, such as solar thermal systems, where the available energy supply and the demand for it are not concurrent. Thus, a storage unit is required in such systems to receive energy from a source when it is available, and allow recovery of that energy upon demand. Rock-beds generally represent one of the most suitable storage units for such systems.

Schumann [1] was one of the earliest to present an analytical solution to the problem of liquid flowing through a porous prism. Extensions of his work were given later by several workers [2,3]. However, these models were rather restrictive for actual applications. More recently, Toovey and Dayan [4] provided a more developed analytical moded by including the axial heat conduction in solid, but the model ignored the transient term in the fluid equation. On the other hand, numerical treatment of the problem was considered by Mumma et al [5], Hughes et al [6], and Coutier and Farber [7]. Torab and Beasley [8] used a numerical model that ignored the conduction heat transfer in solid for an optimization study of a packed bed storage unit.

---

* Faculty of Engineering, Cairo University, Guiza, Egypt.
+ National Research Center, Dokki, Guiza, Egypt.

The present work is concerned with providing a comprehensive study for the thermal characteristics of a rock bed thermal energy storage system during the charging mode. The system is considered perfectly insulated as any proper account for the heat loss to the surroundings would require two-dimensional formulation in space. Conduction in the solid phase is taken into consideration. The effect of bed length, rock diameter, void fraction, inlet air temperature and air mass flow rate on the thermal characteristics of the packed bed is widely investigated.

MATHEMATICAL FORMULATION AND NUMERICAL METHOD

Governing Equations And Boundary Conditions

The packed bed storage system considered in the present work is cylindrical in shape and filled with small spherical rocks. The bed is charged with flowing hot air where thermal interaction occurs between the air and solid rock. In this picture, the system can mathematically be modeled using the appropriate conservation equations applied to the air and solid media separately. The air and solid equations are coupled through the heat exchange process occuring between the air and solid all over the bed.

The set of conservation equations governing the flow of air through the bed is the mass, momentum and energy conservation equations. These equations can generally be written as follows [9].
The continuity equation (mass conservation):

$$\frac{\partial \rho}{\partial t} = - (\underline{\nabla} \cdot \rho \underline{v}) \tag{1}$$

The momentum equation for constant viscosity fluid, μ :

$$\frac{\partial (\rho \underline{v})}{\partial t} = - (\underline{\nabla} \cdot \rho \underline{v} \underline{v}) - \underline{\nabla} p + \mu \nabla^2 \underline{v} + \rho \underline{g} \tag{2}$$

The energy equation for constant thermal conductivity, k :

$$\frac{\partial (\rho c_p T)}{\partial t} = - (\underline{\nabla} \cdot \rho c_p T \underline{v}) + k \nabla^2 T + S \tag{3}$$

The air is moving through the packed bed at low Mach number, and as a consequence its flow can be considered incompressible. The continuity equation reduces in this case to

$$\underline{\nabla} \cdot \underline{v} = 0 \tag{4}$$

Using cylindrical coordinates and noting that the air flows axially through the bed, which is axially symmetric, equation (4) can be integrated to yield

$$\rho_a (v_a)_z = G = \text{constant} \tag{5}$$

Since the air flows axially through the bed, the only equation to be considered in the set of momentum equations (2) is the one in the axial direction, z. Ignoring the viscosity of air and assuming that no body forces exist, the z-direction of the momentum equation reduces to the following form:

$$\rho_a (v_a)_z \frac{\partial (v_a)_z}{\partial z} = - \frac{\partial p}{\partial z} \tag{6}$$

For low speed flow, the inertial term in equation (6), the left hand side, is so small that it can be ignored and upon integration, equation (6) yields the following

$$p \simeq constant \tag{7}$$

Making use of equation (5) and neglecting the heat transfer by conduction in the air field, the energy equation (3) can be reduced to the following form:

$$\varepsilon \rho_a c_{p_a} \frac{\partial T_a}{\partial t} + G c_{p_a} \frac{\partial T_a}{\partial z} = S_a \tag{8}$$

where ε is the void fraction, which is defined as the ratio of the volume occupied by the air in the bed to the total bed volume.

As mentioned before, the source or sink term in the energy equation is produced due to the thermal interaction between the air and solid in the bed. During the thermal charging of the bed, heat is lost from the air to the solid and S_a in equation (8) would be a sink term, which can be formulated as follows:

$$S_a = - h_v (T_a - T_s) \tag{9}$$

where h_v is the volumetric heat transfer coefficient between the air and solid in the bed. h_v is related to the air mass flow rate and diameter of solid spheres in the bed according to the following equation, ref. [7],

$$h_v = 0.7 \left(\frac{G}{D_s} \right)^{0.75} \tag{10}$$

Regarding the solid field, the relevant conservation equation which is applicable to it is the energy equation. Noting that no convective terms exist, the energy equation is deduced from equation (3), and takes the following form

$$(1-\varepsilon) \rho_s c_{p_s} \frac{\partial T_s}{\partial t} = k \frac{\partial^2 T_s}{\partial z^2} + S_s \tag{11}$$

S_s is the source term resulting from the transfer of heat from the air to the cooler solid. It is written in the following form

$$S_s = h_v (T_a - T_s) \tag{12}$$

The boundary condition applicable to the air phase reflects the fact that

533

the bed is charged with hot air at constant inlet temperature, T_h, i.e.

$$\text{at} \quad z = 0 \qquad T_a = T_h \qquad \text{for all } t \qquad (13)$$

The initial condition of air temperature expresses the physical situation in which the air has a temperature equal to the ambient temperature, T_o, throughout the field except at the boundary where its temperature is T_h. Therefore, the initial condition is described by the following step function

$$\text{at} \quad t = 0 \qquad T_a = T_h \qquad \text{for } z = 0$$
$$\qquad\qquad\qquad\qquad T_a = T_o \qquad \text{for } z \neq 0 \qquad (14)$$

Concerning the solid phase, the appropriate upstream and downstream boundary conditions are specified by the solid temperature which satisfy the energy equation at the upstream and downstream boundaries, respectively, i.e.,

$$\text{at} \quad z = 0 \qquad \text{for all } t$$

$$(1-\varepsilon)\,\rho_s\,c_{ps}\frac{\partial T_s(0,t)}{\partial t} \;=\; k\frac{\partial^2 T_s(0,t)}{\partial z^2} + h_v\,[T_h - T_s(0,t)] \qquad (15)$$

$$\text{at} \quad z = L \qquad \text{for all } t$$

$$(1-\varepsilon)\,\rho_s\,c_{ps}\frac{\partial T_s(L,t)}{\partial t} \;=\; k\frac{\partial^2 T_s(L,t)}{\partial z^2} + h_v\,[T_a(L,t) - T_s(L,t)] \qquad (16)$$

The initial condition of the solid temperature field is described as follows:

$$\text{at} \quad t = 0 \qquad T_s = T_o \qquad \text{for all } z \qquad (17)$$

Numerical Method

The thermal behavior of the packed bed storage system is governed by the two coupled partial differential equations (8) and (11), along with their associated boundary conditions. In the present section, the numerical scheme which is devised to solve the governing equations simultaneously, is described.

The air energy equation is discretized using forward and upwind differencing for the time and space derivatives, respectively. It is to be noted that the use of upwind differencing for the space derivative ensures the stability of solution. Further, a fully implicit scheme is adopted to avoid limiting the time step to unnecessarily small values. Denoting the space index by i and time index by K the air energy equation takes the following form upon discretization

$$\left(\frac{\varepsilon\,\rho_a\,c_{pa}}{\Delta t} + \frac{G\,c_{pa}}{\Delta z} + h_v\right) T_{a_{i,K+1}} = \frac{G\,c_{pa}}{\Delta z}\,T_{a_{i-1,K+1}} + \frac{\varepsilon\,\rho_a\,c_{pa}}{\Delta t}\,T_{a_{i,K}}$$
$$+\; \dot{h}_v\,T_{s_{i,K}} \qquad (18)$$

where Δt and Δz are the time step and space mesh size, respectively. Solving the above equation for $T_{ai,K+1}$ produces the marching equation for solving the air temperature field.

Regarding the solid energy equation, forward and central differencing are used to discretize the transient and conductive terms, respectively. The discretized equation is written as follows:

$$(\frac{-k}{\Delta z^2}) \, T_{si+1,K+1} + (\frac{(1-\epsilon)\,\rho_s \, c_{ps}}{\Delta t} + \frac{2k}{\Delta z^2} + h_v) \, T_{si,K+1}$$

$$+ (\frac{-k}{\Delta z^2}) \, T_{si-1,K+1} = \frac{(1-\epsilon)\,\rho_s \, c_{ps}}{\Delta t} \, T_{si,K} + h_v \, T_{ai,K+1} \quad (19)$$

The above equation constitutes a set of algebraic equations having a tri-diagonal matrix coefficient. This set of equations are solved directly using a tri-diagonal matrix solver, "Thomas Algorithm" [10]. The solution yields the temperature field throughout the solid in the bed at the (K+1) time step. The air and temperature solutions are used again in equations (18) and (19) to obtain the solutions at the next time step and so on.

Having obtained the air and solid temperature fields at each time step, the energy stored in solid at each time step can be obtained by numerically integrating the following integral

$$E = \rho_s \, c_{ps} \, (1-\epsilon) \int_0^L \int_0^{2\pi} \int_0^{D/2} (T_s - T_o) \, r \, dr \, d\theta \, dz$$

$$= \rho_s \, c_{ps} \, (1-\epsilon) \, A \, (\int_0^L T_s \, dz - T_o \, L) \quad (20)$$

The numerical values of c_{ps}, ρ_s and k are taken to be 0.960 Kj/Kg K, 2560 Kg/m^3 and 0.48×10^{-3} Kj/m K s, respectively, while those of c_{pa} and ρ_a, 1.005 Kj/Kg K and 1.185 Kg/m^3, respectively.

RESULTS AND DISCUSSION

The numerical solution of equations (8) and (11) along with their assiciated boundary conditions yields a complete picture for the thermal behavior of packed bed storage system operating under a certain set of conditions during the charging process. The thermal behavior of the bed is generally characterized by the temperature distribution in the air and solid phases, and the energy stored inside the bed at any instant of time. The packed bed considered in the present work is cylindrical in shape with diameter, D, equal to 0.4 m and length, L, equal to 1 m. It is thermally charged with hot air having constant inlet temperature of 80 °C, and flowing at a rate of 0.15 Kg/m^2 s. The bed is packed with spherical rock of diameter, D_s, 0.02 m. The void fraction is strongly dependent on the method by which the container is packed and the ratio of the container diameter to that of the packing particles. Torab and Beasley [8] provided experimental results for the void fraction in a randomly packed bed of uniform spheres in a cylindrical container as a function of D/D_s. For D/D_s equal to 20, the void fraction was approximately given equal to 0.365, which is the value adopted in the present work.

The air temperature field for the aforedescribed bed is shown in Figs. (1) and (2). The first figure depicts the variation of air temperature along the bed length at differnt times. It is noted that the temperature variation is rather steep near the upstream boundary during the early stage of the charging process. This behavior comes as a natural consequence of the step function adopted for the upstream air temperature boundary condition. It is interesting to note the long time that is required for the downstream boundary temperature to rise as more than two hours elapse before any appreciable change in its value takes place. Figure (2) shows the variation of air temperature with time at different positions of the bed length. The figure indicates that the air temperature at the first 25 cm of the bed approaches its asymptotic value after about 6 hours of charging, meaning a much longer time is required for fully charging the bed.

The rock temperature distribution inside the bed is shown in Figs. (3) and (4). The former shows the spatial distribution at different times, and the latter the temporal distribution at different locations. It can be seen that the solid at the upstream boundary, which has an initial temperature of 25 $^{\circ}$C, starts to get heated fast by the hot air at the boundary, and reaches its asymptotic temperature at about one hour. Examining the results of Figs. (1) and (3) one can notice the similarity in the general shape of the air and solid temperature distribution after about 30 minutes, with the air temperature being always higher than that of solid at all locations. Such temperature difference between the air and solid generally decreases with time and distance. It is to be noted that no comparison is made between the results of the present model and the experimental results reported in references [7,8] as the conditions of the present model are not similar to those of the reported experiments.

Figure (5) gives the variation of the total energy stored in bed with time at the stantard length of 1 m as well as L = 3, 2 and 0.5 m. From this figure one can deduce the required time for the bed to be completely charged. For the case of L = 1 m, it is seen that more than 10 hours are required for the bed to be almost completely charged. Naturally, the charging proceeds at a fast rate in the beginning, and then its rate keeps declining with the approach of the full charging. Thus, it may be thought that the full charging situation does not necessarily provide for best economical operation, and design could be based on partial charging.

The effect of bed length on its behavior can be seen in Fig.(5). It is interesting to note that the length has no effect on the energy stored in the bed during the first two hours of charging, which suggests that the solid temperature in only the first 50 cm of the bed is pronouncedly affected by charging during this period. With the increase of time, the shorter beds start getting saturated while the longer ones keep storing more energy. Regarding the effect of bed diameter, which is the second variable representing the bed geometry, on its thermal performance it is thought that it can be better elucidated through a spatial two-dimensional model, and consequently its study will be deferred till further development of the model is carried out.

The next variable to consider is the diameter of the rock sphere, D_s. Solutions of the modeling equations were obtained for D_s equal to 0.01, 0.03 and 0.04 m in addition to the stantard case of 0.02 m, while keeping all other parameters fixed. Figure (6) depicts the variation of the energy stored in the packed bed with time, and it is clear that changing D_s increases remarkably the rate of charging without affecting the storage capacity of the bed. This occurs

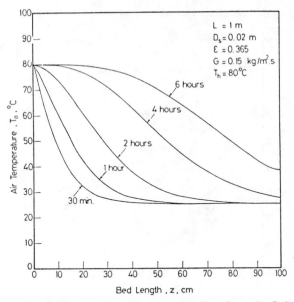

Fig.(1) Variation of Air Temperature, T_a, along the Bed
Length, z, at Different Times.

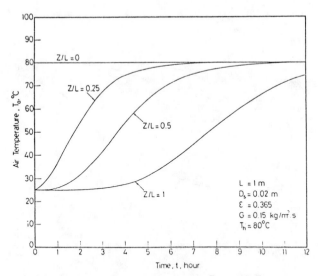

Fig.(2) Variation of Air Tempareture, T_a, with Time, t, at
Different Axial Locations.

537

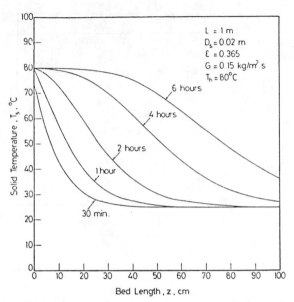

Fig.(3) Variation of Solid Tempareture, T_S, along the Bed
Length, z, at Different Times.

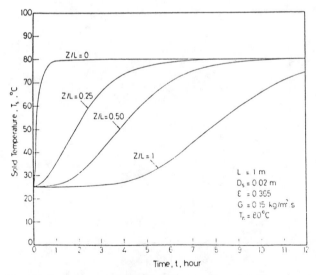

Fig.(4) Variation of Solid Temperature, T_S, with Time, t, at
Different Axial Locations.

538

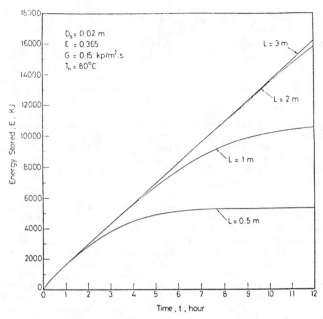

Fig.(5) Variation of Energy Stored, E, with Time, t, for Different Bed Lengths.

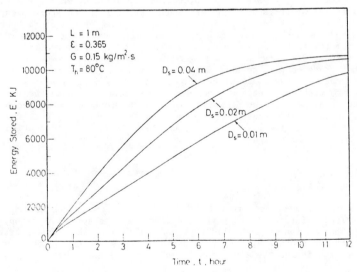

Fig.(6) Effect of Rock Diameter, D_s, on Energy Stored in Bed, E.

as a result of the increase of volumetric heat transfer coefficient, which increases with the rock sphere diameter.

An important variable which may have a significant effect on the bed performance is the void fraction, ε. Although for spherical rock, its value changes in a narrow range it is possible to have in practice a wider range of ε variation as rock would not be perfectly and homogeneously spherical. In addition to the stantard ε of 0.365, values of 0.40 and 0.50 were also studied. Figure (7) shows the variation of energy stored in bed with time at different ε. As indicated in the figure, the void fraction affects the amount of stored energy at all times as a result of the increase that takes place in bed storage capacity with the decrease of void fraction. However, it is interesting to note that the difference in stored energy at any time among the cases shown in Fig.(7) is not generally that large especially during the first five hours of charging. Regarding the rate of charging, it increases a little with the decrease of ε. The total charging time, however, is not affected by the void fraction, as the curves of different ε in Fig.(7) seem to approach their corresponding asymptotic values at about the same time.

the hot air charging the packed bed may have different inlet temperatures depending on the solar collection system. The effect of such temperatures on the bed storage behavior is depicted in Fig.(8). It is noted that all curves have similar trend, and approach their corresponding asymptotic values after about 12 hours of charging. This suggests a linear variation of the energy stored in the packed bed with inlet air temperature, which is born out in Fig.(9). In this figure, all staight lines should intersect the abscissa at 25 °C, which means that the energy stored in bed at any time is directly proportional to the difference between the inlet and ambient air temperatures. The proportionality constant, which is the slope of the stored enery-inlet air temperature relation, increases a little with time as indicated in Fig.(9).

The final parameter considered in the present study is the air mass flow rate per unit area, G. Figure (10) depicts the variation of stored energy with time at different G. It is noticed that G does not affect the storage capacity of the bed, but has a significant influence on the rate of charging. Whereas at $G = 0.15$ Kg/m^2 s the time required to almost fully charge the bed is about 12 hours or more, this time is reduced to about 6 hours by increasing the air mass flow rate to 0.8 Kg/m^2 s. This takes place due to the increase in convective heat transfer rate to the air phase from one side and the increase in volumetric heat transfer coefficient between the air and solid from another side. However, with the continual increase in air mass flow rate the reduction in charging time becomes less dramatic. This can be clearly elucidated by plotting the charging time versus the air mass flow rate, Fig.(11). This figure indicates that beyond a value of G equal to about 1 Kg/m^2 s the reduction in charging time will be insignificant.

CONCLUSIONS

In view of what has been presented the following conclusions can be drawn.
1. The numerical solution of the partial differential equations representing the energy conservation in the solid rock and air fields, which constitute a thermal packed bed storage system, provides a complete picture of the temperature distribution inside the air and solid phases, and the total energy stored

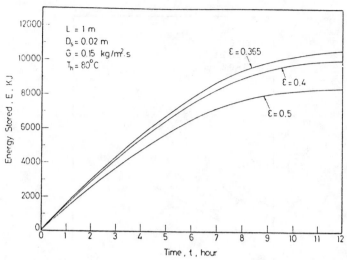

Fig.(7) Effect of Void Fraction, ε, on Energy Stored in Bed, E.

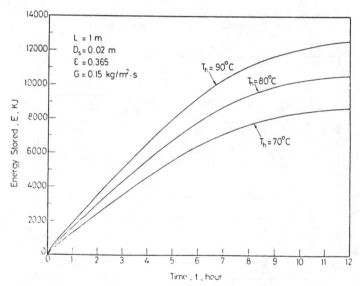

Fig.(8) Effect of Inlet Air Temperature, T_h, on Energy Stored in Bed, E.

541

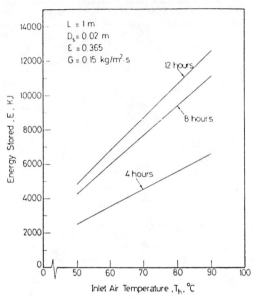

Fig.(9) Variation of Energy Stored, E, with Inlet
Air Temperature, T_a, at Different Times.

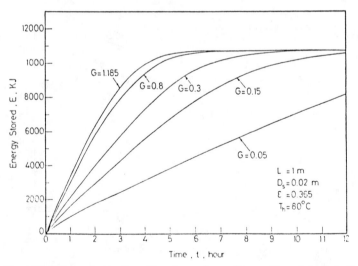

Fig.(10) Effect of Air Mass Flow Rate, G, on Energy Stored in Bed, E.

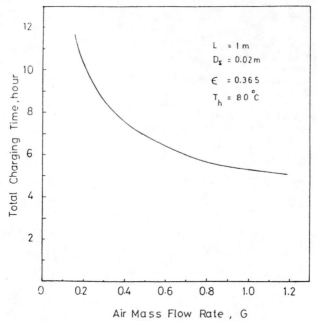

Fig.(11) Effect of Air Mass Flow Rate, G, on Total Charging Time.

inside the bed during charging.
2. Increasing the bed length increases the storage capacity of the bed, and does not affect the charging rate. Thus a longer time is required to fully charge a longer bed.
3. The increase in spherical rock diameter increases the charging rate without affecting the storage capacity of the bed.
4. The void fraction exerts a remarkable effect on the packed bed thermal performance where its increase increases the charging time for a particular required energy stored. Further, the increase in void fraction decreases the storage capacity of the bed.
5. The total energy stored in the packed bed at any particular time increases linearly with the increase of the inlet air temperature. On the other hand, the inlet air temperature increase decreases the time required for charging the bed with a certain amount of energy.
6. The increase of air mass flow rate pronouncedly enhances the heat transfer rate between the air and solid, and consequently decreases markedly the time required for fully charging the bed. It does not, however, affect the total storage capacity of the packed bed.

NOMENCLATURE

A   cross sectional area of bed, m^2
c_p specific heat at constant pressure, $Kj/Kg\,K$
D   diameter of packed bed, m
D_s diameter of spherical rock solid, m
E   energy stored in packed bed, Kj
G   mass flow rate of air per unit area, Kg/m^2s
g   gravitational acceleration, m/s^2
h_v volumetric heat transfer coefficient, $Kj/m^3\,s\,K$
k   thermal conductivity, $Kj/m\,K\,s$
L   bed length, m
p   pressure, N/m^2
S   heat source or sink, $Kj/m^3\,s$
T   temperature, $^{\circ}C$
t   time, s
v   velocity, m/s
z   axial direction
ε void fraction
ρ density, Kg/m^3

Subscripts

a   air
h   hot
p   ambient condition
s   solid

REFERENCES

1. Schumann,T.E.W., J. Franklin Instit., 208, 405–416 (1929).
2. Furnas,C.C., Ind. Engng. Chem., Part 1: 22(1), 26–31, Part 2: 22, 721–729 (1930).
3. Riaz,M., Solar Energy, 21, 123–128 (1978).
4. Toovey,I. and Dayan,J., Journal of Heat Transfer, 107, 713–716 (1985).
5. Mumma,S.A. and Marvin,W.C., ASME paper No 76-HT-73.
6. Hughes,P.J., Klein,S.A. and Close,D.J., ASME Trans. J. Heat, 336–337 (1976).
7. Pascal Coutier,J. and Farber,E.A., Solar Energy, 29 (6), 451–462 (1982).
8. Torab,H. and Beasley,D.E., Trans. ASME, Journal of Solar Engineering, 109, (1987).
9. Bird,R.B., Stewart,W.E, and Lightfoot,E.N., "Transport Phenomena", Wiley, (1960).
10. Roach,P., "Computational Fluid Dynamics", Hermosa Publishers, (1976).

Measurement of Nonuniformity
of Spray Water Deposition
in a Counterflow Cooling Tower

H. A. BRITTAIN and S. C. KRANC
College of Engineering
University of South Florida
Tampa, Florida 33620, USA

Abstract

Counterflow cooling towers operate by throwing a coarse spray of water down over the tower fill and into an upward flowing stream of air. Because the water spray is generated by an array of low head nozzles, the deposition of water over the plan area of the tower is not necessarily uniform. Nonuniformity contributes to a degradation of the thermal performance of the tower and increased operational expense. Detailed measurement and characterization of the nonuniformity of the spray deposition can be used to indicate the thermal performance of the tower and eventually to improve design.

Tests were performed in an experimental counterflow cooling tower in order to measure and characterize the nonuniformity of deposition for a particular type of nozzle. For this purpose an automated sampling system was developed and used to measure the patterns produced by both single nozzles and arrays. The effect of the upward air velocity on the spray was also examined. The results of these experiments were analyzed and compared to existing theory relating thermal performance to characterization of nonuniformity of spray deposition.

1. INTRODUCTION

Counterflow cooling towers are constructed with a volume of fill elevated above a plenum region at the base of the tower. Water is pumped to a manifold of nozzles located near the top of the tower and cascades down on the fill under the influence of gravity. Air enters at the base of the tower and flows upward and out the top, against the water flow. The air may be driven by natural convection or by forced draft. The fill may consist of splash bars or sheets of cellular media to promote filming of the water.

Frequently the distribution of water from the manifold is accomplished by an arrangement of regularly spaced, low head nozzles. These nozzles may produce a spray by a series of splatter plates or turning vanes which throw the water over a broad area. An alternative is to use an internal, stationary

545

vane arrangement which swirls the water flow, resulting in breakup and spreading at the exit. Regardless of design, individual nozzles deposit water in a pattern which usually has circular symmetry but exhibits a distinct radial dependence. For example, water may be deposited in a conical pattern, heavy at the center, or the pattern may have an annular form, with only modest flow in the central region. In order to produce a uniform downward flow of water the patterns caused by individual nozzles must be strategically spaced and possibly overlapped to smooth the deposition which results from arrangements of nozzles.

Design and operation of the tower is based on the assumption of uniform water flow down through the fill region. Nonuniform deposition of water can degrade tower performance in several ways. First, local areas which are not loaded sufficiently are underperforming although the correct approach temperature may be achieved or even exceeded. This is because less water is actually being cooled. Areas which are overloaded do not achieve the design approach. Overloading is compounded if the tower flow and head are increased to compensate for underwatered area since power demands are increased and a larger fraction of the distribution area becomes overloaded. Finally, overloaded regions may retard air flow and "channel" the air to more lightly loaded regions moving the operating point further from the design point.

The effort reported here is concerned with an experimental measurement of nonuniformity in a counterflow geometry. An automated sampling system was used to determine the patterns produced by both single nozzles and by square arrangements of these nozzles. Improvements in uniformity which can result from correct spacing and overlap can result in reduced power consumption and improved performance for the tower.

2. CHARACTERIZATION AND EFFECT OF NONUNIFORM DEPOSITION

While a uniform deposition of water is easily identified, nonuniformity is difficult to characterize and quantify. This problem is not unique to the cooling tower industry but also occurs in such diverse fields as agriculture, fire protection, land spreading of wastewater and fish preservation [1,2]. Typically, a physical measurement of areal distribution produced by water sprays is conducted by placing capture containers at regular intervals in the sample area and measuring the quantity of water collected after a time interval has elapsed. Some statistical method is then applied to the data set and an index of uniformity is derived. One such approach is to compute the standard deviation [3]. For this effort, a related measure of nonuniformity called the Christiansen coefficient [4] will be borrowed from the agricultural irrigation literature. The Christiansen coefficient is based on the average deviation from the mean and is defined in the following manner:

$$Cu = 1 - \left(\sum_i \frac{|x_i - \bar{x}|}{(N\bar{x})} \right)$$ [1]

The use of the Christiansen coefficient to characterize the uniformity of deposition in cooling towers has been discussed and analyzed in References 5 and 6. It was found that the use of this coefficient produced a somewhat more sensitive correlation than a similar coefficient based on the the standard deviation. If the thermal performance P, of the tower is defined as the ratio of actual range to design range and expressed in terms of temperature differences then

$$P = \frac{(T_1 - T_d)}{(T_1 - T_2)}$$ [2]

The correlation between performance and the Christiansen coefficient Cu, may be expressed as [6]

$$P = 0.324 Cu + 0.712$$ [3]

No significant dependence on the skew or kurtosis of the distribution of data was indicated. In principle, the Christiansen coefficient is less sensitive to wild data points. Combined with the mean deposition or loading factor then, the Christiansen coefficient serves to quantify nonuniformity of deposition but does not characterize its cause. It should be noted that the word "distribution" is used here to denote the statistical distribution of deposition data, while "deposition" refers to the actual wetting of a horizontal plane.

In the situation encountered here, the pattern resulting from an array of nozzles is more deterministic in nature rather than dominated by random effects. If the information from capture containers is displayed as an array, distinct geometric patterns can often be discerned. The Christiansen coefficient cannot separate this type of nonuniformity from an equivalent but totally random pattern. From the standpoint of improving tower performance any method which indicates nonuniformity will suffice. If patterns can be seen then improvement is possible. In principle, all methods for characterizing nonuniformity should converge in the limit of uniform deposition.

3. MEASUREMENT OF NONUNIFORMITY

A small experimental tower (Figure 1) was constructed for the purpose of making uniformity measurements. Either a single nozzle, located centrally, or a square arrangement of four nozzles on 0.914 M (3 FT) center could be accommodated. Head on the manifold was measured directly with a standpipe. Air flow was induced by a fan located above the manifold, and tortuous path drift eliminator placed on top of the manifold acted as a flow straightener. Air flow through the central area of the

547

tower was measured to be a nearly uniform 3.3 M/S (650 FPM). All tests reported here were conducted without fill using cold water and no attempt was made to place a thermal load on the tower. Water was captured in a sump and recirculated.

An automated sampling system was developed to measure the water flow in the tower at discrete locations with minimal disturbance to the upward flowing air. The principal element of this system was a cup sampler mounted on a pipe so that the vertical position could be adjusted for measurements in several horizontal planes. Water captured in the cup flows down the pipe which terminates in a small orifice. The flow backs up in the pipe to create a head behind the orifice so that a jet of water emerges from the bottom of the probe. This stream drives a small paddle wheel producing a voltage pulse which can be detected and counted remotely. In order to achieve the proper range for the probe a series of holes were cut in the pipe at various vertical positions. These holes change the response of the probe so that high flow rates can be measured without changing the orifice. To enhance the overall performance a base flow was introduced at the side of the pipe so that very low flows could be detected. The probe was calibrated against a flow meter before a test series. As an additional precaution, the probe was moved through the data position sequence with no flow except the base to ensure that no bias was introduced as a result of sampler positioning.

The probe was moved to a particular sample position by means of a two dimensional traversing mechanism, as shown in Figure 2. Position was indicated by permanent magnets located on the tracks and detected by Hall effect sensors. The entire apparatus was interfaced to a microcomputer system for control and data acquisition. A program was developed which moved the sample probe to a particular point, ordered a pause to allow equilibration at the new flow rate, averaged the flow as indicated by voltage pulses for a period of time, then moved the probe to a new position.

4. EXPERIMENTS

A series of experiments were conducted using commercial nozzles of the internal swirling vane type, designed to produce a circularly symmetric spray of 120 degrees included angle. All tests were conducted at a pressure of 20.7 KPA (3 PSIG) for which the manufacturer indicates a flow rate of 0.82 KG/S (13 GPM).

The first set of tests were designed to measure the pattern produced by an individual nozzle centered in the experimental tower described above. The automated collection system was programmed to measure a 49 point square array of sample points centered on the nozzle. The data obtained were reduced using calibration information obtained previously. Data for all test series are shown in Table I. For single nozzle data the total flow acts as one test of reproducibility and indicates that while the flow is somewhat higher than the manufacturer reports, the results are consistent form test to test. There are two possible causes for this discrepancy, the method of measuring head above the nozzle may use a slightly different datum or possibly the total capture may tend to overestimate flow since a large area is included at the periphery of the deposition, where the flow is light. Thus small errors tend to be magnified. Table I also includes an indication of the pattern radius (estimated from the manufacturer's data for spray angle) at the sample plane with no air flow.

As a first examination of the data taken for single nozzles, an array presentation was constructed by indicating points lying above or below a band centered about the mean. In general, a strong radial pattern is indicated as shown in Figure 3. A more quantitative and useful measure of the pattern can be obtained by correlating radial position with measured deposition. Accordingly, the same array data are plotted as a function of distance from the center. In general the features of nozzle deposition pattern were consistently reproduced even though the data at a particular position could vary indicating a random component. Since several sample points have the same radial position, an average at each point was computed. As shown in Figure 4, an annular artifact emerges, confirming the visual pattern exhibited in the array presentation.

Arrays of data were obtained at three sample planes to determine the spread of the pattern with distance downward from the nozzle exit. All data with air were taken at 3.3 M/S (650 FPM) upward flow. Figure 4 shows the broadening and flattening of the pattern moving down from the nozzle exit. As would be expected, the pattern radius is slightly larger than the estimate obtained from the manufacturer's data for no air flow. An attempt was also made to determine the effect of upward air flow on the pattern. At the 0.264 M (10.4 IN) sample plane measurements were repeated with the fan off. A comparison with data taken at 3.3 M/S (650 FPM) shows that air flow tends to broaden the pattern as would be expected. The pattern radius determined from the data taken without air flow compare favorably with the estimates from spray angle.

Measurements of deposition patterns were made at the same sample planes for square arrangements of four nozzles. The side of the square arrangement was 0.914 M (3 FT), with the sample planes selected so that the degree of overlap was varied strategically, the intermediate sample plane corresponding to the point where patterns from adjacent nozzles were just touching with no overlap. The data was analyzed for uniformity in the manner outlined in the previous section. Since the choice of forty-nine evenly spaced data points means that some data points lie on the border of the sample area, some of the points must be weighted in order to obtain correct values for the mean deposition and the Christiansen coefficient. Accordingly, the four corners were weighted 0.25 and all other border points were weighted 0.5 to correct for the area differences at these sample points. The results of these tests are summarized in Table II. Again, array presentation of the data shows distinct geometric patterns on the test area (Figure 3). The Christiansen coefficient was computed for each data set and the effect of spacing can be immediately seen. In the plane closest to the nozzles, the patterns from adjacent nozzles are separated and the uniformity is low. In the lowest sample plane there is some overlap and the uniformity is considerably improved.

A comparison of uniformity with and without upward air flow indicates an improvement when air flow is present. This may be due in part to improved overlap which results from an increase in pattern radius. In fact, the uniformity for the array without air is quite close to the uniformity for data with air at the plane closer to the nozzle, just as the single nozzle patterns would indicate. A second probable cause is the improved randomization of the patterns which occurs when air flow interacts with the spray.

5. SYNTHESIS OF ARRAY DATA FROM SINGLE NOZZLE DATA

Unfortunately, the determination of proper spacing and overlap for nozzle arrangements can be a time consuming and expensive undertaking if carried out as a series of experiments. Although an automated sampler such as that described here could be an effective method for reducing the effort involved, it is still difficult to vary the spacing of nozzles on the manifold as a parameter of the problem. As a potential solution to this problem, the construction of nozzle arrangements by a numerical experiment has been investigated. The underlying assumption of this technique requires that the patterns produced by individual nozzles can be superposed without interaction. Experimental observations show that this is not completely true but still reasonable as a first assumption. The radial distribution pattern was modeled using the correlation for each nozzle. The averages computed above were interpolated by a cubic spline and incorporated in a larger program which performed the superposition. The program was constructed so that the deposition at any point could be determined.

Initially, the same forty-nine data stations used in the experimental work were examined and the data set obtained in this manner was subjected to the same analysis. The results of four numerical experiments, one at each sample plane and one without air are compared in Table III. In general, the results are very similar to the data for the actual arrays. Pattern presentations are much more geometric because the results are completely deterministic.

To further investigate the comparison, the superposition program was rerun using 1296 data station on a rectangular array (also shown in Table III). These results indicate a deposition and coverage virtually the same as the spline fit for the single nozzle representations. Presumably, using this many points gives much more accurate values for the parameters of interest. Comparing the Christiansen coefficient for the various conditions shows similar discrepancies, although the fundamental conclusions based on either the experimental data or the synthesized data would not be altered. The problem lies in the relatively small number of data stations used in accumulating data. There is bound to be some correspondence between the array of stations and the deterministic pattern produced by the nozzle arrangement. Thus, a disproportionate number of stations may lie in regions of high or low deposition and in the same manner the discrepancy in the total flow rate observed is explained. Although generally impractical, one obvious solution is to increase the number of data stations.

6. CONCLUSIONS

Once a measurement of the degree of nonuniformity in the deposition pattern has been made, an estimate of the reduction in thermal performance can be made by utilizing the correlation of Reference 6. Figure 5 shows this correlation with the experimental uniformity coefficients observed here plotted as data. In all cases substantial improvements in performance could be obtained by improving uniformity.

Although an automated scanning system can provide an effective method of gauging the uniformity of water distribution in a cooling tower, acquisition of large amounts of data is impractical even with such a system, therefore some cautions must be exercised in interpretations drawn from small data sets. The total effort involved in determining optimum spacing and overlap for nozzle arrangements can be reduced by using preliminary models for single nozzles combined with numerical experiments to indicate array performance.

As indicated from the results above, uniformity of deposition is not perfect for the array sampled. Considering the trends of the data observed in this work, it is likely that lowering the sample plane further will increase the uniformity at the expense of increasing the total head required to drive the nozzle. The mean deposition or loading will remain constant. A simple extrapolation could easily be made to estimate how much lower the sample plane should be to improve uniformity. The value of improved uniformity for a particular nozzle can be estimated using the methods of Reference 6, as discussed above. The increase in thermal performance for the tower must be weighted against the increase in power needed to produce the distribution.

A second method for improving the uniformity is to change the spacing in the nozzle arrangement. With the results above for the lowest sample plane several numerical experiments were conducted using the same density of sample points. A spacing of 0.51 M (20 IN), which is more than 100% overlap raised the uniformity coefficient to .84. Improving the uniformity in this manner does not require an increase in head but the loading is increased along with the number of nozzles. These increases may, in turn, cause a change in the air flow and power requirements. Power requirements are changed due to the increase in flow.

One final comment concerns the method of obtaining the data and fit for a single nozzle representation. In this work the radial distribution was deduced from the data taken with the automated sampling system. A more practical method may be to arrange for samples to be taken from stations set on a radial pattern spacing. These data could be taken conveniently close if the deposition is varying rapidly. Using the rectangular array of the sampler the radial positions are fixed and not necessarily well chosen.

7. NOMENCLATURE

Cu - Christiansen coefficient of uniformity
P - thermal performance
T - temperature
N - number of samples
x - deposition
$\bar{x}$ - mean deposition

subscripts

a - achieved
i - datum index
1 - entering
2 - leaving

REFERENCES

1. Pair, C. H., et al, eds., Sprinkler Irrigation, The Irrigation Association, Silver Springs, MD., 1975

2. Kolbe, E. R., "Spray Head Design on Fishing Vessels Using Sprayed Refrigerated Seawater", Transactions of the American Society of Heating, Refrigeration and Air-Conditioning, Vol. 86, part 2, No. 2, pp. 281-289, 1980

3. Fay, H.P., and Hesse, G., 1985, "Application of Upspray Type Water Distribution Systems in Cooling Towers", TP No. 85-9, Cooling Tower Institute

4. Christiansen, J. E., "Irrigation by Sprinkling", Bulletin 670, California Agricultural Experiment Station, Univ. of California, Berkeley, CA., 1942

5. Kranc, S. C., "The Effect of Nonuniform Water Distribution on Cooling Tower Performance", J. Energy, Vol. 7, No. 6, Nov.-Dec. 1984, pp. 636-639

6. Kranc, S. C., "Characterization and Influence of Water Deposition in Counterflow Cooling Towers", <u>ASME</u> <u>Maldistribution</u> <u>of</u> <u>Flow</u> <u>and</u> <u>its</u> <u>Effect</u> <u>on</u> <u>Heat</u> <u>Exchanger</u> <u>Performance</u>, HTD Vol. 75, pp. 113-118, Aug. 1987

TABLE I: SINGLE NOZZLE PERFORMANCE

TEST	AIRFLOW M/S	PLANE M	RADIUS M	FLOW KG/S
1	3.3	0.21	0.37	1.03
2	3.3	0.26	0.46	0.99
3	3.3	0.32	0.54	1.00
4	0.0	0.26	0.46	0.98

TABLE II: PERFORMANCE OF NOZZLE ARRANGEMENTS

TEST	AIRFLOW M/S	PLANE M	MEAN KG/S M^2	UNIFORMITY Cu
1	3.3	0.21	0.89	.30
2	3.3	0.21	0.85	.28
3	3.3	0.26	1.01	.36
4	3.3	0.26	1.01	.38
5	3.3	0.32	0.88	.54
6	3.3	0.32	0.92	.53
7	0.0	0.26	0.88	.27
8	0.0	0.26	0.86	.27

TABLE III: SYNTHESIZED DATA FOR NOZZLE ARRANGEMENTS

AIR FLOW M/S	SAMPLE M	MEAN KG/S M^2		UNIFORMITY Cu	
		49 PT	1296 PT	49 PT	1296 PT
3.3	0.21	0.92	1.26	.32	.20
3.3	0.26	0.91	1.20	.41	.54
3.3	0.32	0.88	1.20	.56	.69
0.0	0.26	0.89	1.19	.26	.25

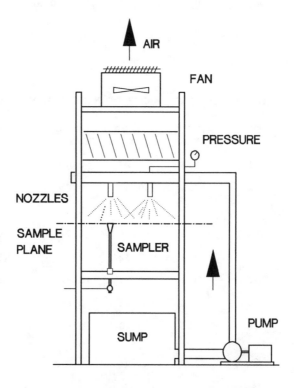

Fig. 1. - Experimental Tower with Automated Sampling System.

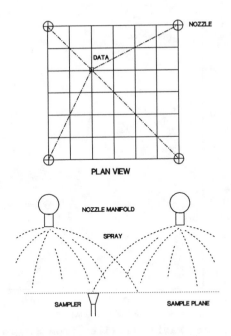

PLAN VIEW

Fig. 2. - Manifold Arrangement of Nozzles with Overlap and Sampling Locations.

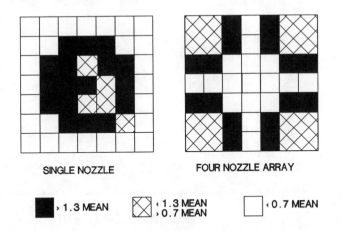

SINGLE NOZZLE   FOUR NOZZLE ARRAY

■ › 1.3 MEAN   ⊠ ‹ 1.3 MEAN › 0.7 MEAN   ☐ ‹ 0.7 MEAN

Fig. 3. - Array Presentation of Sample Data for Single Nozzles and a Square Arrangement of Nozzles.

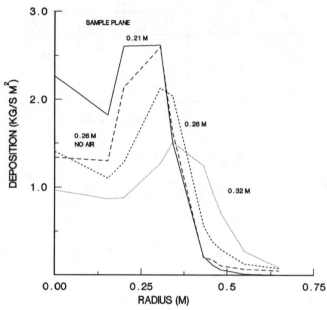

Fig. 4. - Averaged Radial Profiles from Single Nozzles at
Various Sample Planes.

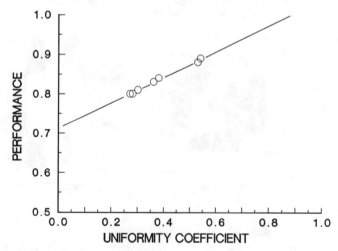

Fig. 5. - Thermal Performance as Predicted from Data for
Nozzle Arrangements (cf. Ref. [6]).

556

A Transient Technique for Measuring the Rates of Heat and Mass Transfer to a Sphere in a Humid Air Flow

I. OWEN
Heat and Mass Transfer Laboratory
Department of Mechanical Engineering
The University of Liverpool
P.O. Box 147, Liverpool, L69 3BX, UK

Abstract

This paper describes a simple experimental technique for measuring condensing and evaporating heat transfer to a body exposed to humid air flow. By measuring the changing temperature of the body it is possible to calculate the instantaneous heat transfer coefficient on the basis of Newtonian heating. To illustrate the validity of the technique, a heated copper sphere was placed in an air stream and the sensible convective heat transfer coefficient was obtained for Reynolds numbers between 7800 and 48000. The results are seen to agree with standard steady state formulations.

By cooling the sphere to about 4°C and placing it in a humid air flow, the heat transfer was seen to be accompanied by condensation and then, as it warmed, by evaporation. The total heat transfer coefficient, which was not constant, was thus obtained and was observed to approach the steady state value as the sphere dried. From the difference between the heat transfer rates with and without phase change it is possible to calculate the latent enthalpy flux and hence the mass flux.

1. INTRODUCTION

The majority of studies relating to convective heat transfer are concerned with steady conditions. In many circumstances, however, convective heat transfer takes place under non-steady conditions. A particular example is when a body is experiencing heat transfer with condensation or evaporation. The situation of interest in the present study is that of a chilled body placed in a warmer, humid air flow. When the body surface temperature is below the dew point of the air, condensation will take place. As the body surface temperature rises above the dew point, condensation will cease and the condensate on the body will begin to evaporate. In this process the sensible heat transfer coefficient between the air stream and the body will remain constant whereas the latent heat transfer will be governed by the rate and the direction of the phase change and the total heat transfer coefficient will therefore not be constant. The heat transfer rates cannot be

measured by steady-state techniques and neither can they be
calculated by the recognised formulations which are also
restricted to the steady-state.

In this paper a technique first proposed by Ahmed et al [1]
has been used to measure the transient heat and mass transfer
rates between a cooled body and a warmer, humid air flow. For
convenience the present study uses a sphere although there is
no reason why a body with a more complex shape should not be
used. To demonstrate the validity of the technique it is shown
how, when used for steady convective heat transfer without
phase change, the measured heat transfer rates agree with
accepted formulae.

2. TRANSIENT CONDENSATION AND EVAPORATION

Figure 1 shows how the temperatures at the interface between
the humid air flow and the body might be expected to vary with
time. It is assumed that the body has a uniform temperature
throughout, i.e. a Biot number much less than unity, this is
easily achieved in practice by using a copper body. It is
further assumed that the temperature of the condensate film is
the same as that of the body. Since the condensate film will
be relatively thin and of good thermal conductivity, this too
is a sound assumption. The dry bulb and dew point temperatures
of the air flow remain constant whilst the temperature of the
condensate film and the body, T, increases through Newtonian
heating from an initial value of T_i as time progresses.

Initially, the temperature of the body, T, is below that of
the dew point and condensation will occur. As T approaches the
dew point, the rate of condensation will decrease and will
cease when both temperatures are equal. The temperature of the
body is still below the dry bulb temperature of the air and
heat transfer will continue but it will now be accompanied by
evaporation since the temperature of the condensate will be
above the dew point. During these stages the heat transfer
from the air stream accounts for the latent heat of
evaporation/condensation and for the temperature rise of the
body. After the condensate film has evaporated completely, the
heat transfer will go solely to increasing the temperature of
the body and this will continue until the system reaches
thermal equilibrium.

The enthalpy fluxes associated with these events are shown in
Fig. 2. Applying the first law of thermodynamics to the
control volume and assuming that the internal energy stored in
the condensate film is negligible compared with that stored in
the body, it can be shown [1] that:

$$h_T \, A(T_\infty - T) = C\rho V \frac{dT}{dt} \qquad (1)$$

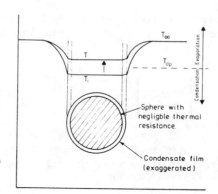

Fig. 1 Temperatures in

and around Sphere

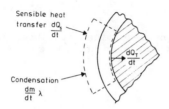

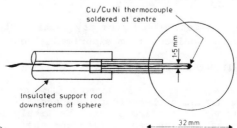

Fig. 2 Heat and Mass

Transfers to Sphere

Fig. 3 Copper Sphere

559

where h_T is the total heat transfer coefficient combing both sensible and latent heat transfers.

Thus,

$$\int h_T \, dt = \int \frac{C\rho V}{A} \frac{dT}{(T_\infty - T)} \tag{2}$$

The sensible heat transfer coefficient, h_S, will be constant and in the absence of phase change equation (2) can be integrated to give:

$$\ln \left[\frac{T_\infty - T}{T_\infty - T_i} \right] = - \left[\frac{h_s A}{C\rho V} \right] t \tag{3}$$

When phase change is also present, h_T will not be a constant and in this case equation (2) will have to be represented in finite difference form,

$$h_T = \frac{C\rho V}{A} \frac{\Delta T}{(T_\infty - T)\Delta t} \tag{4}$$

The rate of mass transfer can be determined from the difference between the total heat transfer and the sensible heat transfer.

Thus,

$$\frac{1}{A} \frac{dm}{dt} \lambda = (h_T - h_s) \, (T_\infty - T) \tag{5}$$

h_S can be determined from dry tests with no phase change and h_T can be calculated from experimental data using equation (4).

3. EXPERIMENTAL INVESTIGATION

Tests were carried out using a 32mm copper sphere with a fine copper-constantan thermocouple embedded at its centre, Fig. 3. Calculation [2] confirmed that the likely temperature differences between the surface and the centre of the sphere would be negligible. The sphere was mounted into the working section of an open wind tunnel. A small steam generator fed steam to the tunnel intake to vary the humidity levels.

Humidity was measured using a psychrometer mounted in the working section. The free stream temperature was measured using a thermocouple which, together with the thermocouple from the sphere, was connected to a chart recorder. The system was capable of resolving temperatures to $0.1°C$. This was important since in the later stages of each test the difference between the free stream and sphere temperatures, $(T_\infty - T)$, became very small and greater accuracy was needed.

For the dry tests, to measure the sensible heat transfer coefficients and to provide a comparison with established correlations, the humidity of the air was left at ambient laboratory conditions. The sphere was heated to about $50°C$ and allowed to cool in the air stream. The flow rate was varied to give a series of results for Reynolds numbers from about 7800 to 48000.

Figure 4 shows a specimen set of results plotted in accordance with equation (3). It can be seen that, as anticipated, they produce straight lines the slopes of which are used to calculate the sensible heat transfer coefficient, h_S. In Fig. 5 the Nusselt number is shown as a function of the Reynolds number and is compared with a body of data collected by McAdams [3], the agreement is seen to be reasonable.

For the wet tests the sphere was cooled in a refrigerator to a temperature of about $4°C$. By varying the amount of steam injected at the wind tunnel intake the humidity could be varied, although it also depended on the air flow through the tunnel. The sphere was inserted into the air flow and the temperatures carefully recorded as the sphere warmed. It was possible to observe the condensation forming on the sphere and then evaporating back into the air steam. Figure 6 shows some typical data for three different Reynolds numbers and dew points. As the sphere temperature passes through the dew point temperature there is a discontinuity where the total heat transfer coefficient is influenced by the change in direction of the mass transfer. Using equation (4), taking the finite difference values for the temperature and the time from the chart recorder trace, it was possible to calculate the total heat transfer coefficient for each test. The results are shown in Fig. 7. Initially the rate of heat transfer is high where the sensible and latent heat fluxes are in the same direction. As the sphere temperature reaches the dew point the condensation ceases and at this point the total heat transfer coefficient is the same as the sensible heat transfer coefficient. As the temperature of the sphere, with the condensate deposited upon it, rises above the dew point so evaporation begins; slowly at first but then at a greater rate as the temperature difference between the sphere and the dew point increases. During this stage the latent heat flux is away from the sphere and in the opposite direction to the sensible heat flux: h_T falls to its minimum value. As the

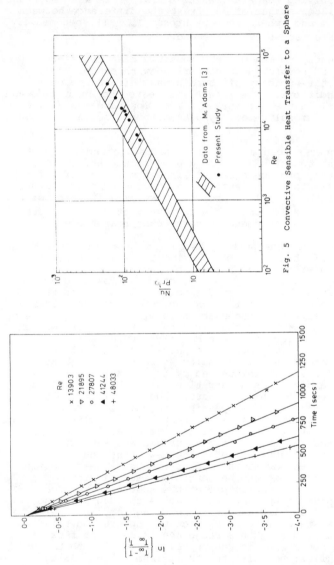

Fig. 5 Convective Sensible Heat Transfer to a Sphere

Fig. 4 Sphere Cooling by Sensible Heat Transfer

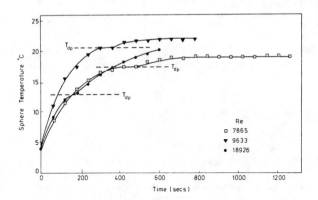

Fig. 6   Temperature of Cold Sphere in Humid Air Flow

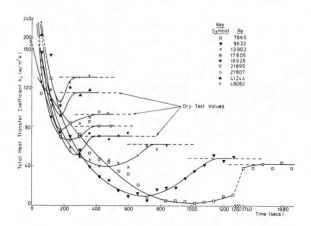

Fig. 7   Total Heat Transfer Coefficient of Cold Sphere in Humid
Air Flow

surface of the sphere beings to dry, the effect of evaporation reduces and h_T increases again until the sphere is totally dry at which point it asymptotes to the dry test value, h_S.

Having thus determined both h_T and h_S it is now possible, using equation (5), to calculate the mass transfer. The results are shown on Fig. 8 where condensation is positive mass transfer and evaporation is negative. The stages of condensation, evaporation and dry-out described earlier are clearly seen.

A mass transfer coefficient h_m can be defined on the basis of the water vapour concentrations in the free stream and at the surface:

$$\frac{1}{A} \frac{dm}{dt} = h_m (c_\infty - c_s)$$

The free stream concentration, c_∞, is calculated from the partial pressure of the water vapour whilst the surface concentration, c_s, can be calculated by assuming the conditions at the surface to be saturated at the temperature of the body. However, when the mass transfer and the differences in concentration both approach zero together, h_m will be indeterminate and therefore the validity of this exercise is questionable. For the sake of completeness and to provide a comparison with a known steady state mass transfer correlation, h_m can be calculated for the remainder of the transient period.

Figure 9 shows how the calculated values for h_m vary with time. It can be seen that in the early stages (which for the higher Reynolds numbers constitutes a substantial part of the transient period) the mass transfer coefficient is considerably higher than the steady state value calculated from the Froessling equation [4]. However, for the evaporation period for the lower Reynolds numbers, where the process has time to settle to the steady state, it can be seen that is not too different from the theoretical value. As the surface of the sphere dries the area available for mass transfer vanishes and h_m appears to fall to zero. It should be stated that the errors which will have accumulated during the calculation of h_m ensure that the values are no better than estimates.

4. CONCLUSIONS

It has been demonstrated that it is possible to measure the instantaneous rates of heat and mass transfer between a body and a humid air stream by using a simple transient technique based on Newtonian heating. The validity of the technique has been confirmed by measuring the rates of heat transfer for a

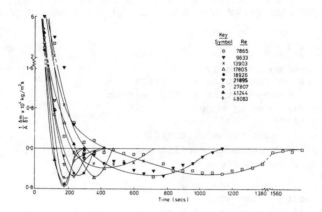

Fig. 8   Mass Transfer between Cold Sphere and Humid Air Flow

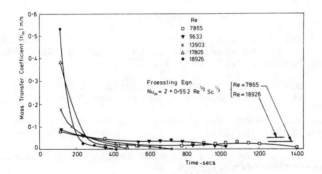

Fig. 9   Mass Transfer Coefficient

565

sphere experiencing forced convection heat transfer without phase change and comparing these with standard correlations.

It has been shown how the total heat transfer coefficient for a body whose surface is experiencing condensation or evaporation varies with time and cannot therefore be adequately described by steady state predictions.

The technique described is simple to use and is a useful tool for studying convective heat transfer, with or without phase change.

5. NOMENCLATURE

A   surface area of body
c_s vapour concentration at surface
c_∞ vapour concentration in free stream
C   specific heat of body
h_m mass transfer coefficient
h_s sensible heat transfer coefficient
h_T total heat transfer coefficient
m   mass of condensate
t   time
T   temperature of body
T_{dp} dew point temperature
T_i initial temperature of body
T_∞ dry bulb temperature of free stream
V   volume of body
λ latent heat of evaporation
ρ density of body

REFERENCES

1. Ahmed, I.Y., Barrow, H. & Dixon, S.L. "An experiment on transient heat and mass transfer in humid air flow" Heat Recovery Systems, 2, 1, p. 3-11, 1982.

2. Carslaw, H.S. & Jaeger, J.C. "Conduction of heat in solids" 2nd edition, Oxford University Press, New York, 1959.

3. McAdams, W.H. "Heat transmission" 3rd edition, McGraw-Hill, New York, 1954. (p. 266).

4. Froessling, N. "The evaporation of falling drops" Gerlands Beitr. Geophys., 52, 170-216, 1939.

Freezing in the Boundary Layer over a Flat Plate

L. CARLOMUSTO, P. D'AGOSTINO, and A. PIANESE
Facolta di Ingegneria
Universita di Cassino
via Zamosch 43, Cassino (FR), Italy

L. M. DE SOCIO
Dipartimento di Meccanica e Aeronautica
Universita di Roma "La Sapienza"
via Eudossiana 18, 00184 Roma, Italy

Abstract

An effective numerical method is proposed in order to evaluate the characteristics of a liquid flow which freezes in the boundary layer over a flat plate. The temperature of the incoming liquid is above the freezing point whereas the temperature of the flat plate is below.

The governing equations are relative to the liquid and to the solid region and correspond to the classic boundary layer equations for the fluid (continuity, equilibrium, energy) and to the heat conduction equation for the solid in contact with the plate. The boundary conditions at the interface (velocity vector equal to zero, temperature equal to the melting temperature) are given in correspondence with an unknown surface to be determined from the local energy balance between heat flux and latent heat.

The solid and liquid domains are divided into a finite number of tracts separated by the unknown boundary, and the values of the dynamical and thermal fields in the nodes are taken as main unknowns, subjected to state-type equations. These are solved by means of the differential quadrature method. The equation for the free boundary is also expressed as a state-type equation and its solution enables one to determine the extents of the two adjacent regions.

A numerical experiment was performed to show the practicality of the proposed procedure and to show the influence of the Prandtl and Reynolds numbers of the fluid on the characteristics of the freezing boundary layer.

1. ANALYSIS

A variety of analytical and numerical methods has been recently applied to the solution of Stefan problem, since increasing interest has been devoted to freezing and melting questions related with such situations as food processing, metal casting, aircraft wings icing and many others. Stefan problem is a non linear free or moving boundary problem which, depending on the particular phenomenology under investigation, is governed by partial differential equations of the parabolic, elliptic or hyperbolic type. In the majority of cases of practical applications, the fundamental equations to be dealt with are parabolic and, in this circumstance, a very efficient method of solution is the differential quadrature method, as proposed, for the first time, in [1].

In this paper the development of a frozen layer near a solid surface from the flow of a condensable fluid is considered. In particular, let $(Y = 0, X > 0)$ represent a semi-infinite flat plate in a uniform incompressible flow, whose unperturbed upstream velocity is q. Let ρ and T be the density and the temperature, respectively, and let c be the specific heat, k the heat conductivity, μ the viscosity and C the latent heat of melting. Furthermore U and V are the X and Y components of the velocity and T_M is the freezing temperature. The temperature of the isothermal plate is T_W and the temperature of the unperturbed fluid is T_0.

The boundary between the solid and the liquid phase is represented by the equation $S = S(X)$.

It is assumed that the flow satisfies the hypotheses leading to the boundary layer approximation of the full Navier-Stokes equations, whereas in the solid phase the heat conducted in the X direction is negligibly small when compared to the heat transferred along the Y direction.

Introducing the dimensionless quantities $x=X/L$, $s=S/L$, $y=Y/L$, $u=U/q$, $v=V/q$, $t=(T-T_M)/(T_0-T_M)$ leads to the following set of non dimensional governing equations, in the boundary layer approximation, for the fluid region $0 < x < \infty$, $s < y < \infty$

Continuity equation

$$u_x + v_y = 0 \tag{1}$$

Momentum equation

$$uu_x + vu_y = (1/Re)\,u_{yy} \tag{2}$$

Energy equation

$$ut_x + vt_y = (1/Pr\,Re)t_{yy} \qquad (3)$$

where Re is the Reynolds number and Pr is the Prandtl number.

In the solid phase region $0 < x < \infty$, $0 < y < s$ the energy equation reduces to

$$t_{yy} = 0 \qquad (4)$$

To the system (1-4) appropriate boundary conditions are:

$$
\left.
\begin{aligned}
&\text{for } x = 0,\ 0 < y < \infty \\
&\quad u = 1,\ t = 1, \\
&\text{for } 0 < x < \infty,\ y = s \\
&\quad u = v = 0,\ t = 0, \\
&\text{for } y \to \infty \\
&\quad u = 1,\ t = 1, \\
&\text{for } 0 < x < \infty,\ y = 0 \\
&\quad t = -1
\end{aligned}
\right\} \qquad (5)
$$

and the energy balance at the interface gives

$$ds/dx = A(t_y)_+ - B(t_y)_- \qquad (6)$$

where A and B are two dimensionless quantities equal to $[c(T_0 - T_M)/C]$ and $[ck_s(T_0 - T_M)/Ck]$, respectively, with k_s the thermal conductivity of the solid phase.

The temperature distribution in the solid phase is immediately found in terms of the unknown s,

$$t = (y/s) - 1 \qquad\qquad x > 0,\ 0 < y < s.$$

The semi-infinite fluid region is then transformed into a strip by means of the change of variables

$$\xi = x\ ;\ \eta = 1 - e^{-ay} \qquad (7)$$

which leads to a set of transformed partial differential equations corresponding to the original system (1-3,6) and related boundary conditions.
At this stage the new integration range along η, $s' < \eta < 1$, (where s' is
the value of s trasformed by (7)) is divided into N-1 equal tracts by
means of N nodal points and for each of them the following relation
holds

$$\eta_i = (i - 1)/(N - 1) + [(N - i)/(N - 1)] y_f$$

with $y_f = s'$. In each internal node the basic set of equation now is

$$u_\xi + a(1-\eta)\left\{\frac{d}{d\xi}\left[-\frac{1}{a}\ln(1-\eta)\right]\right\}u_\eta + v_\eta\, a(1-\eta) = 0$$

$$uu_\xi + a(1-\eta)\left\{u\frac{d}{d\xi}\left[-\frac{1}{a}\ln(1-\eta)\right]+v\right\}u_\eta = \frac{a^2}{Re}\left[-(1-\eta)u_\eta + (1-\eta)^2 u_{\eta\eta}\right] \quad (8)$$

$$ut_\xi + a(1-\eta)\left\{u\frac{d}{d\xi}\left[-\frac{1}{a}\ln(1-\eta)\right]+v\right\}t_\eta = \frac{a^2}{RePr}\left[-(1-\eta)t_\eta + (1-\eta)^2 t_{\eta\eta}\right]$$

whereas the appropriate boundary conditions immediately give the values
of velocity and temperature in y_f, and eq. 6 becomes

$$\frac{dy_f}{d\xi} = a^2(1-y_f)^2\left\{A(t_\eta)_+ - B(t_\eta)_-\right\} \quad (9)$$

in $y_f = s'$. When the Differential Quadrature Method [1,2] is applied, the
set (8-9) reduces to a system of ordinary differential equations which
can be solved by a standard Runge-Kutta procedure.

2. RESULTS

A numerical experiment was carried out relative to a liquid flow
which freezes over a flat plate. In all the cases a was taken equal to 1.

At the first step of the integration procedure, in order to overcome the initial singularity, a linear approximation was assumed

$$y_f = 2\lambda(\xi)^{\frac{1}{2}}$$

where λ is the root of a trascendental equation [3].

The following ranges of physical characteristics of the fluid were considered: $Pr = 1; 10; 100; Re = 100; 1000; 10,000; 100,000$. Furthermore $A = 5 \, 10^{-6}$ and $B = 5 \, 10^{-5}$.

The integration process was applied until separation occurred of the boundary layer from the frozen surface, at a distance x_f, to which the thickness y_f of the frozen layer corresponds. The condition assumed at x_f for separation was $[u_y(y_f)] = 0$.

The table I reports the values of x_f and y_f at different Pr and Re, whereas Fig. 1 show the shape of the liquid-solid interface and Figs. 2,3 and 4 show the axial velocity and temperature profiles in some signifi_ cant situations.

As one would expect, the results show that the separation point x_f moves downstream at increasing Re, at fixed Pr, while the thickness x_f also increases. Pr has a relatively little effect on the same quantities at given Re. In any case, as Pr increases, the frozen region, for x_f practically uninfluenced, is a little thinner, due to the increased thermally insulating properties of the fluid.

Figure 1 is self-evident. Fig.1a) and 1b) show that the rate of growth of the frozen layer is higher close to the leading edge, where it should be theoretically infinite, and then decreases as one moves downstream.

Fig. 2 shows the (coincident) profiles of u and t at some selected stations along x versus η, for $Pr = 1$ and $Re = 10,000$.

Fig. 3 gives the results obtained for $Re = 100,000$ and $Pr = 1$. Comparison with the data in Fig. 2 immediately shows the great influence of the Reynolds number on both the velocity and the temperature profiles.

Figs. 4a) and 4b) finally, report u and t, respectively, for the same $Re = 10,000$ but $Pr = 10$. The u-graph does not really much differ from that in Fig. 2 and he influence of the change in Pr practically affects the temperature distribution only.

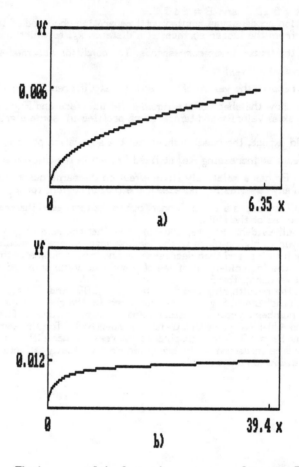

Fig. 1. - Thickness y_f of the frozen layer versus x for some Re and Pr values: a) Pr=10; Re=10,000. b) Pr=1; Re=100,000.

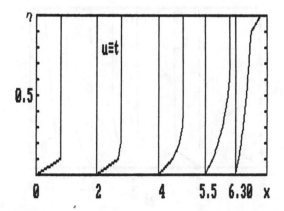

Fig. 2 - Profiles of u≡t at some selected stations along x versus η
for Pr = 1 and Re = 10,000

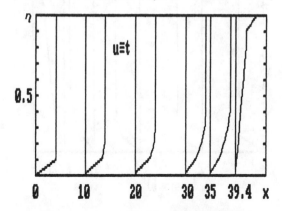

Fig. 3 - Profiles of u≡t at some selected stations along x versus η
for Pr = 1 and Re = 100,000

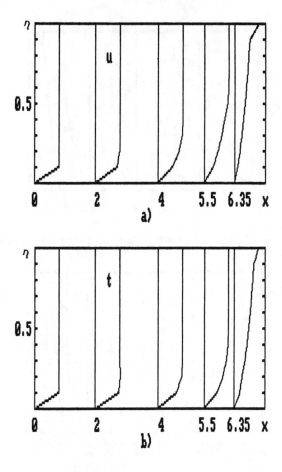

Fig. 4 - Profiles of u (a)) and t (b)) at some selected stations
along x versus η for Pr = 10 and Re = 10,000

574

Re = \ Pr =	1	10	100
100	$y_f = 8.05 \ 10^{-4}$ $x_f = 0.076$	$y_f = 8.02 \ 10^{-4}$ $x_f = 0.076$	$y_f = 8.018 \ 10^{-4}$ $x_f = 0.076$
1000	$y_f = 2.36 \ 10^{-3}$ $x_f = 0.702$	$y_f = 2.33 \ 10^{-3}$ $x_f = 0.702$	$y_f = 2.33 \ 10^{-3}$ $x_f = 0.702$
10,000	$y_f = 6.41 \ 10^{-3}$ $x_f = 6.308$	$y_f = 6.18 \ 10^{-3}$ $x_f = 6.353$	$y_f = 6.14 \ 10^{-3}$ $x_f = 6.359$
100,000	$y_f = 0.012$ $x_f = 39.45$	⸺	⸺

Table I: SEPARATION DISTANCES (x_f) AND THICKNESS OF THE FROZEN LAYER CORRESPONDENTS (y_f) FOR SOME Re AND Pr VALUES.

575

Acknowledgements

This work was partially supported by the Italian Ministry for Education through a MPI 60% grant.

REFERENCES

1. L. M. de Socio, G. Gualtieri: A Bellman's Procedure for a Phase Change Problem, Math. Modelling 8 (1987) 24-28.

2. Bellman R., B.G. Kashelf and J. Costi: Differential Quadrature "A Technique for the rapid solution of nonlinear partial differential equations", J. Comp. Phys 10,40,52 (1972)

3. Ozisik M.N., Heat Conduction, Wiley, N.Y. 1980.

Simulation and Design of Direct Contact Evaporators

P. L. C. LAGE and C. M. HACKENBERG
COPPE - Federal University of Rio de Janeiro
Chemical Engineering Department
C.P. 1191
20000 Rio de Janeiro, RJ, Brazil

Abstract

A model is developed to simulate the continuous direct contact evaporator. It consists of the overall energy and mass balances in the evaporator at the transient regime. These balances originate an ordinary differential system of equations. which is solved numerically. A complementary model that considers the heat and mass transfer processes inside the bubbles is applied to provide the necessary temperature and concentration profiles. This complementary model needs the evaluation of two parameters, which are estimated to specific experimental conditions. Experimental data and simulated results are in good agreement. From the outcome of the simulations we can conclude that the two parameters described above are functions of the continuous phase temperature.

1. INTRODUCTION

The direct contact evaporators can be defined as the equipment used to concentrate solutions by the bubbling of superheated gases. Every bubbling process can be divided in two stages : the bubble formation and the bubble ascension to the liquid surface. Although it is very fast, the former is very important because it mainly determines the bubble volume and shows the largest temperature and concentration gradients. Moreover the inital ascension conditions (volume, temperature and vapor concentration) are those of the detachment of the bubble.

During the bubble ascension, one usually considers that the volume and the shape of the bubbles are constant. Then the single phenomenum consists of a sphere of superheated gas moving upward within a continuous phase (a liquid), changing heat and mass with it. The bubles residence time determines the extension of these changes. Thus the bubble hidrodynamic is important to define the whole process.

The volume of a forming bubble has been widely studied when there are no transfer processes or when they can be neglected. KUMAR & KULOOR [1] showed a comprehensive review of models that had been developed before 1969. On the other hand, the studies that took the tranfer phenomena into account are very few. Most

577

of them dealt with the nucleated boiling as the works of DARBY [2] and SCRIVEN [3]. The bubble formation of superheated water vapor in saturated water was studied by SCHIMDT [4,5]. He was able to derive empirical dimensionless relations which can be used to determine the average heat transfer coefficient and the final bubble area. It was noted that 40-60% of the total heat was transferred during the bubble formation.

In order to understand the bubble ascension one has to know the bubble hydronamic behavior and the heat and mass transfer process between the two phases.

There are three basic hydrodinamic models for an isolated drop or bubble in an infinite medium : the completely mixed drop model, where there are no internal gradients; the rigid drop model, where the bubble behaves like a solid sphere; the internal circulation drop model, where the internal velocity profile has already been established completely. The analitical determination of the bubble ascension velocity is possible in the range of low Reynolds numbers. At the steady state LEVICH [6] gave the classical solutions, while HACKENBERG [7,8] supplied the transient regime solution for the inviscid sphere. The later can evaluate the time the bubble ascension takes to reach the steady state. Hence, it is possible to classify the bubbling processes as fast or slow. The fast processes have its bubble residence time smaler than the time necessary to set the bubble internal circulation profile, so the rigid drop model can be applied. That is the case of bubbles in a liquid medium. On the other hand, the residence time of the slow processes is large enough to develop the internal velocity profile, so the internal circulation drop model should be used. Although these conclusions do not have an analitical background outside the range of low Reynolds numbers, they can be used as a guide to choose a model for a general process.

The behavior of a set of bubbles can be attained from the single bubble behavior by a cellular model. Several cellular models have already been developed, as the ones of HAPPEL [9,10] and RUCKENSTEIN [11], which are described by HACKENBERG [12].

In a bubbling process, the heat transfer process occurs in both the continuous phase and the dispersed one. On the other hand, the evaporation produces a heat flux descontinuity at the bubble surface, so the mass transfer only occurs in the dispersed phase. Therefore one can define the heat transfer coefficient for both phases but the mass transfer coefficient can be defined only for the dispersed phase.

The continuous phase heat transfer coefficient can be calculated by the analitical equations given by LEVICH [6] or by the empirical relations developed by CALDERBANK & MOO-YOUNG [13]. HACKENBERG [12] and RUCKENSTEIN [11] derived an analitical equation similar to the CALDERBANK & MOO-YOUNG [13] equation for small bubbles. When they were studying the heat transfer process without evaporation, ANDRADE [14] and ANDRADE & HACKENBERG [15] used the modified Basset theorem, proposed by HACKENBERG [7], to solve the heat diffusion equation in terms of the bubble surface

temperature. The results were excellent. ANDRADE & HACKENBERG
[16] developed the transient analogies between the heat and mass
transfer. When studying the superheated gas bubbling process,
HACKENBERG & ANDRADE [17,18] utilized these analogies to
determine the simultaneous heat and mass transfer coefficients
of the dispersed phase as a function of the average bubble
temperature and concentration. HACKENBERG & ANDRADE [17,18] also
added the evaporation flux to the heat diffusion equation and he
solved it to get the necessary defining equations to the average
bubble temperature and concentration.

QUEIROZ & HACKENBERG [19] proposed a rigid drop model for
the heat and mass transfer in the superheated bubbles processes.
Their model consists of the heat and mass diffusion equations
with the latent heat of evaporation included in the surface heat
flux condition. They also supposed that there is mass
equilibrium at the bubble surface, so they assumed that the
Antoine equation holds. This model was completely solved by the
functional method of HACKENBERG [7], although two parameters
remained without determination.

The present work develops a model for the direct contact
evaporator with which this kind of equipment can be simulated
and designed. The simulation method is also implemented. The
rigid drop model solved by QUEIROZ & HACKENBERG [19] is used.
The continuous phase heat transfer coefficient is calculated by
HACKENBERG [12] equation. The cellular model described in
HACKENBERG [12] is also employed. Batch transient experimental
data from HACKENBERG & ANDRADE [17,18] are simulated to test the
proposed algorithm. The resulting conclusions validate the
developed model and the algorithm.

2. MODELS

Hidrodynamic model and the continuous phase heat transfer
coefficient

The gas-holdup H_g is the volume fraction of dispersed phase
contained in the total liquid volume in the evaporator. It can
be evaluated by the ratio between the superficial velocity, v_s,
and the ascension velocity, v_a :

$$v_s = W_g/A_b \qquad (1)$$

$$H_g = v_s/v_a \qquad (2)$$

where W_g is the incoming gas flow rate and A_b is the evaporator
cross section area.

From the cellular model developed by RUCKENSTEIN [11] we
have :

$$\Xi_1(\eta) = \frac{3 - 4,5\eta + 4,5\eta^5 - 3\eta^6}{3 + 2\eta^5} \qquad (3)$$

$$\Xi(\eta) = \frac{3(1 - \eta^5)}{2 - 3\eta + 3\eta^5 - 2\eta^6} \qquad (4)$$

where

$$\eta = H_g^{1/3} \qquad (5)$$

These functions estimate the ascension velocity and the Nusselt number for a group of bubbles :

$$v_a = v_a^\circ \; \Xi_1(\eta) \qquad (6)$$

$$Nu = [2/3 \; \Xi(\eta) \; \Xi_1(\eta)]^{1/3} \; Nu^\circ \qquad (7)$$

where

$$Nu = \frac{2ha}{k_c} \qquad (8)$$

where a is the bubble radius, h is the continuous phase heat transfer coefficient and k_c is the continuous phase thermal condutivity. The ascension velocity v_a°, and the Nusselt number Nu°, for a bubble in a infinite medium is given by HACKENBERG [12] for fast and low Reynolds processes :

$$v_a^\circ = \frac{ga^2(\rho_c - \rho_d)}{9\mu_c} \qquad (9)$$

$$Nu^\circ = 0,6282 \left[\frac{(\rho_c - \rho_d)ga^3}{\rho_c v_c^2} \right]^{1/3} Pr^{1/3} \qquad (10)$$

where

$$Pr = \frac{Cp_c \mu_c}{k_c} \qquad (11)$$

is the Prantl number and μ_c is the dynamic viscosity, v_c is the kinematic viscosity, ρ_c is the density, Cp_c is the specific heat at contant pressure for the continuous phase, ρ_d is he dispersed phase density, and g is the gravity acceleration.

If the Reynolds number

$$Re = \frac{2av_a}{v_c} \qquad (12)$$

is greater than 1, equation (9) does not hold. Thus it is necessary to use the empirical equation attained by DAVIES & TAYLOR as described in DAVIDSON & HARRISON [20] :

$$v_a = 0,711 \; (2 \; g \; a)^{1/2} \qquad (13)$$

Equation (10) is identical to CALDERBANK & MOO-YOUNG [13]

equation for small bubbles, if the correction factor is calculated for the typical 5% gas-holdup value.

The direct contact evaporator

This equipment consists of a tank with a cross section area A_b. It is fed by a solution stream with mass flow rate W_c, temperature T_c and solid fraction ρ. A superheated gas is bubbled from its base at a volume flow rate W_g (and mass flow rate W_g), temperature T_i and vapor concentration C_i. The tank also has an outlet through which the concentrated solution leaves the evaporator with mass flow rate W_s, temperature T_s and solid fraction $\mathcal{P}$. The mixture of the injected gas and produced vapor W_a finds its way out at the top of tank. Moreover there is a heat loss Q_p through an area A_p to an ambient at temperature T_a. Some simplifying hypotheses can be made : the gas and the continuous phase are completely immiscible; the continuous phase inside the evaporator is completely mixed, so its conditions are the same of the leaving stream (the vigorous agitation that is induced by the bubbling process supports this assumption); there is no continuous phase in the withdrawal gas; the flow rates W_c, W_s, W_g and the incoming conditions T_c°, ρ, T_i, C_i are constant during the process.

The overall mass balances and the continuous phase heat balance are listed below. The continuous phase volume V_c, its temperature T_c, and its solid fraction $\mathcal{P}$ are the dependents variables. The time t is the independent one, which starts at the beginning of the evaporator operation.

The mass balances are :

$$\frac{d}{dt}(V_c\rho_c) = W_c - W_s - W_a \tag{14}$$

$$\frac{d}{dt}(V_c\rho_c\mathcal{P}) = \rho W_c - \mathcal{P}W_s \tag{15}$$

$$\frac{d}{dt}[V_c\rho_c(1 - \mathcal{P})] = (1 - \rho)W_c - (1 - \mathcal{P})W_s - W_a \tag{16}$$

And the continuous phase energy balance is :

$$\frac{d}{dt}\left[\rho_c V_c C_{pc}(T_c - T_c^\circ) \right] = W_g\overline{C_{pd}}(T_i - \langle T \rangle_f) -$$

$$- W_a \left[\overline{C_{pc1}}(\overline{T_s} - T_c^\circ) + L + \overline{C_{pa}}(\langle T \rangle_f - \overline{T_s}) \right] -$$

$$- W_s\overline{C_{pc2}}(T_c - T_c^\circ) - Q_p(T_c, T_a, hp, A_p, l) \tag{17}$$

where l is the continuous phase height in the evaporator, Q_p is the heat loss rate, $\overline{T_s}$ is the time mean temperature of the bubble surface, $\langle T \rangle_f$ is the volumetric mean of the bubble temperature when the bubble leaves the continuous phase, hp is the internal wall—continuous phase heat transfer coefficient, $\overline{Cpa}$ is the vapor mean specific heat at contant pressure between $<T>f$ and $\overline{Ts}$, $\overline{Cpci}$ is the continuous phase mean specific heat at contant pressure between Tc° and $\overline{Ts}$, $\overline{Cpcz}$ is the continuous phase mean specific heat at constant pressure between Tc and Tc°, $\overline{Cpd}$ is the dispersed phase mean specific heat at constant pressure between Ti and $<T>f$, L is the continuous phase latent heat of evaporation at $\overline{Ts}$.

The equations (14), (15) and (17) are a linearly independent system that represents the evaporator process. Although they could be integrate directly, it is advantageous to manipulate these equations to get the system :

$$\frac{d\mathcal{P}}{dt} = \frac{\rho Wc - \mathcal{P}(Wc - Wa)}{\rho c Vc} \qquad (18)$$

$$\frac{dTc}{dt} = \left\{ \ Wg\overline{Cpd} \ (Ti - <T>f) - Ws\overline{Cpcz} \ (Tc - Tc^{\circ}) - \right.$$

$$- \ Wa \ \left[\ \overline{Cpci} \ (\overline{Ts} - Tc^{\circ}) + L + \overline{Cpa} \ (<T>f - \overline{Ts}) \ \right] -$$

$$- \ Qp(Tc, Ta, hp, Ap, l) - Cpc(Tc - Tc^{\circ})(Wc - Ws - Wa) -$$

$$- \ (Tc - Tc^{\circ}) \ \frac{\partial Cpc}{\partial \mathcal{P}} \ \left[\ \rho Wc - \mathcal{P}(Wc - Wa) \ \right] \left. \right\} \ \bigg/$$

$$\bigg/ \ \left\{ \ \rho c Vc Cpc + \rho c Vc \ (Tc - Tc^{\circ}) \ \frac{\partial Cpc}{\partial Tc} \ \right\} \qquad (19)$$

$$\frac{dVc}{dt} = \left[\ Wc - Ws - Wa - Vc \ \left(\ \frac{\partial \rho c}{\partial Tc} \frac{dTc}{dt} + \frac{\partial \rho c}{\partial \mathcal{P}} \frac{d\mathcal{P}}{dt} \ \right) \ \right] \ / \ \rho c$$

$$(20)$$

The Vc volume can be rewritten in terms of the total volume V and the gas-holdup. From equations (1) and (2), we have :

$$Vc = V(1 - Hg) \qquad (21)$$

Since $V = Abl$, we have :

$$Vc = l(Ab - \mathcal{W}g/va) \qquad (22)$$

Since Ab and $\mathcal{W}g$ are constant, Vc is a function of the liquid level inside the evaporator and the ascension velocity. Hence, the initial conditions necessary to the numerical integration are :

$$Tc = Tc^{\circ}, \ \mathcal{P} = \rho, \ Vc = l^{\circ} \ (Ab - \mathcal{W}g/va) \qquad (23)$$

582

Of course, a complementary model that considers the heat and mass transfer process for a single bubble is necessary to provide the correct values of $\langle T \rangle_f$, T_s and W_a. We also need to determine the time volumetric mean of the bubble temperature $\langle \overline{T} \rangle$ and the time volumetric mean of the bubble vapor concentration $\langle \overline{C} \rangle$ with which we can evaluate the dispersed phase properties.

The heat and mass transfer model for a single bubble

QUEIROZ & HACKENBERG [19] and HACKENBERG & ANDRADE [17,18] solved a rigid drop model for the heat and mass transfer in a spherical bubble with a first order aproximation. The temperature T and concentration C solutions are :

$$\frac{T(r,t) - T_i}{T_i} = \frac{1}{(1 + \lambda) T_i} \left\{ -\frac{L}{h} \frac{dw}{dt} - b\lambda \times \right.$$

$$\times \left[(T_i - T_c) e^{-bt} + \frac{L}{h} w(t) - \frac{bL}{h} w(t) * e^{-bt} \right] \right\} *$$

$$* \left[1 - \frac{2a}{\pi r} \sin\left(\frac{\pi r}{a} \right) \exp\left(-\frac{a\pi^2 t}{a^2} \right) \right] +$$

$$+ \frac{1}{(1 + \lambda) T_i} \left[-(T_i - T_c) - \frac{L}{h} w(0) \right] \times$$

$$\times \left[1 - \frac{2a}{\pi r} \sin\left(\frac{\pi r}{a} \right) \exp\left(-\frac{a\pi^2 t}{a^2} \right) \right] \qquad (24)$$

$$C(r,t) - C_i = \left[\frac{a}{2\mathcal{D}} \frac{dw}{dt} + \frac{\pi^2}{2a} w \right] *$$

$$* \left[1 - \frac{2a}{\pi r} \sin\left(\frac{\pi r}{a} \right) \exp\left(\frac{\mathcal{D}\pi^2 t}{a^2} \right) \right] + \left[C_s(0) - C_i \right] \times$$

$$\times \left[1 - \frac{2a}{\pi r} \sin\left(\frac{\pi r}{a} \right) \exp\left(-\frac{\mathcal{D}\pi^2 t}{a^2} \right) \right] \qquad (25)$$

where

$$\lambda = \frac{2kd}{ha} \qquad (26)$$

$$b = \frac{a\pi^2}{(1 + \lambda) a^2} \qquad (27)$$

where r is the radial coordinate, w is the evaporated mass flux, C_s is vapor concentration at the bubble surface, kd is the thermal condutivity, α is the thermal diffusivity, $\mathcal{D}$ is the mass diffusivity for the dispersed phase. And the asterisk represents the convolution integral.

QUEIROZ & HACKENBERG [19] supposed that the proccess occurs at such conditions that it is possible to consider the bubble surface saturation. Considering an ideal gas behavior for the dispersed phase, an equilibrium equation could be used to relate the temperature and the concentration at the bubble surface. The Antoine equation was choosen, as in REID et al.[22], and it is utilized in a linearized form. They attained the following results for the evaporation flux and the bubble surface temperature :

$$w(t) = \frac{w(0)}{p'' - p'} \left[\left(p'' - \frac{\Pi}{\Gamma} \right) e^{-p't} - \left(p' - \frac{\Pi}{\Gamma} \right) e^{-p''t} \right] +$$

$$+ \frac{\left[T_s(0) - T_c \right] b}{(p'' - p') \Gamma} \left(e^{-p''t} - e^{-p't} \right) \qquad (28)$$

$$T_s(t) - T_c = \frac{w(0)}{p'' - p'} \frac{\Gamma}{b} \left(p' - \frac{\Pi}{\Gamma} \right) \left(p'' - \frac{\Pi}{\Gamma} \right) \times$$

$$\times \left[e^{-p't} - e^{-p''t} \right] + \frac{T_s(0) - T_c}{p'' - p'} \times$$

$$\times \left[\left(p'' - \frac{\Pi}{\Gamma} \right) e^{-p''t} - \left(p' - \frac{\Pi}{\Gamma} \right) e^{-p't} \right] \qquad (29)$$

where

$$\frac{\Pi}{\Gamma} = \frac{(\pi^2 h + 2abAL) \mathcal{D} (1 + \lambda)}{[ha (1 + \lambda) + 2LA\mathcal{D}] a} \qquad (30)$$

$$\Gamma = \frac{ha (1 + \lambda) + 2LA\mathcal{D}}{2hA\mathcal{D} (1 + \lambda)} \qquad (31)$$

where p' and p" are the moduli of the roots of

$$\frac{\pi^4 h^2 \alpha \mathcal{D}}{a(ha + 2kd)} + \left[\frac{\pi^2 h^2 a(\mathcal{D} + \alpha) + 2\pi^2 h\mathcal{D} (kd + AL\alpha)}{ha + 2kd} \right] p +$$

$$\left\{ ha^2 \left[1 + \frac{2LA\mathcal{D}}{ha + 2kd} \right] \right\} p^2 = 0 \qquad (32)$$

The model parameters are $w(0)$ and $T_s(0)$ (or $\Delta T_s^o = T_s(0) -$

584

- Tc). They are the initial mass flux and the initial bubble surface temperature, respectively, in the bubbling process. These parameters try to take into account the heat and mass transfer at the bubble formation.

Clearly, such parameters must be evaluated by a model for the bubble formation step, which may also provide a best aproximation for the initial temperature inside the bubble. Here, this temperature is considered to be equal to the incoming gas temperature. The parameters should be determined from experimental data.

All the necessary variable values can be determined from the above solutions and the following definitions :

$$\bar{w} = \frac{1}{\zeta} \int_0^\zeta w(t) \, dt \qquad (33)$$

$$\overline{T_a} = \frac{1}{\zeta} \int_0^\zeta T_a(t) \, dt \qquad (34)$$

$$\langle T \rangle (t) = \frac{3}{a^3} \int_0^a r^2 T(r,t) \, dr \qquad (35)$$

$$\langle T \rangle_f = \langle T \rangle (\zeta) \qquad (36)$$

$$\langle C \rangle (t) = \frac{3}{a^3} \int_0^a r^2 C(r,t) \, dr \qquad (37)$$

$$\langle C \rangle_f = \langle C \rangle (\zeta) \qquad (38)$$

$$\langle \bar{T} \rangle = \frac{1}{\zeta} \int_0^\zeta \langle T \rangle (t) \, dt = \frac{3}{\zeta a^3} \int_0^\zeta \int_0^a r^2 T(r,t) \, dr \, dt \qquad (39)$$

$$\langle \bar{C} \rangle = \frac{1}{\zeta} \int_0^\zeta \langle C \rangle (t) \, dt = \frac{3}{\zeta a^3} \int_0^\zeta \int_0^a r^2 C(r,t) \, dr \, dt \qquad (40)$$

where ζ is the bubble hydrodynamic residence time in the continuous phase.

If we suppose that all the bubbles have the same volume, we can calculate the evaporated flow rate, W_a by :

$$W_a(t) = \frac{3 V_g \zeta}{a} \bar{w} \qquad (41)$$

It is important to note that all these values are recalculated during the numerical integration of the evaporator model. The linearization of Antoine equation is also recalculated during the integration.

TABLE I.1, I.2 : THE CALCULATED VALUES OF w(0), Hg AND va FOR
THE ANDRADE [21] EXPERIMENTS

Δt (min)	Tc (K)	Tg (K)	w(0) (10^4 kg/m^{2}s)	Hg	va (m/s)
0 - 15	311,9	417,8	2,290	0,1411	0,3409
15 - 30	324,5	406,5	2,812	0,1411	0,3316
30 - 45	326,0	410,5	8,493	0,1411	0,3349
45 - 60	326,0	410,0	7,598	0,1411	0,3345
60 - 75	326,0	408,0	7,166	0,1411	0,3329
75 - 90	326,0	405,5	7,694	0,1411	0,3308

Table I.1 - Experiment 1

Δt (min)	Tc (K)	Tg (K)	w(0) (10^4 kg/m^{2}s)	Hg	va (m/s)
0 - 15	309,4	466,3	4,238	0,1411	0,3804
15 - 30	320,0	449,5	4,325	0,1411	0,3667
30 - 45	321,5	450,0	10,19	0,1411	0,3671
45 - 60	322,0	450,5	10,30	0,1411	0,3675
60 - 75	322,0	449,0	13,42	0,1411	0,3663
75 - 90	322,0	448,0	11,10	0,1411	0,3659

Table I.2 - Experiment 4

586

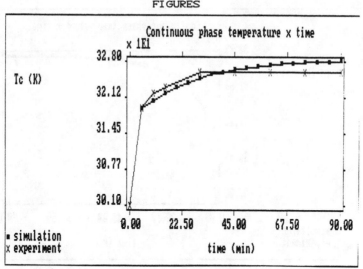

Fig.1.1 - Simulation of experiment 1

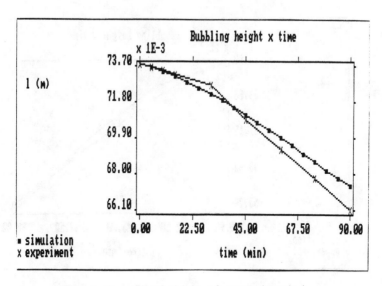

Fig.1.2 - Simulation of experiment 1

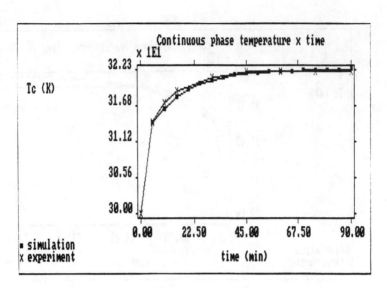

Fig.2.1 - Simulation of experiment 4

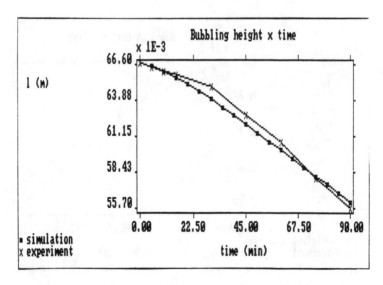

Fig.2.2 - Simulation of experiment 4

3. ANALYSIS OF THE EVAPORATOR SIMULATIONS

The parameters determination

The analized experimental data were obtained by ANDRADE [21]. They consist of twelve batch runs which utilized water as the continuous phase and air as the superheated gas. The gas was fed at two temperature levels. The data is given by the values of T_g, T_c, l and V_c, which were taken down at every 15 minutes, along the 90 minutes of each experiment. There are also the values of T_c at 5 and 10 minutes. We choose the first six experiments to be simulated, since they have the largest changes of T_c and l.

At first, we tried to use the mean values of the initial parameters that were determined by QUEIROZ & HACKENBERG [19], but the simulated results, during the heating of the continuous phase, were far from the experimental data, showing that the initial values must be determined for each gas and continuous phase temperature, as should be expected. Then we determined the values of the initial parameters directly from the experiments. The values of v_a and H_g were also calculated experimentally, so that the inaccuracies of the other models did not affect our results. We utilized the value 0.6 cm for the bubble radius, as reported by ANDRADE [21]. Since there are not enough data to evaluate $T_s(0)$ and w(0) along each experiment, we have taken $T_s(0) = T_c$. The calculations results for the first experiment and the fourth one of ANDRADE [21] are at tables (I.1) and (I.2), respectively.

Simulations

After some time simulations of the heating of the continuous phase, we noticed that it is impossible to simulate the first five minutes of each experiment, due to equipment reversible heating. Thus, all the simulations started at 5 minute initial time. For each experiment, we used mean values of H_g and v_a from the values at tables (I.1) and (I.2), and we also neglected heat loss to the ambient.

The results obtained with constant values of w(0) did not fit the experimental data. Futhermore, the W_a value decreased with increasing T_c values, for an almost constant value of l. Therefore, we concluded that w(0) must be an increasing function of T_c. The function that gave the best results is the exponential :

$$w(0) = k_0 \exp(k_1 T_c) \qquad (42)$$

This fitting led to increasing W_a values with increasing T_c values. Figures (1) and (2), utilizing an exponential fitting for w(0), show that the simulation results (the worst and the best ones) fit the experimental data.

4.   CONCLUSIONS

The good results validate the proposed model and the developed algorithm. They also confirm the applicability of QUEIROZ & HACKENBERG [19] model. The used continuous phase heat transfer coefficient equations (HACKENBERG [12] and CALDERBANK & MOO-YOUNG [13]) are also valid.

In addition, we can conclude that the parameters of the QUEIROZ & HACKENBERG [19] model are increasing functions of the continuous phase temperature. The best fitting is the exponential one. This dependence should be verified, since it includes the effect of $T_s(0)$ which was made equal to T_c. How the gas temperature affects these parameters are not elucidated and so it should be investigated.

REFERENCES

1.   KUMAR, R. & KULOOR, N. R.. The formation of bubbles and drops. In : DREW, T. B. ED.; COKELET, G. R. ED.; HOOPES JR., J. W. ED.; VERMEULEN, T. ED. Advances in Chemical Engineering, 8, pp. 255-368. New York, Academic Press, 1970.

2.   DARBY, R.. The dynamics of vapor bubbles in nucleate boiling. *Chem. Engng. Sci.*, 19, pp. 39-49, 1964.

3.   SCRIVEN, L. E.. On the dynamics of fase growth. *Chem. Engng. Sci.*, 10, pp. 1-13, 1959.

4.   SCHMIDT, H.. Bubble formation and heat transfer during dispersion of superheated steam in saturated water - I : bubble size and bubble detachment at single orificies. *Int. J. Heat Mass Transfer*, 20, pp. 635-646, 1977.

5.   SCHMIDT, H.. Bubble formation and heat transfer during dispersion of superheated steam in saturated water - II: heat transfer from superheated steam bubbles to saturated water during bubble formation. *Int. J. Heat Mass Transfer*, 20, pp. 635-646, 1977.

6.   LEVICH, V. G.. Physicochemical hidrodynamics. Englewood Cliffs, N. J., PrenticeHall Inc., 1962. 700 pp..

7.   HACKENBERG, C. M.. On the unsteady resistance of spherical bodies. Ph.D. Thesis, University of Florida, 1969. 325 pp..

8.   HACKENBERG, C. M.. On the unsteady resistance of an inviscid fluid sphere. V Interamerican Congress of Chemical Engineering, Rio de Janeiro, 1973.

9. HAPPEL, J.. *J. Trans. N. Y. Acad. Sci.* 20, p. 404, 1958.

10. HAPPEL, J.. *A. I. Ch. E. Journal* 4, p. 197, 1958.

11. RUCKENSTEIN, E.. On the mass transfer in the continuos phase from spherical bubbles or drops. *Chem. Engng. Sci.*, 19, p. 131-146, 1964.

12. HACKENBERG, C. M.. Transferência de calor em processos de borbulhamento. VI Congresso Interamericano de Engenharia Química, Caracas, Venezuela, 1975.

13. CALDERBANK, P. H. & MOO-YOUNG, M. B.. The continuos phase heat and mass transfer properties of dispersions. *Chem. Engng. Sci.*, 16, p. 39-54, 1961.

14. ANDRADE, A. L.. Transferência de calor transiente em processos de borbulhamento. Tese de M. Sc., COPPE/UFRJ, Rio de Janeiro, 1972. 52 p..

15. HACKENBERG, C. M. & ANDRADE, A. L.. Transferência de calor transiente em processos de borbulhamento : determinação da temperatura superficial. VI Congresso Interamericano de Engenharia Química, Caracas, Venezuela, 1975.

16. ANDRADE, A. L. & HACKENBERG, C. M.. Sistemas bifásicos : a analogia da transferência de calor e massa em regime transiente. *Rev. latinoam. transf. cal. mat.*, 9 : 59-73, 1985.

17. HACKENBERG, C. M. & ANDRADE, A. L.. A temperatura superficial transitória de uma bolha superaquecida. Anais do II Congresso Latinoamericano de Transferência de Calor e Matéria, vol. I, p. 544-554, São Paulo, 1986.

18. HACKENBERG, C. M. & ANDRADE, A. L.. Transient Surface Temperature of Superheated Bubbles. Publ. in Particulate Phenomena and Multiphase Transport, Ed. T. N. Veziroglu, Hemisphere Publ. Corp. - Springer - Verlag, New York, Vol. I, p. 377, 1988.

19. QUEIROZ, E. M. & HACKENBERG, C. M.. O equilíbrio interfacial e a previsão da taxa de evaporação em bolhas superaquecidas. Anais do I Simpósio Brasileiro de Transferência de Calor e Massa, p. 448-455, Campinas,

SP, UNICAMP, 1987.

20. DAVIDSON, J. F. & HARRISON, D.. Fluidised Particles. Cambridge, Cambridge University Press, 1963.

21. ANDRADE, A. L.. Transferência de calor em bolhas superaquecidas. Tese de D.Sc., COPPE/UFRJ, Rio de Janeiro, 1985. 146 p..

22. REID, R. C.; PRAUSNITZ, J. M.; SHERWOOD, T. K.. The Properties of Gases and Liquids. 3th Edition. New York, McGraw-Hill Book Company, 1977. 688 p..

Study on the Second-Law Efficiency of a Heat Exchanger

W. H. HUANG
Department of Mechanical Engineering
Tatung Institute of Technology
Taipei, Taiwan, ROC

Abstract

The thermodynamic efficiency of the heat exchanger is traditionally based on the first law of thermodynamics. The second-law efficiency is less discussed by other authors. Other papers relating to the second-law efficiency of the heat exchanger is only considering the temperature difference without pressure drop. This paper presents the second-law efficiency of the heat exchanger brought about both the temperature difference and pressure drop, associated with inlet temperature ratio, ratio of thermal capacity, effectiveness, and fluid properties.

1. INTRODUCTION

Scientists, engineers and economists take efficiency as a measure to gauge the performances of systems and processes. Engineering efficiency is useful for maximizing an output for a given input, where economic efficiency is utilized to find the minimum cost for a given product value. At any rate, efficiency is defined as

$$\eta = \frac{\text{Output}}{\text{Input}}$$

Therefore, η frequently has been used as an index to identify how near the output approaches the input. In addition, it serves as a inputs for improving a system or process.

The first-law efficiency, η_1, is defined by the ratio of energy output to energy input as

$$\eta_1 = \frac{\text{Energy output}}{\text{Energy input}}$$

So long as the first law is concerned

$$(\text{Energy input}) = (\text{Energy output}) + (\text{Loss})$$

593

For a heat transfer process or heat exchange, η_1 is 100% if heat loss from the system to the surroundings is negligible, since the total energy "input" of the hot fluid is "gained" (output) by the cold fluid. Hence η_1 can not be used as a useful performance index for a heat transfer process.

The second-law efficiency η_2* may be defined by the ratio of exergy output to exergy input as

$$\eta_2 = \frac{\text{Exergy output}}{\text{Exergy input}}$$

For the second law,

(Exergy input) = (Exergy output) + (Exergy unavoidable and intrinsic undesired output)

Because exergy is not conserved when there is unavoidable entropy production which can not be reduced to zero for a real process, where loss in the first law can be diminished in the ideal case by some "external" methods such as insulation or recovery for most of thermal systems. Losses are understood to include such things as waste, improper or unable to use that might be gained to another by means of spending another useful energy or method. Energy loss does not mean a disappearance of energy. Rather it results from inefficient transformation. However, irreversibility is the inevitable destruction of available energy. Irreversibility can not be diminished or recovered but could be minimized to some extent by certain means. Aside from identifying the location, type and magnitude of such waste and destruction by entropy production, second-law efficiency gives a finer understanding of performance than η_1. η_2 is intended to complement but not to replace η_1 in order to give better judgement of the performance of a process. The second-law analysis is good for diagnosis that can be understood from previous studies.

η_2 characterizes the degree of imperfection of the processes which the working fluid undergoes, but η_1 fails to reflect the degree of imperfection.

---

*η_2 is also called differently as "effectiveness", "exergetic efficiency", "rational efficiency" or "thermodynamic efficiency ratio" by different investigators.

The higher η_2 indicates a greater quantity of energy available in the working fluid to perform a useful function. Two surfaces A and B with $(\eta_1)_A > (\eta_1)_B$ and $(\eta_2)_A < (\eta_2)_B$, if the selection of these two surfaces were to be made for some application based solely on η_1, as is typically the case, then the selection would undoubtedly be in favor of the one that would deliver the less availability that ultimately would result in the performance of less useful work. It is not sufficient that a surface merely captures the incoming energy. What is important is how much of the captured energy is available for use. The second-law approach demonstrates that to quantify the performance of a thermal process lends much better insight into the total performance than does the conventional solely applied first-law approach.

The η_1 is necessary only for comparing systems which have like (equal grade) inputs and like outputs, or for comparing two models of a device. It is the η_2, not the η_1, which measures how well a system is performing, compared to the possible performance. In other words, $(1-\eta_2)$, not $(1-\eta_1)$ is a measure of the potential for improvement. η_2 could be called the energy utilization factor, a degree of thermodynamic perfection.

Proposed in the many studies are entropy productions and their ratios to the reference surface as criteria. But the entropy production is sometimes ambiguous in amount. With the aids of η_2 it will be useful to show the goodness of a process. Since the second law shows

$$Ex,_{in} - Ex,_{out} = I$$

and for any system, substance or process η_2 is $1 > \eta_2 \geq 0$, where

$$\eta_2 = 1 - \frac{I}{Ex,_{in}} = 1 - \frac{T_o \dot{S}_p}{Ex,_{in}} \tag{1}$$

The fraction $(1-\eta_2)$ is being lost (destructed) due to irreversible processes causing an entropy production $\dot{S}_p$. The irreversibility in any process is taken as the decrease in the available energy or as the increase in the unavailable energy. The definition of Equation (1) is based on the definition of the irreversibility rate $I = W_{rev} - W_{act}$, with W_{act} equal to zero in a heat exchanger, so that $I = W_{rev}$ for heat exchanger. According to the second law, since $I > 0$ and $W_{rev} < Q$, η_2 lies between zero and unity.

L. C. Witte, and N. Shamsundar [1] indicated that the second-law efficiency, η_2, has certain value which is not zero as $\varepsilon=0$ by defining $\eta = 1 - I/Q$ in the absence of pressure drop. And, by defining $\eta_2 = 1 - W_{rev}/Ex,in$, it initially decreases to the minimum and then increases as ε increases. These properties make these definitions are unsuitable for the second-law efficiency. Golem [3] derived η_2 without considering pressure drop. This paper defines the second-law efficiency η_2 as (Exergy output)/(Exergy input) and considering the temperature difference and pressure drop.

2. ANALYSIS

For the purpose of making the heat exchanger designer predict the loss of available energy in a given heat exchanging operation condition, the second-law efficiency of a counter-flow heat exchanger as an example in this study. Because all losses in available energy translate into increase consumption, the design of a heat exchanger for minimum available energy loss or maximum second-law efficiency is an energy conservation measure with obvious economic implications.

For each fluid of a counter-flow heat exchanger inlet temperature, $T_{h,i}$, $T_{c,i}$, inlet pressure $P_{h,i}$, $P_{c,i}$, heat capacity rates of $(\dot{m} C_p)_{min}$, $(\dot{m} C_p)_{max}$ are given. The rate of entropy production of a heat exchanger can be calculated as

$$\dot{S}_p = \dot{m}_h s_h + \dot{m}_c s_c \tag{2}$$

where heat flow across the outer wall of the heat exchanger is neglected. The subscript h in above equation represents for the hot fluid and c for the cold fluid. Assuming the ideal gas equation is applied, Equation (2) becomes

$$\dot{S}_p = C_h [\ln \frac{T_{h,o}}{T_{h,i}} - (\frac{R'}{C_p})_h \ln \frac{P_{h,o}}{P_{h,i}}] + C_c [\ln \frac{T_{c,o}}{T_{c,i}} - (\frac{R'}{C_p})_c \ln \frac{P_{c,o}}{P_{c,i}}]$$

$$= \dot{S}_{\Delta T} - \dot{S}_{\Delta P} \tag{3}$$

where

$$\dot{S}_{\Delta T} = C_h \ln(\frac{T_{h,o}}{T_{h,i}}) + C_c \ln(\frac{T_{c,o}}{T_{c,i}}) \tag{4a}$$

$$\dot{S}_{\Delta P} = C_h (\frac{R'}{C_p})_h \ln(\frac{P_{h,o}}{P_{h,i}}) + C_c (\frac{R'}{C_p})_c \ln(\frac{P_{c,o}}{P_{c,i}})_c \tag{4b}$$

and subscript i represents inlet fluid, o represents outlet fluid. R' is a
gas contant.

From the energy blance and the definition of the effectiveness gives

$$C_h \ (T_{h,i} - T_{h,o}) = C_c \ (T_{c,o} - T_{c,i})$$

$$\varepsilon = \frac{C_h (T_{h,i} - T_{h,o})}{C_{min} (T_{h,i} - T_{c,i})} \tag{5}$$

Combining the above equations and eliminating $T_{h,i}$ and $T_{c,o}$ the following
equation is given for R<1 as

$$\dot{S}_p = C_h \ ln[1 + \varepsilon (\frac{T_{c,i}}{T_{h,i}} - 1)] + C_c \ ln[1 - \frac{C_h}{C_c} \varepsilon (1 - \frac{T_{h,i}}{T_{c,i}})]$$

$$- C_h \ (\frac{R'}{C_p})_h \ ln[1 - (\frac{\Delta P}{P_{h,i}})_h] - C_c (\frac{R'}{C_p})_c \ ln[1 - (\frac{\Delta P}{P_{c,i}})_c] \tag{6}$$

Similary for R>1

$$\dot{S}_p = C_h \ ln[1 - \frac{\varepsilon}{R} (1 - \frac{T_{c,i}}{T_{h,i}})] + C_c \ ln[1 + \varepsilon (\frac{T_{h,i}}{T_{c,i}} - 1)]$$

$$- C_h \ (\frac{R'}{C_p})_h \ ln[1 - (\frac{\Delta P}{P_{h,i}})_h] - C_c (\frac{R'}{C_p})_c \ ln[1 - (\frac{\Delta P}{P_{c.i}})_c] \tag{7}$$

where

$$\frac{\Delta P}{P,i} = \frac{P,i - P,o}{P,i}$$

The pressure drop in the channel is given as

$$\Delta P = f \ \rho L \ \frac{2 \ u^2}{D_h}$$

The efficiency formula developed is more useful if the effects of the
heat transfer area on the efficiency are assessed. Since both increase
together, a larger area implies higher fixed costs, and higher η_2 implies
lower operating costs. As a result, a compromise is necessary to fix the
design values. However, the relationship between η_2 and area depends on
the type of heat exchanger, heat transfer coefficient, etc.. Consequently,
instead of an area, a quantity reflecting the extent of the heat transfer
surface is used that yields quantitative results which are not restricted to

597

any type of heat exchanger. Since $\varepsilon = f$ (NTU, C_h/C_c), the $T_{h,o}$, $T_{c,o}$ of outlet temperatures would depend on the heat exchanger design whereas the inlet temperatures, $T_{c,i}$ and $T_{h,i}$, are almost specified. Assuming the fluids remain a single phase or a double phase between inlet and outlet and C_p = constant, then η_2 can be obtained.

Assuming fluids are ideal gases with $p = \rho R'T$, the above $(\frac{R'}{C_p}\frac{\Delta P}{P,i})$ terms becomes $\Delta p/(\rho T C_p)_{in}$ for each fluid. From $\ln(1+x) \doteq x$, and NTU=UA/C_{min}, then $\dot{S}_p$ or η_2 can be obtained, combined with the data of U,f, and ε. The entropy production rate becomes, for R<1

$$\dot{S}_p \doteq C_h\, \varepsilon(\frac{T_{c,i}}{T_{h,i}} -1) + C_c\, \varepsilon R(\frac{T_{h,i}}{T_{c,i}} -1) + (\frac{A_c L\, \mu^3 fRe^3}{2\rho^2 D_h^4 T_{h,i}})_h + (\frac{A_c L\, \mu^3 fRe^3}{2\rho^2 D_h^4 T_{c,i}})_c$$

(8)

and for R>1

$$\dot{S}_p = C_h\, \frac{\varepsilon}{R}(\frac{T_{c,i}}{T_{h,i}} -1) + C_c\, \varepsilon(\frac{T_{h,i}}{T_{c,i}} -1) + (\frac{A_c L\, \mu^3 fRe^3}{2\rho^2 D_h^4 T_{h,i}})_h$$
$$+ (\frac{A_c L\, \mu^3 fRe^3}{2\rho^2 D_h^4 T_{c,i}})_c$$

(9)

If the entropy production rate due to flow friction is negligible and there is no heat transfer to the surroundings from the wall of the heat exchanger, then the entropy production per unit capacity rate will be

$$N_s = \frac{\dot{S}_p}{C_h} = \ln[1+ (\frac{T_{c,i}}{T_{h,i}} -1)] + \frac{C_c}{C_h}\ln[1- \frac{C_h}{C_c}\varepsilon(1- \frac{T_{h,i}}{T_{c,i}})]$$

(10)

The maximum $(\dot{S}_p/C_h)$ can be found by differentiating Equation (10) with respect to ε as

$$(\frac{\dot{S}_p}{C_h})_{max} = (1+\frac{C_c}{C_h})\ln\frac{\frac{T_{h,i}}{T_{c,i}} + \frac{C_h}{C_c}}{1+\frac{C_h}{C_c}} - \frac{C_c}{C_h}\ln\frac{T_{h,i}}{T_{c,i}}$$

(11)

and

$$(\varepsilon)_{opt} = \frac{1}{1+\frac{C_h}{C_c}}$$

(12)

If $C_c = C_h$, then

$$(\dot{S}_p/C_h)_{max} = \ln[(\frac{T_{h,i}}{T_{c,i}} +1)^2/(\frac{T_{h,i}}{T_{c,i}})]$$ (13)

at $\varepsilon = 0.5$, are obtained.

The inlet exergy of heat, $Ex,_{in}$, is

$$Ex,_{in} = Q_{in}(1- \frac{T_o}{T_{h,lm}}) = C_h(T_{h,i}-T_{h,o}) (1- \frac{T_o}{\frac{T_{h,i} - T_{h,o}}{\ln \frac{T_{h,i}}{T_{h,o}}}})$$ (14a)

where $T_{h,lm}$ is a mean logarithm temperature of the hot fluid as

$$T_{h,lm} = (T_{h,i} - T_{h,o})/\ln(T_{h,i}/T_{h,o})$$

$$Ex,_{in} = C_h (\Delta T_h - T_o \ln \frac{T_{h,i}}{T_{h,o}})$$

$$= C_h T_{h,i}[\varepsilon(1 - \frac{T_{c,i}}{T_{h,i}}) + \frac{T_o}{T_{h,i}} \ln[1-\varepsilon(1 - \frac{T_{c,i}}{T_{h,i}})]$$ (14b)

For R<1, the hot fluid has a smaller capacity rate, then

$$\varepsilon = \frac{C_h [T_{h,i}-T_{h,o}]}{C_h [T_{h,i}-T_{c,i}]} = \frac{1}{R} \frac{(\frac{T_{c,o}}{T_{c,i}}) - 1}{(\frac{T_{h,i}}{T_{c,i}}) - 1}$$

and for R>1

$$\varepsilon = \frac{C_h [T_{h,i}-T_{h,o}]}{C_c [T_{h,i}-T_{c,i}]} = R \frac{[1-(T_{h,o}/T_{h,i})]}{[1-(T_{c,i}/T_{h,i})]} = \frac{(T_{c,o}/T_{c,i})-1}{(T_{h,i}/T_{c,i})-1}$$

are obtained. Utilizing the values for ε, R, $T_{h,i}/T_{c,i}$, yields η_2 as follows.

(a) For R < 1

$$\eta_2 = 1- \frac{T_o \dot{S}_p}{Ex,_{in}} = 1- \frac{T_o}{T_{h,i}} \frac{\ln[1+\varepsilon(\frac{T_{c,i}}{T_{h,i}} -1)] + \frac{1}{R} \ln[1-R \varepsilon(1- \frac{T_{h,i}}{T_{c,i}})]}{[\varepsilon(1- \frac{T_{c,i}}{T_{h,i}})] + \frac{T_o}{T_{h,i}} \ln[1-\varepsilon(1- \frac{T_{c,i}}{T_{h,i}})]}$$

$$- (\frac{R'}{C_p})_h \ln[1-(\frac{\Delta p}{P_{h,i}})_h] - \frac{1}{R} (\frac{R'}{C_p})_c \ln[1-(\frac{\Delta p}{P_{c,i}})_c]$$

(15a)

(b) For $R > 1$, Eq. (B. 14a) becomes

$$Ex,_{in} = C_h T_{h,i} [\frac{\varepsilon}{R}(\frac{T_{c,i}}{T_{h,i}} -1) + \frac{T_o}{T_{h,i}} \ln[1+ \frac{\varepsilon}{R}\frac{T_{c,i}}{T_{h,i}} -1)]$$

then,

$$\eta_2 = 1 - \frac{T_o}{T_{h,i}} \frac{\ln[1+ \frac{\varepsilon}{R}(\frac{T_{c,i}}{T_{h,i}} -1)] + \frac{1}{R} \ln[1-\varepsilon(\frac{T_{h,i}}{T_{c,i}})]}{\frac{\varepsilon}{R}(\frac{T_{c,i}}{T_{h,i}} -1) + \frac{T_o}{T_{h,i}} \ln[1+ \frac{\varepsilon}{R} (\frac{T_{c,i}}{T_{h,i}} -1)]}$$

$$-(\frac{R'}{C_p})_h \ln[1-(\frac{\Delta p}{P_{h,i}})_h] - \frac{1}{R} (\frac{R'}{C_p})_c \ln[1-(\frac{\Delta p}{P_{c,i}})_c]$$

(15b)

(c) For $R = 0$ (boiler, evaporator), $T_o > T_{c,i}$, $\Delta p_c = 0$, T_c = constant

$$= 1 - \frac{T_o}{T_{h,i}} \frac{\ln[1+\varepsilon(\frac{T_{c,i}}{T_{h,i}} -1)] - \varepsilon(1-\frac{T_{h,i}}{T_{c,i}}) - (\frac{R'}{C_p})_h \ln[1-(\frac{\Delta p}{P_{h,i}})_h]}{\varepsilon(1-\frac{T_{c,i}}{T_{h,i}}) + \frac{T_o}{T_{h,i}} \ln[1-\varepsilon(1-\frac{T_{c,i}}{T_{h,i}})]}$$

assuming $\ln(1+x) \doteq x$.

(15c)

(d) For $R = \infty$ (condenser), $\Delta p_h = 0$, T_h = constant

$$Ex,_{in} = Q_{in} (1-\frac{T_o}{T_h}) = C_c[T_{c,o} - T_{c,i}] (1-\frac{T_o}{T_h})$$

$$= C_c T_{c,i} [\varepsilon(\frac{T_{h,i}}{T_{c,i}} -1)] (1-\frac{T_o}{T_h})$$

$$\eta_2 = 1 - \frac{\ln[1+\varepsilon(\frac{T_{h,i}}{T_{c,i}} -1)] - \varepsilon(1-\frac{T_{c,i}}{T_{h,i}}) - (\frac{R'}{C_p})_c \ln[1-\frac{\Delta p}{P_{c,i}})_c]}{\varepsilon (\frac{T_{h,i}}{T_{c,i}} -1) (1-\frac{T_o}{T_h})}$$

(15d)

The results of η_2 versus ε for various C_h/C_c without pressure drops for

$\dfrac{T_{h,i}}{T_{c,i}} = 1.5$, $\dfrac{T_{h,i}}{T_{c,i}} = 2.0$ and $\dfrac{T_{h,i}}{T_{c,i}} = 3$ are shown in Figure 1, Figure 2 and

Figure 3, respectively.

$T_{h,i}/T_{c,i}=1.5$ pertains to most liquid -- liquid, liquid-air and gas-air heat exchangers, while $T_{h,i}/T_{c,i} = 2.0$ corresponds to most economizers, air pre-heaters and waste heat recovery units and $T_{h,i}/T_{c,i}=3$ is the range for cooler, medium pressure steam boilers or pressurized water reactors [1]. However, it should be noted here that different value of $T_{h,i}/T_{c,i}$ for the same value of $T_{h,i}/T_{h,i}$ or $T_{c,o}/T_{c,i}$ give different effect for each value of $T_{h,i}/T_{c,i}$ as shown below.

(1) For R < 1

    (a) If $T_{h,i}/T_{h,o}$ = constant, then

$$\varepsilon = [1-(T_{h,o}/T_{h,i})]/[1-(T_{c,i}/T_{h,i})]$$

$$\varepsilon_{t_i^* = 1.5} = 3[1-(T_{h,o}/T_{h,i})], \quad \varepsilon_{t_i^* = 2} = 2^{[1-(T_{h,o}/T_{h,i})]}$$

$$\varepsilon_{t_i^* = 3.0} = \frac{3}{2}[1-(T_{h,o}/T_{h,i})]$$

$$\therefore \varepsilon_{t_i^* =3.0} = \frac{1}{2}\,\varepsilon_{t_i^* = 1.5}, \quad \varepsilon_{t_i^* = 2} = 2 = \frac{2}{3}\,\varepsilon_{t_i^* = 1.5} \text{ are obtained,}$$

    where t_i^* defines as the ratio of inlet temperatures of the hot fluid to the cold fluid, $T_{h,i}/T_{c,i}$.

    (b) If $T_{c,o}/T_{c,i}$ = constant, then

$$\varepsilon = \frac{1}{R}[1-(T_{c,o}/T_{c,i})]/[1-(T_{h,i}/T_{c,i})]$$

    thus,

$$\varepsilon_{t_i^* = 2.0} = \frac{1}{2}\,\varepsilon_{t_i^* = 1.5} \text{ and } \varepsilon_{t_i^* = 3} = \frac{1}{4}\,\varepsilon_{t_i^* = 1.5} \text{ are obtained.}$$

(2) For R > 1

    (a) If $T_{h,i}/T_{h,o}$ = constant, then

$$\varepsilon = R[1-(T_{h,o}/T_{h,i})]/[1-(T_{c,i}/T_{h,i})]$$

$$\varepsilon_{t_i^* = 2.0} = \frac{2}{3}\,\varepsilon_{t_i^* = 1.5} \text{ and } \varepsilon_{t_i^* = 3} = \frac{1}{2}\,\varepsilon_{t_i^* = 1.5} \text{ are obtained.}$$

    (b) If $T_{c,o}/T_{c,i}$ = constant, then

$$\varepsilon = [1-(T_{c,o}/T_{c,i})]/[1-(T_{h,i}/T_{c,i})]$$

    thus,

601

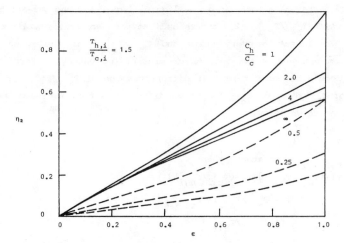

Figure 1 Second-law efficiency, η_2, against heat exchanger effectiveness, ε, at $T_{h,i}/T_{c,i}=1.5$ for various C_h/C_c if pressure drops are neglected.

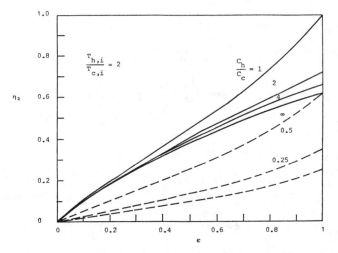

Figure 2 Second-law efficiency, η_2, against heat exchanger effectiveness, ε, at $T_{h,i}/T_{c,i}=2$ for various C_h/C_c if pressure drops are neglected.

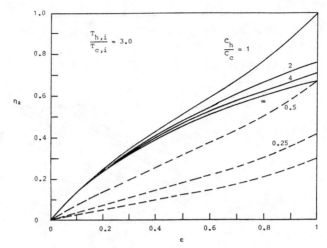

Figure 3   Second-law efficiency, η_2, against heat exchanger
effectiveness, ε, at $T_{h,i}/T_{c,i}=3.0$ for various C_h/C_c
if pressure drops are neglected.

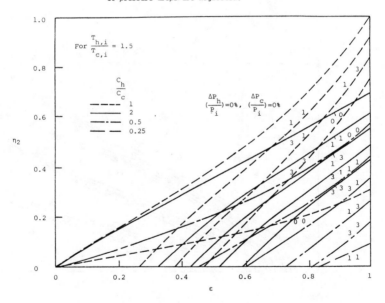

Figure 4   Second-law efficiency, η_2, versus heat exchanger effectiveness, ε, at
$T_{h,i}/T_{c,i}=1.5$ with pressure drops of various C_h/C_c.

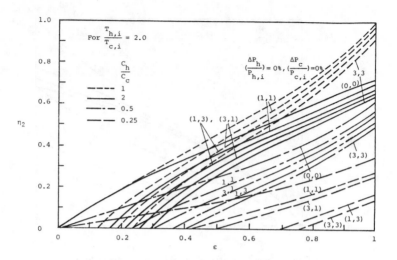

Figure 5 Second-law efficiency, η_2, versus heat exchanger effectivenss, ε, at $T_{h,i}/T_{c,i}=2$ with pressure drops of various C_h/C_c.

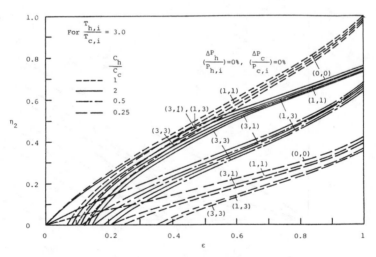

Figure 6 Second law efficiency, η_2, versus heat exchanger effectiveness, ε, at $T_{h,i}/T_{c,i}=3$ with pressure drops of various C_h/C_c.

$$\varepsilon_{t_i^*} = 2.0 = \frac{1}{2} \quad \varepsilon_{t_i^*} = 1.5 \quad \text{and} \quad \varepsilon_{t_i^*} = 3 = \frac{1}{4} \quad \varepsilon_{t_i^*} = 1.5 \quad \text{are obtained.}$$

Plots of η_2 against effectiveness for various combinations of parameters (C_h/C_c), $(T_{h,i}/T_{c,i})$ are shown in Figures 1 to 3. The following results are drawn from the figures:

(1) Increased ε always results an increase in η_2, because increasing the areas of a heat exchanger reduces the temperature difference.

(2) As the temperature ratio $T_{h,i}/T_{c,i}$ increases, the η_2 decreases at the same ε which is correct from the second-law basis, because of the increase of temperature difference between the hot fluid and the cold fluid.

(3) The highest η_2 is obtained when the capacity rates are equal. When the capacity rates are not equal, better η_2 is obtained when the hot fluid has the larger capacity rate. In constrast to the effectiveness, which is unaffected by interchanging C_c and C_h, the η_2 is strongly dependent on which one of the two capacity rates is greater. An example of the effect of this is that boilers and eveporators (R=0) are inherently less efficient than condenser (R=∞). The η_2 decreases as the R different from unity. This decrease is due to more heat being transfered through a higher temperature difference between the wall and the fluid.

(4) The η_2 reaches the maximum value of unity when R=1 and the ε is equal to unity. Thus, an infinitely large counterflow heat exchanger has a second-law efficiency of unity. As the ε approaches zero, the η_2 is also approached zero, regardless of the value of the capacity rate ratio R.

(5) For small temperature difference, the η_2 is generally high. Consequently, in design with large $T_{h,i}/T_{c,i}$, competing effects of η_2 and ε on the operating cost must be taken into account.

If pressure drops exist in heat exchanger, either on hot fluid side or cold fluid side or both, they will decrease second-law efficiency significantly which are shown in Figures 4 to 6 for various $T_{h,i}/T_{c,i}$, and combinations of C_h/C_c, $(\Delta P/P_i)$. In some situations as pressure drop increases, even though ε is not equal to zero, the pressure drop reduces the second-law efficiency and approaches zero which indicates that the fluid which does not transfer any available energy.

For a counter-flow heat exchanger the relationship between ε, NTU, and R is shown in the textbook. When NTU, R, the inlet fluid temperatures and pressure are fixed, the second-law efficiency of the exchanger is only a function of the mass flow rate. As the mass flow rate increases, the presure drop also increases, causing a decrease in the η_2 for the same effectiveness. In case of low value of ε (low NTU), the pressure drop has significant effect on the η_2, since no energy is transferred effectively. As NTU is increased, more energy is transferred between the hot & cold streams, then η_2 also is increased. However, further increase in ε results in reducing the temperature difference and in a decrease in irreversibility in heat transfer, this giving a higher η_2 with an increase in effectiveness.

The exchanger effectiveness is defined as

$$\varepsilon = \frac{\dot{Q}}{\dot{Q}_{max\ possible}}$$

It provides a comparison of the actual heat transfer rate, $\dot{Q}$, to the thermo-dynamically limited maximum possible heat transfer rate. In a regenerator, ε can be called the degree of regeneration or regenerator fraction or temperature efficiency without considering pressure effects. From Equation (4a), if only the entropy production rate resulting from heat transfer is taken consideration, assuming $C_h = C_c$, then

$$\dot{S}_{\Delta T} = C\ [\ln(\frac{T_{out}}{T_{in}})_c - \ln(\frac{T_{in}}{T_{out}})_h\ ]$$

$$= \dot{m}\ C_p\ (T_{out}-T_{in})_c\ [\frac{\ln(\frac{T_{out}}{T_{in}})_c}{(T_{out}-T_{in})_c} - \frac{\ln(\frac{T_{in}}{T_{out}})_h}{(T_{in}-T_{out})_h}] \qquad (16)$$

where $\qquad C_c(T_{out}-T_{in})_c = C_h(T_{in}-T_{out})_h$

then $\qquad S_{\Delta T} = \dot{m}\ C_p\ (T_{out}-T_{in})_c\ [\ \frac{T_{h,lm}-T_{c,lm}}{T_{h,lm} \times T_{c,lm}}\ ] \qquad (17)$

is obtained, where

$$T_{lm} = \frac{\int dH}{\frac{\int dH}{T}} = \frac{\int dT}{\frac{\int dT}{T}} = \frac{T_{out}-T_{in}}{\ln \frac{T_{out}}{T_{in}}}$$

Substituting the definition of effectiveness ε into Equation (17) then

$$\dot{S}_{\Delta T} = \dot{m}\ C_p\left(\frac{(T_{out}-T_{in})_c}{T_{h,i}-T_{c,i}}\ \frac{(\frac{T_{h,lm}-T_{c,lm}}{T_{h,i}-T_{c,i}})\ (T_{h,i}-T_{c,i})^2}{T_{h,lm}\cdot T_{c,lm}}\right)$$

$$= \dot{m}\ C_p\ \varepsilon\left(\frac{T_{h,lm}-T_{c,lm}}{T_{h,i}-T_{c,i}}\right)\ \left(\frac{(T_{h,i}-T_{c,i})^2}{T_{h,lm}\cdot T_{c,lm}}\right)$$

can be obtained. Assuming $\ln(1+x) \fallingdotseq x$, then

$$\frac{T_{h,lm}-T_{c,lm}}{T_{h,i}-T_{c,i}} = 1 - \varepsilon$$

and thus the heat transfer irreversibility can be given as

$$I_{\Delta T} = T_o\ [\dot{m}\ C_p \varepsilon(1-\varepsilon)]\ \frac{(T_{h,i}-T_{c,i})^2}{T_{h,lm}\cdot T_{c,lm}} \tag{18}$$

By differentiating Equation (18) with respect to ε, it may be found that $I_{\Delta T}$ is maximum for $\varepsilon=0.5$, and

$$(I_{\Delta T})_{max} = T_o\ \frac{\dot{m}\ C_p}{4}\ \frac{(T_{h,i}-T_{c,i})^2}{T_{h,lm}\cdot T_{c,lm}} \tag{19}$$

The result is the same as Equation (13) at $C_h/C_c=1$ for $(\dot{S}_p/C_h)_{max}$ with $\varepsilon_{opt}=0.5$.

It may be shown that the temperature difference needed for heat transfer is approximately

$$\dot{Q} \fallingdotseq \dot{m}\ C_p(1-\varepsilon)(T_{h,i}-T_{c,i})$$

and this term is the numerator of the heat transfer irreverisbility component and this is the factor on which the designer should focus his attention.

3. CONCLUSION

Second-law efficiency of a heat exchanger contains following features:
(a) It indicates how close the heat transfer reaches its input available energy and establishes a yardstick for the heat transfer to be improved.
(b) It illustrates how good the process is in value which is clear to identify the goodness of the process.
(c) For the case that the entropy production is due only to heat transfer, it is possible to combine ε-NTU or LMTD method to compare preformance of

a heat exchanger or a heat transfer surface with or without pressure
drop more reasonable.

NOMENCLATURE

A = heat exchanger area

C = thermal capacity rate = $\dot{m}\, C_p$

C_p = specific heat

D_h = hydraulic diameter

Ex = exergy

f = friction coefficient

I = rate of irreversibility production

L = length of channel

$\dot{m}$ = mass flow rate

Ns = entropy production rate per thermal capacity

NTU= number of transfer units, NTU = U A / C_{min}

P = pressure

R = ratio of thermal capacities = C_h/C_c

R' = gas constant

Re = Reynold number

$\dot{S}$ = rate of entropy production

T = absolute temperature

t* = definition of $T_{h,i}/T_{c,i}$

T_{lm}= log-mean temperature

$\dot{Q}$ = heat transfer rate

U = overall heat transfer coefficient

W = work

ρ = density

η_1 = the first-law efficiency

η_2 = the second-law efficiency

ε = heat exchanger effectiveness

μ = Viscosity of fluid

SUBSCRIPTS

c = cold fluid

h = hot fluid

i = inlet

in = inlet

o = outlet

p = pressure, production term

rev= reversible

Δp = due to pressure drop

ΔT = due to temperature difference

REFERENCE

1. L. C. Witte, and N. Shamsunder, "A Thermodynamic Efficiency Concept for Heat Exchange Devices", ASME Journal Engineering for Power, Vol.105, 1983, pp.199-203.

2. S. Sarangi and K. Chowdhury, "On the Generation of Entropy in a Counter-Flow Heat Exchanger", Cryogenics, February, 1982, pp.63-65.

3. P. J. Golem, and T.A. Brzustowski, "Second-law Analysis of Energy Processes. Part 2: The Performance of Simple Heat Exchangers.", Transactions de la SCGM, Vol. 4, No.4, 1976-1977, pp.219-226.

4. W. H. Huang, et al., "The Second-Law Efficiency of a Heat Exchanger", Proc. of the 4th National Conference of CSME, pp.645-653, 1987.

Electron-Phonon Thermal Resistance between Solid-Solid Interfaces in an Energy Conversion System

M. M. KAILA
Department of Applied Physics
Papua New Guinea University of Technology
Lae, PNG

ABSTRACT

Thermal energy produced by any energy conversion method often includes interfaces for onward transmission to utilising end. Surfaces involved may be highly conducting to start with but slowly change into insulating films, over the period of time. The resulting heat transfer process consequently changes from purely electronic to electron–phononic. This paper starting from the theory of overall effective electronic conductance asperity incorporates modifications produced by electron–phonon interactions at the interface. These heat paths are arranged in paralled and series combinations to satisfy the heat balance equation. Previously studied amorphous silicon films is taken as the representative insulating surface between metallic copper plates.

1. INTRODUCTION:

When a small amount of heat is allowed to flow through a limited area of any material, by physical contact, a localised change in temperature results. This change becomes stable in a few seconds. In the case of thermally insulating materials, interstitial fluid conductance and convection, if present, also become competitive modes of heat transfer. Metal–insulator contact has atom vibrations in the insulator side of the interface. This results in electron–phonon interactions taking up the heat transfer. The fraction of heat transferred from a narrow rounded hot probe across the contact can be written as a mathematical identity [1] in terms of various thermal resistances present.

2. THEORY:

The fractional change in temperature of a hot probe on contact with a cold substrate can be expressed in terms of ratios of probe resistance and series and parallel combinations of intervening thermal resistances. The identity in present case is

$$\frac{V}{(V_0 - V)} = \frac{R_0}{R_c} + \frac{R_0}{(R_1 + R_2)} \qquad (1)$$

The meaning of various terms is as follows:

V_0 = Temperature of the hot probe.

611

V	=	Temperature of the hot probe on contact.
R_0	=	Thermal resistance of the hot probe.
R_c	=	Thermal resistance to heat flow by convection.
R_1	=	Thermal resistance as a result of mechanical contact (probe).
R_2	=	Thermal resistance as a result of mechanical contact (substrate).
R_1 and R_2		containing constriction and phonon–electron [2] heat transfer components are expressed as follows.

$$R_1 = \frac{1}{4ak_1} + e/X_1 \pi a^2 k_{ph1}^{-X_1 x} \qquad\qquad (2)$$

$$R_2 = \frac{1}{4ak_2} + e/X_2 \pi a^2 k_{ph2}^{-X_2 x} \qquad\qquad (3)$$

Suffix 1, 2 refer to probe and contact materials i.e. copper and silicon in our case. Meaning of new terms is as follows.

a [1]	=	Effective radius of the mechanical contact.
k	=	Thermal conductivity.
X	=	Inverse characteristic length of electron–phonon heat transfer at the contact.
k_{ph}	=	Phonon thermal conductivity.
x	=	Thickness from the interface, inside insulator.

In view of the limited knowledge about the degrees of involvement of various heat transfer processes at the interfaces and of X (taken approximately $X_1 = X_2 = 10^{-6}$/m[2], the following estimate [3] of k_{ph} is considered a good approximation.

$$k_{ph} = 4 \left(\frac{k}{h}\right)^3 \frac{M\delta\theta_D}{\gamma^2 T} \qquad\qquad (4)$$

Here M is average atomic mass of the material, δ average interatomic distance, θ_D Debye temperature, γ Grüneissen constant of the material and k Boltzmann constant and h the Planck's constant.

Use of equations (1) –(4) for C_u –Si interface result in estimate of contact thermal resistance as in figure 1. On treating the interface, merely as an effective electron transport path of thickness d, one can also estimate [4] electron–phonon component by $\frac{d}{\pi a^2 k}$, k being thermal conductivity of the equivalent insulating Si [5] film The results due to this classical [4] approach are also included in figure 1 for comparison.

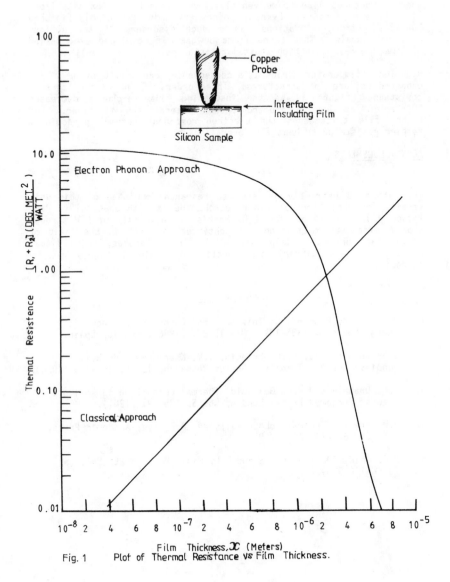

Fig. 1   Plot of Thermal Resistance vs Film Thickness.

613

3. CONCLUSION AND DISCUSSION:

This study is a very good representative of what can happen in high speed structures like space vehicles, and large heat flux situations like nuclear reactors. Even if joints are made from highly finished surfaces, strong vibrations can produce disorders at the surface on atomic scale. Thin oxide films or gaps if developed over period of time can result in high thermal resistances.

High local temperature drops as a consequence, can result in un-expected failures of structures. Three orders of increase of thermal resistance (figure 1) as the insulating film thickness decreases 10^{-5} to 10^{-8} m due to electron-phonon interactions is clear. For higher film thicknesses the electron component becomes predominant keeping resistance to heat flow overall high.

4. ACKNOWLEDGEMENTS

The author is extremely grateful to research Committee of this uni-versity for their keenness in research. He is also grateful to Pro-fessor H.J. Goldsmid and Dr. L.B. Harris of University of NSW through whom he received motivation for continued research in the field of solid state physics. Help extended by Mr. G Farakas, Audio Visual office of Matheson Library in preparation of drawing is highly appreciated.

REFERENCES

1.   M.M. Kaila, Proceedings International Congress on Renewable Energy Sources, 1305-1315, May 18-23, (1986), Madrid, Spain.

2.   L.S. Kokorev, Yu. N.,M Shelagin, V.V. Kharitonou and N.I. Soboleva, Heat - Transfer, Soviet Research, 12, 5, 106-110,(1980).

3.   J.R. Drabble and H.J. Goldsmid, Thermal Conduction in semicond-uctors, Pergamon Press, London, Ch. 5, 135-197 (1961).

4.   R.W. Powell, Thermal Conductivity ed. R.P. Tye, Academic Press, N.Y, Ch. 6, 275-338 (1969).

5.   H.J. Goldsmid, M.M. Kaila and G.L. Paul, Phys. Stat. Sol. (a) 76, K31 (1983).

Utilization of Solar Energy in Nigeria

M. T. OLADIRAN
Department of Mechanical Engineering
University of Ibadan
Ibadan, Nigeria

Abstract

The use of solar energy as a substitute for the non-renewable energy resource in Nigeria is attractive. However climatological data are scanty for many sites. Also, design equations to determine solar radiation levels are not usually available. Consequently solar radiation measurements from 16 different stations are presented and analysed in this paper. Correlation equations based on the form developed by Angstrom are proposed for the three zones. These equations produce results which are in agreement with recorded data. Thus solar radiation at any geographical point can be predicted.

1. INTRODUCTION

The commercial power demand in Nigeria has been supplied mainly from fossil fuel based generation systems. Figure 1 depicts contribution from various sources namely petroleum oil, natural gas, coal and hydro power. Uptil late fifties, coal was the principal energy resource. From that time until now, petroleum oil supplies the bulk energy requirements. The availability of this resource has been guaranteed from local supply as Nigeria produces and exports oil.

However, the population of the country which is almost 100 million continues to increase and the ongoing rapid industrialization imply that the power demand would also increase tremendously. Unfortunately, new oil wells are seldom discovered and the quantity of oil from existing sources is depleting rapidly. To prevent an energy schism in the country, it is thus essential to diversify the power production techniques. The petroleum resource can also be reserved for premium purposes.

Initially hydro power seemed attractive. However, due to incessant national hydrological problems other alternative power production methods have to be assessed and developed. Solar and wind power are considered because of the inherent advantages associated with these resources. Also, over 70% of the population lives in the rural and isolated areas so that alternative technologies can easily be adapted for local applications.

Wind energy has received little attention in Nigeria. However solar power has been developed for low grade energy applications such as hot water heating (Ref.1), drying of agro materials (Ref.2) and water purification (Ref.3).

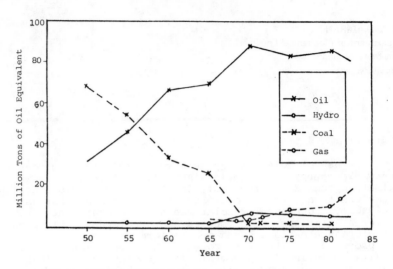

FIG. 1:   CONSUMPTION OF PRIMARY ENERGY IN NIGERIA

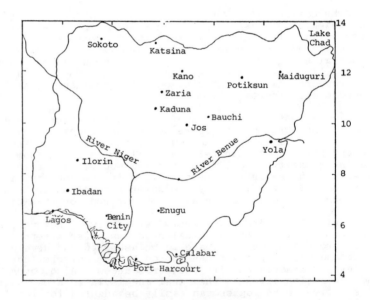

FIG. 2:   MAP OF NIGERIA SHOWING SOME STATIONS MEASURING
METEOROLOGICAL DATA

These devices and appliances are usually located at research
stations and institutions where radiation data are fairly
measured. Oftentimes, these locations are not the sites for
possible solar power stations.

Thus, it is essential to develop correlation relationships
which can be used with some degree of confidence to predict
solar radiation for any particular location where solar driven
artefacts are to be installed.

Solar radiation data using modern pyranometers are very
scarce in Nigeria. However, there are over 75 stations measuring
agrometeorological data. In these stations, the radiation and
the sunshine hours are measured by the Gunn Bellani Integrator
and the Campbell-Stokes recorder respectively. These measurements are adjudged adequate for assessing the solar energy
potentials in the country. In this paper, agrometeorological
data recorded for a period of between three and twelve years for
sixteen locations, (see fig.2) are averaged to obtain correlation
expressions.

2. CORRELATIONS FOR THE GLOBAL SOLAR IRRADIATION.

There are two methods that can be used to predict the
solar irradiation namely the use of the duration of sunshine
hours and opaque cloud cover. The latter model is not adopted
in this study because cloud cover measurements are not recorded
by most of the Nigerian meteorological stations. Also, it has
been suggested that its use in predicting long term solar
irradiation is susceptible to several errors (Ref.4). The
measurements of sunshine hours started several years ago and
it is reliable. Hence, its use in this investigation.

Angstrom (Ref.5) was the earliest climatological researcher
to propose expressions for predicting solar irradiation. The
simple correlation between the monthly average irradiance
received on an horizontal surface and the duration of sunshine hours
is of the form:

$$\frac{Q_\ell}{Q_m} = a+b \, \frac{S_\ell}{S_m} \qquad\qquad --------- \qquad (1)$$

where a and b are constants whose values depend on both the
latitude and some other meteorological variables (e.g.
ambient temperature and relative humidity). Following the work
of Angstrom, many researchers have investigated the availability
of solar resource in various countries (Ref. 6 and 7). For
example, Black et. al. (Ref.8) studied the solar radiation and
the sunshine hours for 32 stations located between Latitude
6.3^OS to 7.0^OS and 18.5^ON to 64.8^ON. A general linear expression similar to equation (1) above was obtained. The values
of a and b were 0.23 and 0.48 respectively.

Also, Ezeilo (Ref.9) proposed a correlation equation to predict the solar radiation for 2 cities in Nigeria. The values of a and b were respectively 0.21 and 0.49. Application of this equation to other locations have not however been tested.

The application of equation (1) depends on the availability of the values of the four variables. Q_ℓ, Q_m, S_ℓ and S_m. The theoretical values of Q_m and S_m for any geographical location have been derived elsewhere (Ref. 4). The extraterrestrial radiation on a horizontal surface can be computed from

$$I = \int E \, I_{sc} Sin\alpha dt \qquad \text{---------} \qquad (2)$$

where

$$Sin \, \alpha = sir\beta \, sin\delta + cos\beta \, cos\delta \, cos \, t_s$$

and

$$E = 1+0.034 \, cos(\frac{360n}{365})$$

The total daily irradiation is also given by

$$I_{Td} = \frac{2I}{\Omega} \qquad \text{---------} \qquad (3)$$

where Ω = angular velocity of the earth

$$= \frac{\pi}{12} \, rad/hr.$$

Integrating equation (2) over the sunshine period, and subtituting into equation (3), we have

$$I_{Td} = \frac{24}{\pi} I_{sc} [cos\beta cos\delta sin t_{sr} + t_{sr} \, sin\beta sin\delta]E \text{ ---} \quad (4)$$

but at sunrise α = o, so that

$$cos t_{sr} = - tan\beta tan\delta \qquad \text{----------} \qquad (5)$$

or $t_{sr} = cos^{-1}[-tan\beta tan\delta] \qquad \text{----------} \qquad (6)$

Substituting equation (5) into (4) gives:

$$I_{Td} = 7.64 I_{sc} E \, sin\beta sin\delta[t_{sr} - tan \, t_{sr}] \qquad \text{------} \qquad (7)$$

Thus, the average daily extraterrestrial radiation can easily be computed for each month. Also, from equation (6), the maximum sunshine hours is

$$S_m = \frac{2}{15} cos^{-1}[-tan\beta tan\delta] \qquad \text{--------} \qquad (8)$$

3. RESULTS AND DISCUSSION

Even when the solar radiation data are not measured, the duration of sunshine hours have been recorded in many Agromet stations in Nigeria. The sunrise and sunset tables (Ref. 10) show that the maximum duration of sunshine varies from eleven to thirteen and a half hours depending on the month of year and other meteorological variables.

The mean monthly solar irradiation for the sixteen stations are presented in Table I and Figs 3, 4 and 5. The curves in these figures are similar. However, the values of radiation appears to be highest in the northern stations which are close to the Sahel desert. For each curve, the radiation increases from January until it reaches a maximum value around March and then decreases to a minimum level around July. It increases again till November and drops off in December.

The climatological observations corroborate these measurements. There are basically two seasons in Nigeria namely, the dry and wet (rainy) seasons. The rainy season extends from around April to October and the dry, (Sunny) season is from November to March. Usually, cold, dry harmattan winds blow across the country between December and February causing a slight reduction in the intensity of solar radiation.

The mean annual insolation for all the stations used in this study is approximately $20MJ/m^2$-day. The intensity of solar radiation at all the stations is adequate to operate solar driven appliances and for power supply by photovoltaic methods. For example, solar devices can be installed in isolated villages which have no access to the national power grid system. The use of solar power is desirable because hitherto, availability and provision of conventional energy even for urban dwellers is reaching a crisis level in many developing countries including Nigeria. In the rural areas, the problem is more acute as they also lack other essential utilities.

A linear regression analysis was carried out on the data for each station by using an IBM computer. The values of the correlation constants a and b are depicted in Table II These values are in agreement with published data. For example, Duffie and Beckman (11) obtained a value of 0.22 and 0.50 for a and b respectively for a station in Massachusets (L 42°N) which experiences rainfall throughout the year. This station is comparable to some stations in the Southern part of Nigeria. The results of Ezeilo (Ref. 9) for Nsukka (L $6^{\circ}54'$N) also falls within the present range of results.

The variation of Q' with respect to S' is shown graphically in figure 6. Even though a linear relationship produced the best fit, it was observed that the use of a single solar radiation model for the whole country produced large

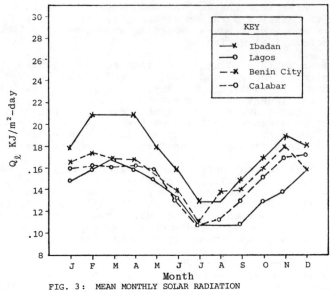

FIG. 3 : MEAN MONTHLY SOLAR RADIATION

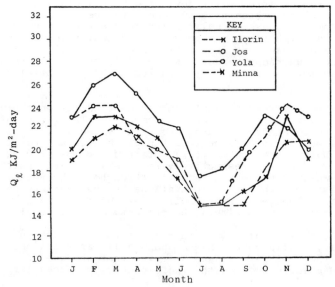

FIG. 4 : MEAN MONTHLY SOLAR RADIATION

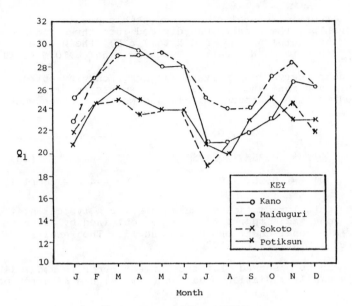

FIG. 5: MEAN MONTHLY SOLAR RADIATION

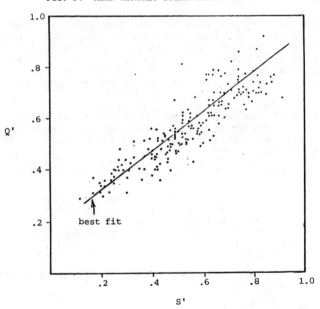

FIG. 6: THE EFFECT OF NORMALISED SUNSHING HOURS
ON NON-DIMENSIONAL SOLAR IRRADIATION
FOR SOME NIGERIAN STATIONS

errors of up to 25%.

In order to determine more acceptable prediction relationships, the country was divided into three zones namely: the Southern zone (L 4^ON to 7.5^ON), the Middle zone (L 7.5^ON to 10^ON) and the Northern zone (L 10^ON to 14^ON).

Linear regression analysis was again carried out on the data for each zone and the following expressions were obtained for the Southern, Middle and Northern zones respectively:

$$Q' = 0.228 + 0.56S' \qquad \text{-------} \qquad (9)$$

$$Q' = 0.235 + 0.49S' \qquad \text{-------} \qquad (10)$$

$$Q' = 0.251 + 0.60S' \qquad \text{-------} \qquad (11)$$

The actual radiation measurements for a typical station in each zone was compared with values obtained by using the appropriate zonal equations 9, 10 and 11. The results are presented in Tables III, IV and V.

It can be observed that the largest error is about 14% Ezeilo's single radiation equation (Ref. 9) has an error of ±30%.

4. CONCLUSION

The mean monthly solar radiation from some stations have been analysed. The use of a single equation based on sunshine hours to predict radiation levels at all points in Nigeria has been found to lead to large errors. Thus the country was divided into three zones and an equation was proposed for each zone. The results of these equations are reliable.

Therefore it can be concluded that the total daily insolation on a horizontal surface located at a known latitude can be estimated fairly readily once the duration of sunshine is available.

5.   NOMENCLATURE

a	=	a constant in the regression equation
b	=	a constant in the regression equation.
E	=	Orbital eccentricity factor.
I	=	The daily intensity of extraterrestrial radiation, kw/m^2.
I_{sc}	=	The solar constant, kw/m^2.
I_{Td}	=	The total daily intensity of extraterrestrial radiation Kw/m^2.

TABLE I: MEAN MONTHLY SOLAR RADIATION (MJ/m^2-day)

Stations Latitude	Bauchi $10^O 1'N$	Enugu $6^O 27'N$	Port Harcourt $4^O 51^O N$	Zaria $11^O .0'N$
Month				
Jan.	22.0	18.0	17.0	22.0
Feb.	26.0	20.0	17.0	24.5
Mar.	26.0	20.0	18.0	24.0
Apr.	20.0	20.0	13.5	20.0
May	23.0	20.0	16.5	23.0
June	19.0	17.5	14.0	19.0
July	15.0	15.0	12.0	16.0
Aug.	20.0	15.0	13.0	15.0
Sept.	23.0	16.0	14.0	22.0
Oct.	24.5	18.0	16.0	22.0
Nov.	23.0	19.0	17.0	23.0
Dec.	23.0	19.5	17.0	21.0
Mean Annual Radiation	18.2	22.0	16.2	21.0

TABLE II: ESTIMATION OF THE CONSTANTS a and b IN THE REGRESSION
EQUATION, Q' = a+bS' FOR VARIOUS STATIONS IN NIGERIA.

Location	Latitude	a	b	Correlation coefficient
Bauchi	$10^{\circ}01'$	0.14	0.75	0.91
Benin City	$6^{\circ}19'$	0.22	0.50	0.86
Calabar	$4^{\circ}58'$	0.23	0.52	0.90
Enugu	$6^{\circ}27'$	0.21	0.62	0.95
Ibadan	$7^{\circ}26'$	0.25	0.58	0.91
Ilorin	$8^{\circ}29'$	0.21	0.49	0.80
Jos	$9^{\circ}52'$	0.13	0.52	0.92
Kano	$12^{\circ}03'$	0.19	0.61	0.91
Lagos	$6^{\circ}20'$	0.21	0.49	0.80
Maiduguri	$11^{\circ}51'$	0.45	0.46	0.79
Minna	$9^{\circ}37'$	0.28	0.47	0.83
Port Harcourt	$4^{\circ}51'$	0.25	0.63	0.89
Potiksun	$11^{\circ}43'$	0.33	0.49	0.85
Sokoto	$13^{\circ}01'$	0.23	0.60	0.83
Yola	$9^{\circ}14'$	0.32	0.48	0.87
Zaria	$11^{\circ}.00$	0.17	0.68	0.89

TABLE III: COMPARISON OF MEASURED AND PREDICTED SOLAR RADIATION
(SOUTHERN ZONE)

STATION: LAGOS

LATITUDE: $6^{\circ}20'N$

Month	Measured Radiation MJ/m²–day	Predicted Radiation MJ/m²–day	% Error based on Measured value
Jan	15.3	16.8	9.8
Feb.	16.6	17.9	7.8
Mar	17.0	17.0	0
Apr.	16.3	18.5	13.5
May	14.9	16.3	9.4
June	13.0	14.5	11.5
July	11.1	12.2	9.9
Aug.	11.1	12.0	8.1
Sept.	11.5	12.8	11.3
Oct.	12.9	14.1	8.5
Nov.	14.3	16.2	13.3
Dec.	16.3	17.4	6.7

TABLE IV: COMPARISON OF MEASURED AND PREDICTED RADIATION (MIDDLE ZONE).

STATION: ILORIN

LATITUDE: $8^O 29'$

Month	Measured Radiation MJ/m^2 –day	Predicted Radiation MJ/m^2–day	% Error based on Measured value
Jan.	19.0	20.4	7.4
Feb.	21.3	21.9	2.8
Mar	22.1	20.8	5.9
Apr.	20.4	20.8	2.0
May	20.8	21.4	2.9
June	17.6	19.4	10.2
July	14.9	16.6	11.4
Aug.	14.9	14.5	2.7
Sept.	14.9	15.4	3.4
Oct.	18.3	18.3	0
Nov.	20.6	20.8	1.0
Dec.	20.4	19.9	2.5

TABLE V: COMPARISON OF MEASURED AND PREDICTED SOLAR RADIATION
(NORTHERN ZONE).

STATION: Maiduguri

LATITUDE: $11^O 51^O N$

Month	Measured Radiation MJ/m^2-day	Predicted Radiation MJ/m^2-day	% Error based on measured value
Jan	22.8	21.6	5.3
Feb.	27.1	27.4	1.1
Mar	28.8	25.4	11.8
Apr.	29.2	26.4	9.6
May	29.2	25.8	11.6
June	28.1	24.8	11.7
July	24.7	21.9	11.3
Aug.	23.8	20.8	12.6
Sept.	24.0	21.6	10.0
Oct.	26.7	24.9	6.7
Nov.	28.5	25.4	10.9
Dec.	26.0	23.8	8.5

n	=	number of days beginning from Jan. 1.
Q_ℓ	=	mean monthly horizontal terrestrial radiation. MJ/m^2-day.
Q_m	=	the mean monthly horizontal extraterrestrial radiation, MJ/m^2-day.
Q'	=	the non-dimensional solar radiation i.e. Q_ℓ/Qm
S_ℓ	=	the mean monthly duration of sunshine, hours
S_m	=	the mean maximum monthly duration of sunshine, hours.
S'	=	the non-dimensional sunshine hours i.e. S_ℓ/S_m
t	=	the hour angle.
t_{sr}	=	sun rise period in hours.
α	=	the solar altitude
β	=	Latitude
δ	=	declination.

Acknowledgements

The author would like to express his gratitude to Dr S. O. Ojo for assistance rendered in the Computer Laboratory of his department.

REFERENCES

[1] Ogunnoiki, K. A.
"Design and Construction of a Solar Hot Water System. Thesis, Mechanical Engrg Dept., University of Ibadan 1988.

[2] Ofi, O.
"Construction and Evaluation of a Solar Dryer" Nigerian Journal of Solar Energy, Vol 2, pg 47, 1982.

[3] Awoniyi, J. A. et al:
Solar Energy Research in Mechanical Engineering Dept; Ahmadu Bello University Zaria." Nigerian Journal of Solar Energy Vol 1, No 1, pg 81, 1980.

[4] Kreith, F and Kreider J.F.
"Principles of Solar Engineering", Mc-Graw Hill Book Company, New York 1978.

[5] Angstrom, A.K.
"Solar and Terrestrial Radiation" Quarterly Journal of Royal Meteorological Society 50, 121, 1924.

[6]   Lewis, G.
        "Irradiance Estimates for Zambia" Solar Energy
        Vol 26, pg 81 No. 1, 1981.

[7]   Chuah, D.G.S. and Lee, S.L.:
        "Solar Radiation Estimates in Malaysia" Solar
        Energy, Vol. 26 No.1 pg. 33 1981.

[8]   Black J.N. et al:
        Solar Radiation and the Duration of sunshine".
        Quarterly Journal of Royal Meteorological Society.
        80, pg 231, 1954.

[9]   Ezeilo, C.C.O.:
        "Solar Radiation Measurements; The Nigerian Experience"
        Proceedings of National Symposium on Solar Energy
        pg 42, 1982.

[10]   Agromet Bulletin published by the Meteorological
        Department, Lagos.

[11]   Duffie, J.A. and Beckman, W.A.:
        "Solar Engineering of Thermal Processes "John Wiley,
        New York, 1980.

An Investigation of Aircooled Condenser with Mathematical Simulation Method

CHOW CHI CHIN and VAN CHANG
Shanghai Institute of Mechanical Engineering
Jun Gong Road
Shanghai, PRC

Abstract

In this paper a mathematical simulation model of aircooled condenser has been given. The model is established by the local analysis method and ε-Ntu method. It has been used to calculate the heat transfer performance and pressure drop of refrigerant R12 in aircooled condensers with different tube set connection types. The calculated results of the mathematical model fairly agree with the experimental results. Five condensers with different tube set connection types have been studied by the mathematical simulation method. The calculated results show that, for "three-row" aircooled condensers, the condenser with "2 way to 1 way" tube set connection type has better heat transfer performance than others.

1, INTRODUCTION

Recently, as the aircooled condensers are widely used in the refrigerating, chemical, cryogenic and other engineering, the problem of enhancing heat transfer in the aircooled condenser becomes more interesting by the people. In the condensing process the thermal parameters (p,v,t,h) are varied in a wide range. For the convenience of calculation the mean parameters method is widely used in the engineering design, but this method can't reflect the real heat transfer process in the aircooled condenser. The mathematical simulating method divides the condenser into several differential units and analyzes the local parameters, this method is applicable to investigation of the heat transfer process in the aircooled condenser in which the thermal parameters are varied in a wide range. Its advantage does not require the experiment of the prototype, on the other hand, this method can also simulate the heat transfer process under the conditions deviated from the design condition.

The local parameters which can't be measured in the experiment can also be calculated. The aircooled condensers of different flow pattern, different tube set connection can be compared by the mathematical simulating method. Because the results can be quickly calculated by the simulating method, less labor is needed. This method, comparatively speaking, is more economical.

The accuracy of the mathematical simulating method depends on the established mathematical model. Some mathematical models of the aircooled condensers have been given by several scholars. N.K. Anand and P.R. Tree [1] investigated a single aircooled condensing tube with plate fin on the air side. He divided the condensing tube into differential heat transfer units by one pitch length. The differential conservation equations in semi-indistinct form were solved by alternate and numerical methods. J.A.R.Parise and W.C.Cartwright[2] adopted the heat transfer units which contain several fin pitches,and introduced Bake's two-phase flow pattern diagram. He adopted different correlations of vapor condensation in the tube to accomplish the mathematical simulation of the aircooled condenser. Raymond,D.Ellision and Fredrick A.Cresuick[3] adopted the method which traces thecomplicate refrigerent circuit, for every heat transfer tube and they used ε-Ntu method to calculate the aircooled condenser with complicate circuit. Furthermore,Soviet scholars N.K.Saveskee[4] and S,D.Goldstein etc. also investigated the simulating calculation of the aircooled condensers in recent years.

This paper based on the former investigated results gives a new mathematical model of the aircooled condenser with local analyses and ε-Ntu method. This model gives fair accuracy and high calculating speed. With this mathematical model we have calculated five aircooled condensers with different tube set connections(figure 1). They are cross flow condenser with 3 parallel-way connection(model 1),cross flow condenser with "2way to 1 way" connection(model 2), counter parallel mixed flow condenser with 3 parallel-way connection(model 3),counter parallel mixed flow condenser with 2 parallel-way connection(model 4),counter parallel mixed flow condenser with 1 way connection(model 5).

2,THE ESTABLISHMENT OF THE MATHEMATICAL MODEL

Basic Hypotheses

The basic hypotheses of the mathematical model are: the inlet parameters of refrigerant vapor and air remain homogeneous and constant;the processing quality of the benders and welding quality of the tube connections are all the same;the mass flow rate in different tube sets are equal;no uncondensed vapor and lubricated oil are contained in the condensing vapor;the heat loss from condenser to ambient air has been neglected; the air is unmixed between the calculating units;the axial heat conduction is neglected;the tube in the subcooled region is full of condensed liquid.

The above hypotheses are almost practical. For example,as the convective term of the air is greatly larger than the diffuser term of the air by the partition of the fin,we may consider that the unmixed model agrees with the practical condition. The heat conduction along the tube wall is much less than the radial heat conduction in the wall,so that the axail heat conduction may be neglected. As these hypotheses are reasonable,the calculated results of the mathematical model have been confirmed by the experiments.

Calculating Method

In order to consider the inhomogeneity of the thermal parameters in the aircooled condenser,we divided the aircooled condenser into several differential basic heat transfer units. The basic heat transfer unit is a part of the condensing tube,the length of unit is variable. Each heat transfer unit contains a tube section and several fins(figure 2). For each heat transfer unit the outlet parameters may be calculated by ε-Ntu method. With ε-Ntu method the alternate time consumption may be decreased and the stability of the calculation increases.

ε -Ntu method is:

the heat transfer coefficient of unit condensing tube may be calculated by equation(1)

$$\frac{1}{Kofo}=\frac{1}{\alpha_i fi}+\frac{1}{2\pi}(\frac{\ln\frac{dm}{di}}{\lambda_{cu}}+\frac{\ln\frac{do}{dm}}{\lambda_{al}})+\frac{Rc}{\pi dm}+\frac{Ri}{fi}+\frac{Ro}{fo}+\frac{1}{\alpha_o fo\eta_s} \tag{1}$$

the thermal resistance of unit length condensing tube is

$$R_w'= 1/Kofo \tag{2}$$

the thermal resistance of thermal unit (Δx) is

$$Rw=1/Kofo\Delta x=R_w'/\Delta x=1/UA \tag{3}$$

the heat capacity of air is

$$Ca=ma\cdot Cpa \tag{4}$$

the heat capacity of the refrigerant in the single phase region is

$$Cr=mr\cdot Cpr \tag{5}$$

Number of thermal units are

$$Ntu=UA/Cmin \tag{6}$$

The minimum heat capacity Cmin is the smaller of Ca and Cr.

The heat transfer effectiveness in the single phase flow region is

$$\varepsilon =1-e^{-Ntu} \tag{7}$$

The heat transfer effectiveness in the single phase flow region depends on the heat capacity of air and refrigerant:

When Ca> Cr

$$\Gamma =1-e^{-Ntu}Cr/Ca \tag{8}$$

$$\varepsilon =1-e^{-\Gamma Ca/Cr} \tag{9}$$

When Cr> Ca

$$\Gamma =e^{-Ntu} \tag{10}$$

$$\varepsilon =Cr/Ca(1-e^{-\Gamma Ca/Cr}) \tag{11}$$

The heat transfer of basic thermal unit is

$$Q= \varepsilon\cdot Cmin (Trin-Tain) \tag{12}$$

The outlet temperature of air is

$$Ta=Tain+Q/Ca \tag{13}$$

633

The outlet temperature and outlet quality of refrigerant are as follows:
in the single phase region,

$$Tro=Trin-Q/Cr \qquad (14)$$

in the two phase region,

$$Tro=Tsat \qquad (15)$$

$$xo=xin-Q/mr \cdot r \qquad (16)$$

According to the heat balance equation the inner wall temperature of the tube is

$$Twi=\frac{Q}{\frac{1}{\alpha_i fi\Delta x}+\frac{Ri}{fi\Delta x}} +Trm \qquad (17)$$

The Calculation of Thermal and Thermal Physical Properties of the Refrigerants

The thermal properties of R12 used in the mathematical model are calculated by Matin and Hou general equation. The thermal physical properties of R12 are calculated by the calculated method given by Soviet literature (7).

The Calculation of Convective Film Coefficient in Refrigerant Side

The local convective film coefficients of vapor condensation have been investigated by most scholars. Many empirical corrlations have been given in the literature[8], [9].

The comparision analyses between literature[9] and experimental results show that the calculated results by Shah Relation fairly agree with experimental results, but in the low Renolds number region Shah Relation must be corrected. J.A.Tichy[10] corrected Shah Relation in the low Renolds number region, but in the Re< 4500 region the tendency of the corrected term is changed. This paper gives the corrected factors inRe < 4500 region. The corrected factors make the tendency of local convective film coefficient unchanged in the low Renolds number region.

The corrected term is

$$Nux=0.023 \ Rel^{0.8}Prl^{0.4}\left[\frac{(1-x)^{0.8}+3.8x^{0.76}(1-x)^{0.04}}{(\frac{ps}{pcr})^{0.38}}\right]A \qquad (18)$$

$$A=4.754 \ (\frac{Rel}{1000})^{-0.712114} \qquad 1500 < Rel< 10^{4}$$

$$A=1 \qquad Rel > 10^{4}$$

other parameters are in the following range:

$$7 < di < 40mm, \ 0 < x < 1, \ 0.002 < Pr < 0.44$$
$$10.833 < Gr < 210 \ Kg/m^{2}s, \ 21^{\circ}C< Tsat < 310^{\circ}C,$$
$$158 < q < 1893000W/m^{2}$$
$$3 < vg < 300 \ m/s, \ 1 < Prl < 13$$

Shah and Corrected Shah Relation are applicable to various flow patterns.

G.C.Hiller and C.R.Glickman's Correlation is adopted to calculate the convective film coefficient in the single phase region,

in the laminar flow region, $Re < 3500$

$$St\ Pr^{2/3} = 1.1064\ Re^{-0.78992} \tag{19}$$

in the transitional flow region, $3500 < Re < 6000$

$$St\ Pr^{2/3} = 3.5194 \times 10^{-7} Re^{1.03804} \tag{20}$$

in the turbulent flow region, $Re > 6000$

$$St\ Pr^{2/3} = 0.018\ Re^{-0.1375} \tag{21}$$

The Calculation of the Convective Film Coefficient on the Air Side

The convective film coefficients on the air side of the flat finned tube aircooled condenser have been measured in the laboratory of Shanghai Institute of Mechanical Engineering (SIME).

The following equation has been correlated by the experimental data [11].

$$\alpha_0 = 27.4144\ Wmax^{0.53773} \tag{22}$$

The air velocity in the minimum area is in the range

$$2.5 < Wmax < 6.8\ m/s$$

The Calculation of Flow Resistance in the Refrigerant Side

The friction coefficient in the single phase flow region can be calculated by Blasius Equation:

$$f = 0.3164\ Re^{-0.25} \tag{23}$$

The pressure drop is

$$\frac{dp}{dz} = -\frac{f}{d}\ \frac{Gr^2}{2\rho}$$

The friction coefficient in the two-phase flow region can be calculated by Lockhart-Martinelli method [12], this method is applicable to the mass flow rate in the range

$$Gr < 1360\ Kg/m^2 s$$

The pressure drop is

$$\frac{dp}{dZ} = \phi_l^2 \left(-\frac{dp}{dZ}\right)_l = \phi_l^2 f1 \frac{Gr^2 (1-x)^2}{2\rho_l d} \tag{24}$$

where

$$\phi_l^2 = \frac{\frac{dp}{dZ}}{\left(\frac{dp}{dZ}\right)_l} = 1 + \frac{C}{\chi} + \frac{1}{\chi^2}$$

coefficient C list in the Table i, where χ is Lockhart-Martinelli parameter

$$\chi^2 = \frac{(\frac{dp}{dZ})_1}{(\frac{dp}{dZ})_g} = \frac{f_1}{f \cdot g} \frac{(1-x)^2}{x^2} \frac{\rho_g}{\rho_1}$$

The pressure drop of the single phase flow passing through the bender is

$$\Delta pb = \zeta \frac{Gr^2}{2\rho} \qquad\qquad (25)$$

$$\zeta = 2(0.131+0.16 \, d/R)^{3.5}$$

The pressure drop of the two-phase flow passing through the the bender may be calculated by Geshom method (12)

$$\frac{\Delta P_b}{\Delta P_{b1}} = 1 + (\frac{\rho_1}{\rho_g} -1)(\frac{2}{\zeta}x(1-x)\Delta(\frac{1}{s}) +x) \qquad (26)$$

where

$$\Delta(\frac{1}{s}) = \frac{1.1}{2+R/d}$$

Δ pb is the pressure drop of two-phase flow passing through the bender supposing that the mass flow rate of two-phase flow is equal to the mass flow rate of liquid phase.

Calculating Procedure

First of all, divide the aircooled condenser into several thermal units, each thermal unit should correspond with I,J,K coordinate and then change the calculating order of I,J,K, finally, the mathematical simulation of aircooled condenser with different tube set connection can be made. The division of thermal units in model 2 is shown in figure 3.

The calculating order is made for the thermal units along the condensing refrigerant. Supposing that the refrigerant mass flow rate of tube sets are equal, the inlet refrigerant parameters of inlet thermal units are equal to the inlet refrigerant parameters of the aircooled condenser. For each thermal unit the outlet parameters can be calculated by ε-Ntu method. The outlet parameters of refrigerant for the former unit is used as the inlet parameters for the later unit. The average value of inlet and outlet parameters for the thermal units may be taken as the reference parameters of air and refrigerant physical properties. As the inlet and outlet parameters are unknown before calculation, several alternate steads must be made. The mixed parameters of individual tube sets are taken as the starting parameters of the gathering tube set. In the case of aircooled condenser with parallel tube set connection, the mixed parameters are taken as the outlet parameters. The flow chart of "several parallel-way to one way" tube set connection is shown in figure 4.

3, THE CALCULATION RESULTS

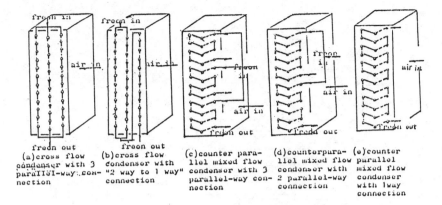

(a)cross flow
condenser with 3
parallel-way con-
nection

(b)cross flow
condenser with
"2 way to 1 way"
connection

(c)counter para-
llel mixed flow
condenser with 3
parallel-way con-
nection

(d)counterpara-
llel mixed flow
condenser with
2 parallel-way
connection

(e)counter
parallel
mixed flow
condenser
with 1way
connection

Figure 1. 5 models with different tube set connections

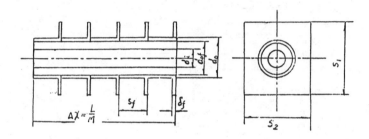

Figure 2. Thermal unit

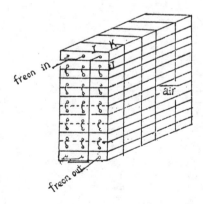

Figure 3. Division of thermal units

637

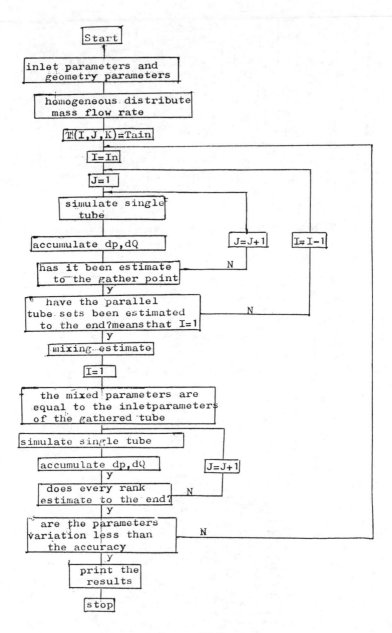

figure 4. Flow chart of several parallel ways
gather to one way tube set connection

638

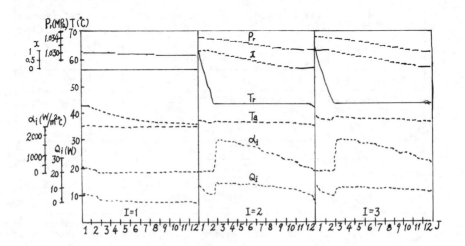

figure 5. The distribution of local parameters
in the model

TABLE 1 THE VALUE OF C IN EQUATION(12)

Rel > 1000 Reg > 1000	Rel ≤ 1000 Reg > 1000	Rel > 1000 Reg ≤ 1000	Rel ≤ 1000 Reg ≤ 1000
20	12	10	5

TABLE 2. THE CHARACTERISTIC PARAMETERS
OF SIMULATION CONDITION

Inlet air pressure Pai mmH₂O	air humidity da g/Kg	face velocity of air Wy m/s	inlet refrigerant pressure P_r kPa
10500	2	2.5	1034.28
mass flow rate of refrigerant mr Kg/h	inlet refrigerant temperature Tri C	heat transfer area Fl m²	refrigerant
25.2	65	6.699	R12

TABLE 3 . THE GEOMETRY PARAMETERS
MODEL 1 AND 2

geometry parameters model	outside diameter dm mm	inside diameter di mm	fin thickness δ_f mm	longitudinal tube spacing s2 mm
1	10.3	9.3	0.2	22.52
2	10.3	9.3	0.2	22.52

transverse tube spacing s1 mm	arrangment	fin pitch sf mm	face area A_f m²
26.0	staggered	2.53	0.333×0.4
26.0	staggered	2.53	0.333×0.4

In addition to the calculation of the average parameters
of the aircooled condenser,the local parameters such as the local
refrigerant temperature,pressure vapor quality,inner wall temperature,air temperature and convective film coefficients in the
refrigerant and the air side can also be calculated. The local
parameters along the tube length in different tube rows are shown
in figure 5.The characteristic parameters of the simulation condition in model 2 are shown in Table 2.

4, VERIFICATION OF MODELS

In order to verify the accuracy of the mathematical models,
the calculated results have been compared with the experimental
results of model 1 and 2.The geometry parameters of the experimental model 1 and 2 are shown in Table 3.
For model 1   22runs data have been compared with the calculated results which indicate that,the mean root deviation of
average outlet air temperature is $0.0328°C$,the mean root deviation
of total heat transfer is 7.23W. The average outlet air temperature and the total heat transfer of the calculated results are
fairly in agreement with the experimental data. The deviation of
pressure drop and outlet temperature of refrigerant between the
calculated and experimental results are a bit greater. The mean
root deviation of refrigerant outlet temperature in model 1 is
$0.43°C$ and in model 2 is $1.12°C$.
The above analyses show that, the mathematical models of
aircooled condensers have been verified in a wide experimental
range. This model may be used to investigate the heat transfer
in aircooled condensers. It is an effective tool to design,simu-
late and optimize the refrigerating,air conditioning and heat
pump systems.

5, DISCUSSION

The mathematical model established in this paper bases on
the same flow rate in different tube sets. If the inhomogeneous
distribution of refrigerant must be considered,the mass flow rate
in individual tube set will be corrected by the flow resistance
distribution. Reference (4) shows that,for the "few-row"(<4 rows)
aircooled condensers which are widely used in the refrigerating
engineering the effect of inhomogeneous distribution may be neglected.
For the five aircooled condensers with different tube set
connection types under the condition listed in Table 2 have been
investigated by the mathematical simulation method,the calculated
results show that,model 2 has the best heat transfer performance,
model 1 is the worst,models 1 and 3 are the same,model 4 is better
than 3,model 5 has the heat transfer performance between model 1,
2 and 4, but its flow resistance increases rapidly. By means of
simulating calculation and experimental investigation, the heat
transfer coefficient of model 2 is higher than that of model 1
by over 10% (13).
The correlation of convective film coefficient for vapor
condensation in the tube given by D.P.Traviss,etc are only appreciable to the annular flow rate range. The correlation given by
this paper(equation 18)can be used to calculate the local convective film coefficient under the low mass flow rate range.

6, CONCLUSION

The present paper offers a mathematical model of aircooled condenser by local analyses and ε-Ntu method. The model has been verified in a wide experimental range, it may be used to investigate the heat transfer in the aircooled condenser, it becomes an effective tool to design, simulate and optimize the refrigerating, air conditioning and heat pump systems.

The calculating results of five 3-row aircooled condensers with different tube set connection type show that, the "2-way to 1 way" tube set connection type is the best type, under the same inlet air temperature and condensing temperature its heat transfer coefficient is 10% higher than that of three-parallel tube set connection type, so that it should be recommended to the public for the production of aircooled condensers.

7, NOMENCLATURE

Ca	heat capacity of air, W/K
Cr	heat capacity of refrigerant, W/K
Cmin	the smaller of Ca and Cr
Cpa	specific heat of air, J/Kg.K
Cpr	specific heat of refrigerant, J/Kg.K
di	inside diameter of heat transfer tube, m
dm	outside diameter of heat transfer tube, m
do	outside diameter of finned tube, m
fi	inner surface area per unit length of thermal units, m^2/m
fo	outer surface area per unit length of thermal units, m^2/m
Gr	mass flow velocity of refrigerant, $Kg/m^2 s$
Ko	heat transfer coefficient for unit outer surface, W/m^2K
ma	mass flow rate of air, Kg/s
mr	mass flow rate of refrigerant, Kg/s
Nux	local Nusselt number
Ntu	number of thermal units
ps	saturated pressure of refrigerant, Pa
pcr	critical pressure of refrigerant, Pa
pr	reduced pressure
Pr	Prandtl number
Prl	liquid Prandtl number
Q	the rate of heat transfer, W
q	the rate of heat transfer per unit area, W/m^2
Re	Reynolds number
Rel	Reynolds number supposing all liquid mass flow rate
r	latent heat of refrigerant, J/Kg
R	radius of bender, m
Rc	contact thermal resistance, $m^2 K/W$
Ri	thermal resistance of scale in the tube, $m^2 K/W$
Ro	thermal resistance of scale on the tube, $m^2 K/W$
Rw'	thermal resistance per unit tube length, m.K/W
Rw	thermal resistance per the length Δx, m.K/W
St	Stanton number
Tro	outlet temperature of refrigerant, $^\circ C$
Trin	inlet temperature of refrigernt, $^\circ C$
Tao	outlet temperature of air, $^\circ C$
Trm	average refrigerant temperature, $^\circ C$
Tsat	saturated temperature of refrigerant, $^\circ C$
UA	the product of heat transfer coefficient and area, W/K

```
vg        velocity of vapor phase,m/s
Umax      air velocity at the minimum area,m/s
χ         Lockhart-Martinelli parameter
x         vapor quality
xo        outlet vapor quality
xin       inlet vapor quality
Z         axial distance from the starting point
Δx        length of thermal unit,m
Δpb       pressure drop of the bender,Pa
Δpbl      pressure drop of the bender,supposing all liquid mass flow,
          Pa
ρl        density of liquid,m³/Kg
ρg        density of gas,m³/Kg
ηs        surface efficiency of the fin
ζ         local drag coefficient
ε         efficiency of thermal unit
αi        convective film coefficient on the refrigerant side,W/m²K
αo        convective film coefficient on the air side,W/m²K
fg        friction coefficient of gas
fl        friction coefficient of liquid
λAl       thermal conductivity of aluminum,W/m.K
λcu       thermal conductivity of copper,W/m.K
```

8, REFERENCES

[1], N.K.Anand,D.R.Tree   "Steady State Simulation of a Single
     Tube Finned Condenser Heat Exchanger",ASHRAE Transaction,
     1982,Vol 88,Part II .p185

[2], .R.Parise,W.C.Cartwright, "Local Analysis of Three-Dimen-
     sional Aircooled Condenser Using A Two-Phase Flow Diagram".

[3], Raymond D.Ellison,Frederick A.Cresuick,etc   "A Computer
     Model For Air Cooled Condensers with Specified Refrigerant
     Circuiting",ASHRAE Transaction,1981,Vol 87, Part I

[4], И.К.САВИЦКИЙ, "РАСЧЕТНО-ТЕОРЕТИЧЕСКОЕ ИССЛЕДОВАНИЕ ВОЗДУШ-
     НЫХ КОНДЕНСАТОРОВ С РАЗЛИЧНЫМИ ДВИЖЕНИЯ ХЛАДАГЕНТА И
     ВОЗДУХА", «Холодильная Техника» 1986, №9

[5], S.D.Goldstein,etc   "A Mathematically Computer Analysis
     of Plate-Fin Heat Exchanger",ASHRAE Transaction,1983.Vol89.
     Part II

[6], Kays,W.M.London.A.L. "Compact Heat Exchanger", M.Graw
     Hill, 1964

[7], МАШИН И ТЕПЛОВЫХ НАСОСОВ   "ТЕРМОДИНАМИЧЕСКИЕ И ТЕПЛОФИ-
     ЗИЧЕСКИЕ СВОЙСТВА"

[8], 藤井哲,本田博司,野津滋, "フロン系冷媒の水平管内凝縮", «冷凍»1980年
     1月刊

[9], M.M.Shah, "A General Correlation For Heat Transfer During
     Film Condensation Inside Pipes", Int.J.Heat & Mass Transfer,
     1979,Vol 22.

[10], J.A.Tichy,N.A.Macken, "An Experimental Investigation of
      Heat Transfer in Forced Convection of Oil Refrigerant
      Mixture",ASHRAE Transaction,1982,Vol 88

[11], Zhao Yan Yun, Chow Chi Chin, "An Investigation on Thermal Performance of Double Side Slit Finned Tube Heat Exchanger", SIME M.S.Thesis

[12], Cheng Zi Hang, "Two Phase Flow and Heat Transfer of Vapor-Liquid Mixture", Mechanical Engineering Publishing House, 1982

[13], Van Chang, Chow Chi Chin, "The Effect of Tube Set Connection Type on the Heat Transfer in Air-Cooled Condenser", SIMEM.S.Thesis, 1988

644

Index